T0136878

Intelligent Systems Reference Library

Volume 118

Series editors

Janusz Kacprzyk, Polish Academy of Sciences, Warsaw, Poland
e-mail: kacprzyk@ibspan.waw.pl

Lakhmi C. Jain, University of Canberra, Canberra, Australia;
Bournemouth University, UK;
KES International, UK
e-mails: jainlc2002@yahoo.co.uk; Lakhmi.Jain@canberra.edu.au
URL: http://www.kesinternational.org/organisation.php

About this Series

The aim of this series is to publish a Reference Library, including novel advances and developments in all aspects of Intelligent Systems in an easily accessible and well structured form. The series includes reference works, handbooks, compendia, textbooks, well-structured monographs, dictionaries, and encyclopedias. It contains well integrated knowledge and current information in the field of Intelligent Systems. The series covers the theory, applications, and design methods of Intelligent Systems. Virtually all disciplines such as engineering, computer science, avionics, business, e-commerce, environment, healthcare, physics and life science are included.

More information about this series at http://www.springer.com/series/8578

Dionisios N. Sotiropoulos · George A. Tsihrintzis

Machine Learning Paradigms

Artificial Immune Systems and their
Applications in Software Personalization

 Springer

Dionisios N. Sotiropoulos
University of Piraeus
Piraeus
Greece

George A. Tsihrintzis
University of Piraeus
Piraeus
Greece

ISSN 1868-4394   ISSN 1868-4408   (electronic)
Intelligent Systems Reference Library
ISBN 978-3-319-83675-1   ISBN 978-3-319-47194-5   (eBook)
DOI 10.1007/978-3-319-47194-5

Printed on acid-free paper

This Springer imprint is published by Springer Nature
The registered company is Springer International Publishing AG
The registered company address is: Gewerbestrasse 11, 6330 Cham, Switzerland

To my beloved family and friends

Dionisios N. Sotiropoulos

*To my wife and colleague, Prof.-Dr. Maria
Virvou, and our daughters, Evina,
Konstantina and Andreani*

George A. Tsihrintzis

Foreword

There are many real-world problems of such high complexity that traditional scientific approaches, based on physical and statistical modeling of the data generation mechanism, do not succeed in addressing them efficiently. The cause of inefficiencies often lies in multi-dimensionality, nonlinearities, chaotic phenomena and the presence of a plethora of degrees of freedom and unknown parameters in the mechanism that generates data. As a result, loss of information that is crucial to solve a problem is inherent in the data generation process itself, making a traditional mathematical solution intractable.

At the same time, however, biological systems have evolved to address similar problems in efficient ways. In nature, we observe abundant examples of high level intelligence, such as

- Biological neural networks, i.e., networks of interconnected biological neurons in the nervous system of most multi-cellular animals, are capable of learning, memorizing and recognizing patterns in signals such as images, sounds or odors.
- Ants exhibit a collective, decentralized and self-organized intelligence that allows them to discover the shortest route to food in very efficient ways.

A third example of a biological system that exhibits high level intelligence is the vertebrate immune system. The immune system in vertebrates is a decentralized system of biological structures and processes within an organism that protects against pathogens that threaten the organism and may cause disease. Even though it is virtually impossible for a system to learn and memorize all possible forms of pathogens that may potentially threaten an organism, the immune system is capable of detecting a wide variety of agents, ranging from viruses to parasitic worms, and distinguishing them from the organism's own healthy tissue.

Dionisios N. Sotiropoulos and George A. Tsihrintzis have authored the book at hand on computational aspects of the vertebrate immune system with the intent to promote the use of artificial immune systems in addressing machine learning problems. Artificial immune systems are a class of intelligent systems, inspired by the vertebrate immune system and capable of learning and memorizing. In the recent years, the discipline of computer science has shown intense research interest into applying artificial immune system techniques in various pattern recognition, clustering and classification problems.

The book at hand is a significant addition to this field. The authors present artificial immune systems from the practical signal processing point of view, emphasizing relevant algorithms for clustering and classification. Additionally, the authors illustrate the use of the proposed algorithms through a number of case studies and application on a variety of real data. Particularly interesting is the authors' proposal to use artificial immune system approaches to tackle classification problems which exhibit high, even extreme, class imbalance (so-called *one-class classification problems*).

A practical application of their approach may be found in the design of recommender systems that require and use only positive examples of the preferences of their users. The authors show, through application on real data, that artificial immune systems may be an efficient way to address such problems.

The book is addressed to any graduate student and researcher in computer science. As such, it is self-contained, with the necessary number of introductory chapters on learning and learning paradigms before specific chapters on artificial immune systems. I believe that the authors have done a good job on addressing the tackled issues. I consider the book a good addition to the areas of learning, bio-inspired computing and artificial immune systems. I am confident that it will help graduate students, researchers and practitioners to understand and expand artificial immune systems and apply them in real-world problems.

Dayton, OH, USA   Prof.-Dr. Nikolaos G. Bourbakis
June 2016   IEEE Fellow, President, Biological and
                                                    Artficial Intelligence Foundation (BAIF)
                                                OBR Distinguished Professor of Informatics and
                                                                     Technology and Director
                                            Assistive Technologies Research Center and Director
                                                   Assistive Technologies Research Center

Preface

In the monograph at hand, we explore theoretical and experimental justification of the use of *Artificial Immune Systems* as a *Machine Learning Paradigm*. Our inspiration stems from the fact that vertebrates possess an immune system, consisting of highly complex biological structures and processes, that efficiently protect them against disease. A biological immune system is capable of detecting a wide variety of agents, including viruses, parasites and cancer cells, and distinguishing them from the organism's own healthy tissue. This is achieved in the *adaptive immune* subsystem of the immune system.

More specifically, the adaptive immune (sub)system continuously performs a self/non-self discrimination process. In machine learning terms, the adaptive immune system addresses *a pattern classification problem with extreme class imbalance*. Over the recent years, classification problems with class imbalance have attracted the interest of researchers worldwide. However, little attention has been paid so far to the use of artificial immune systems in addressing classification problems with a high or extreme degree of class imbalance.

We address the fundamental problems of pattern recognition, i.e. (*clustering*, *classification* and *one-class classification*), by developing artificial immune system-based machine learning algorithms. We measure the efficiency of these algorithms against state of the art pattern recognition paradigms such as *support vector machines*. Particular emphasis is placed on pattern classification in the context of the class imbalance problem. In machine learning terms, we address degenerated binary classification problems where the class of interest to be recognized is known through only a limited number of positive training instances. In other words, the target class occupies only a negligible volume of the entire pattern space, while the complementary space of negative patterns remains completely unknown during the training process. A practical application of this approach may be found in the design of recommender systems that require the use of only positive examples of the preferences of their users. We show, through application on real data, that artificial immune systems address such problems efficiently.

The general experimentation framework adopted throughout the current monograph is an open collection of one thousand (1000) pieces from ten (10) classes of western music. This collection has been extensively used in applications concerning music information retrieval and music genre classification. The experimental results presented in this monograph demonstrate that the general framework of artificial immune system-based classification algorithms constitutes a valid machine learning paradigm for clustering, classification and one-class classification.

In order to make the book as self-contained as possible, we have divided it into two parts. Specifically, the first part of the book presents machine learning fundamentals and paradigms with an emphasis on one-class classification problems, while the second part is devoted to biological and artificial immune systems and their application to one-class classification problems. The reader, depending on his/her previous exposure to machine learning, may choose either to read the book from its beginning or to go directly to its second part. It is our hope that this monograph will help graduate students, researchers and practitioners to understand and expand artificial immune systems and apply them in real-world problems.

Piraeus, Greece   Dionisios N. Sotiropoulos
June 2016   George A. Tsihrintzis

Acknowledgments

We would like to thank Prof.-Dr. Lakhmi C. Jain for agreeing to include this monograph in the Intelligent Systems Reference Library (ISRL) book series of Springer that he edits. We would also like to thank Prof. Nikolaos Bourbakis of Wright State University, USA, for writing a foreword to the monograph. Finally, we would like to thank the Springer staff for their excellent work in typesetting and publishing this monograph.

Contents

Part I
Machine Learning Fundamentals

Chapter 1
Introduction

Abstract In this chapter, we introduce the reader to the scope and content of the book which is a research monograph on the artificial immune system paradigm of machine learning. The book topic falls within the, so-called, biologically motivated computing paradigm, in which biology provides the source of models and inspiration towards the development of computational intelligence and machine learning systems. Specifically, artificial immune systems are presented as a valid metaphor towards the creation of abstract and high level representations of biological components or functions. We summarize the book and outline each subsequent chapter.

One of the primary characteristics of humans is their persistence in observing the natural world in order to devise theories about how the many parts of nature behave. Newton's laws of physics and Kepler's model of planetary orbits constitute two major examples of human nature which for many years now tries to unravel the basic underpinning behind the observed phenomena. The world, however, need not just be observed and explained, but also utilized as inspiration for the design and construction of artifacts, based on the simple principle stating that nature has been doing a remarkable job for millions of years. Moreover, recent developments in computer science, engineering and technology have been determinately influential in obtaining a deeper understanding of the world and particularly biological systems. Specifically, biological processes and functions have been explained on the basis of constructing models and performing simulations of such natural systems. The reciprocal statement is also true, meaning that the introduction of ideas stemmed from the study of biology, have also been beneficial for a wide range of applications in computing and engineering. This can be exemplified by artificial neural networks, evolutionary algorithms, artificial life and cellular automata. This inspiration from nature is a major motivation for the development of artificial immune systems.

In this context, a new field of research emerged under the name of *bioinformatics* referring to principles and methodologies of information technology (i.e. computational methods) applied to the management and analysis of biological data. Its implications cover a diverse range of areas from computational intelligence and robotics to genome analysis [1]. The field of biomedical engineering, on the other hand, was introduced in an attempt to encompass the application of engineering principles to biological and medical problems [5]. Therefore, the subject of biomedical engineering is intimately related to bioinformatics.

D.N. Sotiropoulos and G.A. Tsihrintzis, *Machine Learning Paradigms*,
Intelligent Systems Reference Library 118, DOI 10.1007/978-3-319-47194-5_1

The bilateral interaction between computing and biology can be mainly identified within the following approaches:

1. *biologically motivated computing*, where biology provides sources of models and inspiration in order to develop computational systems (e.g. Artificial Immune Systems),
2. *computationally motivated biology* where computing is utilized in order to derive models and inspiration for biology (Cellular Automata) and
3. *computing with biological mechanisms*, which involves the use of the information processing capabilities of biological systems to replace, or at least supplement,current silicon based computers (e.g. Quantum and DNA computing).

This monograph, however, is explicitly focused on biologically motivated computing within the field of Artificial Immune Systems. In other words, the research presented here does not involve the computational assistance of biology through the construction of models that represent or reproduce biological functionalities of primary interest. On the contrary, the main purpose of this monograph is to utilize biology, and immunology in particular, as a valid metaphor in order to create abstract and high level representations of biological components or functions. A metaphor uses inspiration in order to render a particular set of ideas and beliefs in such a way that they can be applied to an area different from the one in which they were initially conceived. Paton [4] identified four properties of biological systems important for the development of metaphors:

1. *architecture*, which refers to the form or the structure of the system,
2. *functionality*, which corresponds to its behavior,
3. *mechanisms*, which characterize the cooperation and interactions of the various parts, and
4. *organization*, which refers to the ways the activities of the system are expressed in the dynamics of the whole.

The primary contribution of this monograph lies within the field of Pattern Recognition, providing experimental justifications concerning the validity of Artificial Immune Systems as an alternative machine learning paradigm. The main source of inspiration stems from the fact that the Adaptive Immune System constitutes one of the most sophisticated biological systems that is particularly evolved in order to continuously address an extremely unbalanced pattern classification problem by performing a self/non-self discrimination process. The main effort undertaken in this monograph is focused on addressing the primary problems of Pattern Recognition by developing Artificial Immune System-based machine learning algorithms. Therefore, the relevant research is particularly interested in providing alternative machine learning approaches for the problems of *Clustering, Classification* and *One-Class Classification* and measuring their efficiency against state of the art pattern recognition paradigms such as Support Vector Machines. Pattern classification is specifically studied within the context of the Class Imbalance Problem which arises with extremely skewed training data sets. Specifically, the experimental results presented in this book involve degenerate binary classification problems where the class of interest to be recognized is known only through a limited number of positive training instances. In other words, the target class occupies only a negligible volume of the entire pattern space while the complementary space of negative patterns remains completely unknown during the training process. The effect of the Class Imbalance Problem on the performance of the utilized Artificial Immune System-based classification algorithm constitutes one of the secondary objectives of this monograph.

The general experimentation framework adopted throughout the book in order to assess the efficiency of the proposed clustering, classification and one-class classification algorithms was a publicly-available collection of one thousand (1000) pieces from 10 different classes of western music. This collection, in particular, has been extensively used in applications concerning music information retrieval and music genre classification [3, 10].

The following list summarizes the pattern recognition problems addressed in the current monograph through the application of specifically designed Artificial Immune System-based machine learning algorithms:

1. Artificial Immune System-Based music piece clustering and Database Organization [6, 8].
2. Artificial Immune System-Based Customer Data Clustering in an e-Shopping application [9].
3. Artificial Immune System-Based Music Genre Classification [7].
4. A Music Recommender Based on Artificial Immune Systems [2].

The book is structured in the following way:

- Chapter 2 presents an extensive review on the subject of machine learning and extensive references to existing literature. The relevant research is primarily focused on the main approaches that have been proposed in order to address the problem of machine learning and how they may be categorized according to the type and amount of inference. Specifically, the categorization of the various machine learning paradigms according to the type of inference, involves the following approaches:

1. Model Identification or Parametric Inference; and
2. Model Prediction or General Inference.

The general framework of the parametric model, in particular, introduces the principles of Empirical Risk Minimization (ERM) and Structural Risk Minimization (SRM). On the other hand, the Transductive Inference Model is defined as an extension to the original paradigm of General Inference. The categorization of machine learning models according to the amount of inference includes the following approaches:

1. Rote Learning,
2. Learning from Instruction and
3. Learning from Examples.

Specifically, Learning from Examples provides the ideal framework in order to analyze the problem of minimizing a risk functional on a given set of empirical data which constitutes the fundamental problem within the field of pattern recognition. In essence, the particular form of the risk functional defines the primary problems of machine learning, namely:

1. The Classification Problem;
2. The Regression Problem; and
3. The Density Estimation Problem which is closely related to the Clustering Problem.

Finally, this chapter presents a conspectus of the theoretical foundations behind Statistical Learning Theory.

- Chapters 3, 4 deal with a special category of pattern recognition problems arising in cases when the set of training patterns is significantly biased towards a particular class of patterns. This is the so-called Class Imbalance Problem which hinders the performance of many standard classifiers. Specifically, the very essence of the class imbalance problem is unraveled by referring to the relevant literature review. Indeed, there exists a wide range of real-world applications involving extremely skewed data sets. The class imbalance problem stems from the fact that the class of interest occupies only a negligible volume within the complete pattern space. These chapters investigate the particular effects of the class imbalance problem on standard classifier methodologies and present the various methodologies that have been proposed as a remedy. The most interesting approach within the context of Artificial Immune Systems is the one related to the machine learning paradigm of one-class classification. One-Class Classification problems may be thought of as degenerate binary classification problems where the available training instances originate exclusively from the under-represented class of patterns.
- Chapter 5 discusses the machine learning paradigm of Support Vector Machines which incorporates the principles of Empirical Risk Minimization and Structural Risk Minimization. Support Vector Machines constitute a state of the art classifier which is used in this monograph as a benchmark algorithm in order to evaluate the classification accuracy of Artificial Immune System-based machine learning algorithms. Finally, this chapter presents a special class of Support Vector Machines that are especially designed for the problem of One-Class Classification, namely One-Class Support Vector Machines.
- Chapter 6 analyzes the biological background of the monograph, namely the immune system of vertebrate organisms. The relevant literature review presents the major components and the fundamental principles governing the operation of the adaptive immune system, with emphasis on those characteristics of the adaptive immune system that are of particular interest from a computation point of view. The fundamental principles of the adaptive immune system are given by the following theories:

1. Immune Network Theory,
2. Clonal Selection Theory and
3. Negative Selection Theory.

- Chapter 7 introduces the subject of Artificial Immune Systems by emphasizing on their ability to provide an alternative machine learning paradigm. The relevant bibliographical survey was utilized in order to extract the formal definition of Artificial Immune Systems and identify their primary application domains involving:

1. Clustering and Classification,
2. Anomaly Detection/Intrusion Detection,
3. Optimization,
4. Automatic Control,
5. Bioinformatics,
6. Information Retrieval and Data Mining,
7. User Modeling/Recommendation and
8. Image Processing.

Special attention was paid to analyzing the Shape-Space Model which provides the necessary mathematical formalism for the transition from the field of biology to the field of Information Technology. This chapter focuses on the development of alternative machine learning algorithms based on Immune Network Theory, the Clonal Selection Principle and Negative Selection Theory. The proposed machine learning algorithms relate specifically to the problems of:

1. Data Clustering,
2. Pattern Classification and
3. One-Class Classification.

- Chapter 8 applies Artificial Immune System-based clustering, classification and recommendation algorithms to sets of publicly-available real data. The purpose of this chapter is to experiment with real data and compare the performance of Artificial Immune System-based algorithms with related state of the art algorithms and draw conclusions.
- Chapter 9 summarizes the major findings of this monograph, draws conclusions and suggests avenues of future research.

References

1. Attwood, T.K., Parry-Smith, D.J.: Introduction to Bioinformatics. Prentice Hall, New Jersey (1999)
2. Lampropoulos, A., Sotiropoulos, D.G.: (2010). A music recommender based on artificial immune systems. *Multimedia Systems*
3. Li, T., Ogihara, M., Li, Q.: A comparative study on content-based music genre classification. In: SIGIR '03: Proceedings of the 26th Annual International ACM SIGIR Conference on Research and Development in Informaion Retrieval, pp. 282–289. ACM, New York, NY, USA (2003)
4. Paton, R.: Computing with Biological Metaphors. Chapman & Hall, Boca Raton (1994)
5. Schwan, H.P.: Biological Engineering. McGraw-Hill, New York (1969)
6. Sotiropoulos, D., Lampropoulos, A., Tsihrintzis, G.: Artificial immune system-based music piece similarity measures and database organization. In: Proceedings of 5th EURASIP Conference, pp. 5, vol. 254 (2005)

7. Sotiropoulos, D., Lampropoulos, A., Tsihrintzis, G.: Artificial immune system-based music
    genre classification. In: New Directions in Intelligent Interactive Multimedia. Studies in Com-
    putational Intelligence, pp. 191–200, vol. 142 (2008)
8. Sotiropoulos, D., Tsihrintzis, G.: Immune System-based Clustering and Classification Algo-
    rithms. *Mathematical Methods in* (2006)
9. Sotiropoulos, D.N., Tsihrintzis, G.A., Savvopoulos, A., Virvou, M.: Artificial immune system-
    based customer data clustering in an e-shopping application. In: Proceedings, Part I Knowledge-
    Based Intelligent Information and Engineering Systems, 10th International Conference, KES
    2006, October 9–11, 2006, pp. 960–967. Bournemouth, UK (2006)
10. Tzanetakis, G., Cook, P.R.: Music genre classification of audio signals. IEEE Trans. Speech
    Audio Process. **10**(5), 293–302 (2002)

Chapter 2
Machine Learning

Abstract We present an extensive review on the subject of machine learning by studying existing literature. We focus primarily on the main approaches that have been proposed in order to address the problem of machine learning and how they may be categorized according to type and amount of inference. Specifically, the categorization of the various machine learning paradigms according to the type of inference, involves the following two approaches:

- Model Identification or Parametric Inference; and
- Model Prediction or General Inference.

The general framework of the parametric model, in particular, introduces the principles of Empirical Risk Minimization (ERM) and Structural Risk Minimization. On the other hand, the Transductive Inference Model is defined as an extension to the original paradigm of General Inference. The categorization of machine learning models according to the amount of inference includes the following approaches:

- Rote Learning;
- Learning from Instruction; and
- Learning from Examples.

Specifically, Learning from Examples provides the framework to analyze the problem of minimizing a risk functional on a given set of empirical data which is the fundamental problem within the field of pattern recognition. In essence, the particular form of the risk functional defines the primary problems of machine learning, namely:

- The Classification Problem;
- The Regression Problem; and
- The Density Estimation Problem which is closely related to the Clustering Problem.

Finally, in this chapter we present a conspectus of the theoretical foundations behind Statistical Learning Theory.

© Springer International Publishing AG 2017

D.N. Sotiropoulos and G.A. Tsihrintzis, *Machine Learning Paradigms*,

Intelligent Systems Reference Library 118, DOI 10.1007/978-3-319-47194-5_2

2.1   Introduction

The ability to learn is one of the most distinctive attributes of intelligent behavior. An informal definition of the learning process in general could be articulated as: *"The learning process includes the acquisition of new declarative knowledge, the development of new skills through interaction or practice, the organization of new knowledge into general, effective representations, and the discovery of new facts and theories through observation and experimentation"*. The term *machine learning*, on the other hand, covers a broad range of computer programs. In general, any computer program that improves its performance through experience or training can be called a learning program. Machine learning constitutes an integral part of artificial intelligence since the primary feature of any intelligent system is the ability to learn. Specifically, systems that have the ability to learn need not be implicitly programmed for any possible problematic situation. In other words, the development of machine learning alleviates the system designer from the burden of foreseeing and providing solutions for all possible situations.

The study and modelling of learning processes in their multiple manifestations constitute the topic of machine learning. In particular, machine learning has been developed around the following primary research lines:

- *Task-oriented studies*, which are focused on developing learning systems in order to improve their performance in a predetermined set of tasks.
- *Cognitive simulation*, that is, the investigation and computer simulation of human learning processes.
- *Theoretical analysis*, which stands for the investigation of possible learning methods and algorithms independently of the particular application domain.
- *Derivation of machine learning paradigms and algorithms* by developing metaphors for biological processes that may be interesting within the context of machine learning. A typical example is the field of biologically inspired computing which led to the emergence of Artificial Neural Networks and Artificial Immune Systems.

The following sections provide an overview of the various machine learning approaches that have been proposed over the years according to different viewpoints concerning the *underlying learning strategies*. Specifically, Sect. 2.2 provides a more general categorization of the machine learning methodologies based on the particular type of inference utilized while Sect. 2.3 provides a more specialized analysis according to the amount of inference. Finally, Sect. 2.5 gives a theoretical justification of Statistical Learning Theory.

2.2   Machine Learning Categorization According to the Type of Inference

The fundamental elements of statistical inference have existed for more than 200 years, due to the seminal works of Gauss and Laplace. However, their systematic analysis began in the late 1920s. By that time, descriptive statistics was mostly complete since it was shown that many events of the real world are sufficiently described by different statistical laws. Specifically, statisticians have developed powerful mathematical tools, namely distribution functions, that have the ability to capture interesting aspects of reality. However, a crucial question that was yet to be answered concerned the determination of a reliable method for performing statistical inference. A more formal definition of the related problem could be the following: *Given a collection of empirical data originating from some functional dependency, infer this dependency*. Therefore, the analysis of methods of statistical inference signaled the beginning of a new era for statistics which was significantly influenced by two bright events:

1. Fisher introduced the main models of statistical inference in the unified framework of parametric statistics. His work indicated that the various problems related to the estimation of functions from given data (the problems of discriminant analysis, regression analysis, and density estimation) are particular instances of the more general problem dealing with the parameter estimation of a specific parametric model. In particular, he suggested the Maximum Likelihood method as a for the estimation of the unknown parameters in all these models.
2. Glivenko, Cantelli and Kolmogorov, on the other hand, started a general analysis of statistical inference. One of the major findings of this quest was the Glivenko–Cantelli theorem stating that the empirical distribution function always converges to the actual distribution function. Another equally important finding came from Kolmogorov who found the asymptotically exact rate of this convergence. Specifically, he proved that the rate turns out to be exponentially fast and independent of the unknown distribution function.

Notwithstanding, these two events determined the two main approaches that were adopted within the general context of machine learning:

1. *Model Identification* or *particular (parametric) inference* which aims at creating simple statistical methods of inference that can be used for solving real-life problems, and
2. *Model Prediction* or *general inference*, which aims at finding one induction method for any problem of statistical inference.

The philosophy that led to the conception of the model identification approach is based upon the belief that the investigator knows the problem to be analyzed relatively well. Specifically, he/she is aware of the physical law that generates the stochastic properties of the data and the function to be found up to a finite number of parameters. According to the model identification approach, the very essence of the statistical inference problem is the estimation of these parameters by utilizing the available data. Therefore, the natural solution in finding these parameters is obtained by utilizing information concerning the statistical law and the target function is the adaptation of the maximum likelihood method. The primary purpose of this theory is to justify the corresponding approach by discovering and describing its favorable properties.

On the contrary, the philosophy that led to the conception of the model prediction approach is focused on the fact that there is no reliable a priori information concerning the statistical law underlying the problem or the desirable function to be approximated. Therefore, it is necessary to find a method in order to infer the approximation function from the given examples in each situation. The corresponding theory of model prediction must be able to describe the conditions under which it is possible find the best approximation to an unknown function in a given set of functions with an increasing number of examples.

2.2.1   Model Identification

The model identification approach corresponding to the principle of parametric inference was developed very quickly since its original conception by Fisher. In fact, the main ideas underlying the parametric model were clarified in the 1930s and the main elements of theory of parametric inference were formulated within the next 10 years. Therefore, the time period between the 1930 and 1960 was the "golden age" of parametric inference which dominated statistical inference. At that time, there was only one legitimate approach to statistical inference, namely the theory that served the model identification approach. The classical parametric paradigm falls within the general framework introduced by Fisher according to which any signal Y can be modelled as consisting of a deterministic component and a random counterpart:

$$Y = f(X) + \epsilon \tag{2.1}$$

The deterministic part $f(X)$ is defined by the values of a known family of functions which are determined up to a limited number of parameters. The random part ϵ corresponds to the noise added to the signal, defined by a known density function. Fisher considered the estimation of the parameters of the function $f(X)$ as the goal of statistical analysis. Specifically, in order to find these parameters he introduced the maximum likelihood method. Since the main goal of Fisher's statistical framework was to estimate the model that generated the observed signal, his paradigm is identified by the term "*Model Identification*". In particular, Fisher's approach reflects the traditional idea of Science concerning the process of inductive inference, which can be roughly summarized by the following steps:

1. Observe a phenomenon.
2. Construct a model of that phenomenon (*inductive step*).

3. Make predictions using this model (*deductive step*).

The philosophy of this classical paradigm is based upon on the following beliefs:

1. *In order to find a dependency from the data, the statistician is able to define a set of functions, linear in their parameters, that contain a good approximation to the desired function. The number of parameters describing this set is small.*
   This belief was specifically supported by referring to the Weierstrass theorem, according to which any continuous function with a finite number of discontinuities can be approximated on a finite interval by polynomials (functions linear in their parameters) with any degree of accuracy. The main idea was that this set of functions could be replaced by an alternative set of functions, not necessarily polynomials, but linear with respect to a small number of parameters. Therefore, one could obtain a good approximation to the desired function.
2. *The statistical law underlying the stochastic component of most real-life problems is the normal law.*
   This belief was supported by referring to the Central Limit Theorem, which states that under wide conditions the sum of a large number of random variables is approximated by the normal law. The main idea was that if randomness in a particular problem is the result of interaction among a large number of random components, then the stochastic element of the problem will be described by the normal law.
3. *The maximum likelihood estimate may serve as a good induction engine in this paradigm.*
   This belief was supported by many theorems concerning the conditional optimality of the maximum likelihood method in a restricted set of methods or asymptotically. Moreover, there was hope that this methodology would offer a good tool even for small sample sizes.

Finally, these three beliefs are supported by the following more general philosophy:
If there exists a mathematical proof that some method provides an asymptotically optimal solution, then in real life this method will provide a reasonable solution for a small number of data samples.

The classical paradigm deals with the identification of stochastic objects which particulary relate to the problems concerning the estimation of densities and conditional densities.

Density Estimation Problem

The first problem to be considered is the density estimation problem. Letting ξ be a random vector then the probability of the random event $F(\mathbf{x}) = P(\xi < \mathbf{x})$ is called a *probability distribution function of the random vector* ξ where the inequality is interpreted coordinatewise. Specifically, the random vector ξ has a density function if there exists a nonnegative function $p(\mathbf{u})$ such that for all $\mathbf{x}$ the equality

$$F(\mathbf{x}) = \int_{-\infty}^{\mathbf{x}} p(\mathbf{u})d\mathbf{u} \tag{2.2}$$

is valid. The function $p(\mathbf{x})$ is called a *probability density* of the random vector. Therefore, by definition, the problem of estimating a probability density from the data requires a solution of the integral equation:

$$\int_{-\infty}^{\mathbf{x}} p(\mathbf{u}, \alpha)d\mathbf{u} = F(\mathbf{x}) \tag{2.3}$$

on a given set of densities $p(\mathbf{x}, \alpha)$ where $\alpha \in \Lambda$. It is important to note that while the true distribution function $F(\mathbf{x})$ is unknown, one is given a random independent sample

$$\mathbf{x}_1, \ldots, \mathbf{x}_l \tag{2.4}$$

which is obtained in accordance with $F(\mathbf{x})$. Then it is possible to construct a series of approximations to the distribution function $F(\mathbf{x})$ by utilizing the given data set (2.4) in order to form the so-called *empirical distribution function* which is defined by the following equation:

$$F_l(\mathbf{x}) = \frac{1}{l} \sum_{i=1}^{l} \theta(\mathbf{x} - \mathbf{x}_i), \tag{2.5}$$

where $\theta(\mathbf{u})$ corresponds to the step function defined as:

$$\theta(\mathbf{u}) = \begin{cases} 1, & \text{when all the coordinates of vector } \mathbf{u} \text{ are positive,} \\ 0, & \text{otherwise.} \end{cases} \tag{2.6}$$

Thus, the problem of density estimation consists of finding an approximation to the solution of the integral equation (2.3). Even if the probability density function is unknown, an approximation to this function can be obtained.

Conditional Probability Estimation Problem

Consider pairs $(\omega, \mathbf{x})$ where $\mathbf{x}$ is a vector and ω is scalar which takes on only k values from the set $\{0, 1, \ldots, k-1\}$. According to the definition, the conditional probability $P(\omega, \mathbf{x})$ is a solution of the integral equation:

$$\int_{-\infty}^{\mathbf{x}} P(\omega|\mathbf{t})dF(\mathbf{t}) = F(\omega, \mathbf{x}), \tag{2.7}$$

where $F(\mathbf{x})$ is the distribution function of the random vector $\mathbf{x}$ and $F(\omega, \mathbf{x})$ is the joint distribution function of pairs $(\omega, \mathbf{x})$. Therefore, the problem of estimating the conditional probability in the set of functions $P_\alpha(\omega|\mathbf{x})$, where $\alpha \in \Lambda$, is to obtain an approximation to the integral equation (2.7) when both distribution functions $F(\mathbf{x})$ and $F(\omega, \mathbf{x})$ are unknown, but the following set of samples are available:

$$(\omega_1, \mathbf{x}_1), \ldots, (\omega_l, \mathbf{x}_l).$$

$$(2.8)$$

As in the case of the density estimation problem, the unknown distribution functions $F(\mathbf{x})$ and $F(\omega, \mathbf{x})$ can be approximated by the empirical distribution functions (2.5) and function:

$$F_l(\omega, \mathbf{x}) = \frac{1}{l} \sum_{i=1}^{l} \theta(\mathbf{x} - \mathbf{x}_i) \delta(\omega, \mathbf{x}_i),$$

$$(2.9)$$

where the function $\delta(\omega, \mathbf{x})$ is defined as:

$$\delta(\omega, \mathbf{x}) = \begin{cases} 1, & \text{if the vector } \mathbf{x} \text{ belongs to class } \omega, \\ 0, & \text{otherwise.} \end{cases}$$

$$(2.10)$$

Thus, the problem of conditional probability estimation may be resolved by obtaining an approximation to the solution of the integral equation (2.7) in the set of functions $P_\alpha(\omega|\mathbf{x})$ where $\alpha \in \Lambda$. This solution, however, is difficult to get since the probability density functions $F(\mathbf{x})$ and $F(\omega, \mathbf{x})$ are unknown and they can only be approximated by the empirical functions $F_l(\mathbf{x})$ and $F_l(\omega, \mathbf{x})$.

Conditional Density Estimation Problem

The last problem to be considered is the one related to the conditional density estimation. By definition, this problem consists in solving the following integral equation:

$$\int_{-\infty}^{y} \int_{-\infty}^{\mathbf{x}} p(t|\mathbf{u}) dF(\mathbf{u}) dt = F(y, \mathbf{x}),$$

$$(2.11)$$

where the variables y are scalars and the variables $\mathbf{x}$ are vectors. Moreover, $F(\mathbf{x})$ is a probability distribution function which has a density, $p(y, \mathbf{x})$ is the conditional density of y given $\mathbf{x}$, and $F(y, \mathbf{x})$ is the joint probability distribution function defined on the pairs $(y, \mathbf{x})$. The desirable conditional density function $p(y|\mathbf{x})$ can be obtained by considering a series of approximation functions which satisfy the integral equation (2.11) on the given set of functions and the i.i.d pairs of the given data:

$$(y_1, \mathbf{x}_1), \ldots, (y_l, \mathbf{x}_l)$$

$$(2.12)$$

when both distributions $F(\mathbf{x})$ and $F(y, \mathbf{x})$ are unknown. Once again, it is possible to approximate the empirical distribution function $F_l(\mathbf{x})$ and the empirical joint distribution function:

$$F_l(y, \mathbf{x}) = \frac{1}{l} \sum_{i=1}^{l} \theta(y - y_i) \theta(\mathbf{x} - \mathbf{x}_i).$$

$$(2.13)$$

Therefore, the problem is to get an approximation to the solution of the integral equation (2.11) in the set of functions $p_a(y, \mathbf{x})$ where $\alpha \in \Lambda$, when the probability distribution functions are unknown but can be approximated by $F_l(\mathbf{x})$ and $F_l(y, \mathbf{x})$ using the data (2.12).

2.2.2   Shortcoming of the Model Identification Approach

All three problems of stochastic dependency estimation that were thoroughly discussed previously can be described in the following general way. Specifically, they are reduced to solving the following linear continuous operator equation

$$Af = F, f \in \mathcal{F} \tag{2.14}$$

given the constraint that some functions that form the equation are unknown. The unknown functions, however, can be approximated by utilizing a given set of sample data. In this way it possible to obtain approximations to the distribution functions $F_l(\mathbf{x})$ and $F_l(y, \mathbf{x})$. This formulation can reveal a main difference between the problem of density estimation and the problems of conditional probability and conditional density estimation. Particularly, in the problem of density estimation, instead of an accurate right-hand side of the equation only an approximation is available. Therefore, the problem involves getting an approximation to the solution of Eq. (2.14) from the relationship

$$Af \approx F_l, f \in \mathcal{F}. \tag{2.15}$$

On the other hand, in the problems dealing with the conditional probability and conditional density estimation not only the right-hand side of Eq. (2.14) is known only approximately, but the operator A is known only approximately as well. This being true, the true distribution functions appearing in Eqs. (2.7) and (2.11) are replaced by their approximations. Therefore, the problem consists in getting an approximation to the solution of Eq. (2.14) from the relationship

$$A_l f \approx F_l, f \in \mathcal{F}. \tag{2.16}$$

The Glivenko–Cantelli theorem ensures that the utilized approximation functions converge to the true distribution functions as the number of observations goes to infinity. Specifically, the Glivenko–Cantelli theorem states that the convergence

$$\sup_{\mathbf{x}} |F(\mathbf{x}) - F_l(\mathbf{x})| \xrightarrow[l \to \infty]{P} 0 \tag{2.17}$$

takes place. A fundamental disadvantage of this approach is that solving the general operator Eq. (2.14) results in an ill-posed problem. Ill-posed problems are extremely difficult to solve since they violate the well-posedness conditions introduced by Hadamard involving the existence of a solution, the uniqueness of that solution and the continuous dependence of the solution on the empirical data. That is, the solutions of the corresponding integral equations are unstable.

Moreover, the wide application of computers, in the 1960s, for solving scientific and applied problems revealed additional shortcomings of the model identification approach. It was the fist time that researchers utilized computers in an attempt to analyze sophisticated models that had many factors or in order to obtain more precise approximations.

In particular, the computer analysis of large scale multivariate problems revealed the phenomenon that R. Bellman called *"the curse of dimensionality"*. It was observed that increasing the number of factors that have to be taken into consideration requires an exponentially increasing amount of computational resources. Thus, in real-life multidimensional problems where there might be hundreds of variables, the belief that it is possible to define a reasonably small set of functions that contains a good approximation to the desired one is not realistic.

Approximately at the same time, Tukey demonstrated that the statistical components of real-life problems cannot be described by only classical distribution functions. By analyzing real-life data, Tukey discovered that the corresponding true distributions are in fact different. This entails that it is crucial to take this difference into serious consideration in order to construct effective algorithms.

Finally, James and Stein showed that even for simple density estimation problems, such as determining the location parameters of a $n > 2$ dimensional normal distribution with a unit covariance matrix, the maximum likelihood method is not the best one.

Therefore, all three beliefs upon which the classical parametric paradigm relied turned out to be inappropriate for many real-life problems. This had an enormous consequence for statistical science since it looked like the idea of constructing statistical inductive inference models for real-life problems had failed.

2.2.3 Model Prediction

The return to the general problem of statistical inference occurred so imperceptibly that it was not recognized for more than 20 years since Fisher's original formulation of the parametric model. Of course, the results from Glivenko, Cantelli and Kolmogorov were known but they were considered to be inner technical achievements that are necessary for the foundation of statistical inference. In other words, these results could not be interpreted as an indication that there could be a different type of inference which is more general and more powerful than the classical parametric paradigm.

This question was not addressed until after the late 1960s when Vapnik and Chervonenkis started a new paradigm called *Model Prediction* (or predictive inference). The goal of model prediction is to predict events well, but not necessarily through the identification of the model of events. The rationale behind the model prediction paradigm is that the problem of estimating a model is hard (ill-posed) while the problem of finding a rule for good prediction is much easier (better-posed). Specifically, it could happen that there are many rules that predict the events well and are very different from the true model. Nonetheless, these rules can still be very useful predictive tools.

The model prediction paradigm was initially boosted when in 1958 F. Rosenblatt, a physiologist, suggested a learning machine (computer program) called the Perceptron for solving the simplest learning problem, namely the pattern classification/recognition problem. The construction of this machine incorporated several existing neurophysiological models of learning mechanisms. In particular, F. Rosenblatt demonstrated that even with the simplest examples the Perceptron was able to generalize without constructing a precise model of the data generation process. Moreover, after the introduction of the Perceptron, many learning machines were suggested that had no neurobiological analogy but they did not generalize worse than Perceptron. Therefore, a natural question arose:

Does there exist something common in these machines? Does there exist a general principle of inductive inference that they implement?

Immediately, a candidate was found as a general induction principle, the so-called *empirical risk minimization* (ERM) principle. The ERM principle suggests the utilization of a decision rule (an indicator function) which minimizes the number of training errors (empirical risk) in order to achieve good generalization on future (test) examples. The problem, however, was to construct a theory for that principle.

At the end of 1960s, the theory of ERM for the pattern recognition problem was constructed. This theory includes the general *qualitative theory* of generalization that described the necessary and sufficient conditions of consistency of the ERM induction principle. Specifically, the consistency of the ERM induction principle suggests that it is valid for any set of indicator functions, that is $\{0, 1\}$-valued functions on which the machine minimizes the empirical risk. Additionally, the new theory includes the general *quantitative theory* describing the bounds on the probability of the (future) test error for the function minimizing the empirical risk.

The application of the ERM principle, however, does not necessarily guarantee consistency, that is convergence to the best possible solution with an increasing number of observations. Therefore, the primary issues that drove the development of the ERM theory were the following:

1. Describing situations under which the method is consistent, that is, to find the necessary and sufficient conditions for which the ERM method defines functions that converge to the best possible solution with an increasing number of observations. The resulting theorems thereby describe the qualitative model of ERM principle.
2. Estimating the quality of the solution obtained on the basis of the given sample size. This entails, primarily, to estimate the probability of error for the function that minimizes the empirical risk on the given set of training examples and secondly to estimate how close this probability is to the smallest possible for the given set of functions. The resulting theorems characterize the generalization ability of the ERM principle.

In order to address both issues for the pattern recognition problem it is necessary to construct a theory that could be considered as a generalization of the Glivenko–Cantelli–Kolmogorov results. This is equivalent to the following statements:

1. For any given set of events, to determine whether the uniform law of large numbers holds, that is whether uniform convergence takes place.
2. If uniform convergence holds, to find the bounds for the *non-asymptotic* rate of uniform convergence.

This was the theory constructed by Vapnik and Chervonenkis which was based on a collection of new concepts, the so-called capacity concepts for a set of indicator functions. The most important new concept was the so-called VC dimension of the set of indicator functions which characterizes the variability of the set of indicator functions. Specifically, it was found that both the necessary and sufficient conditions of consistency and the rate of convergence of the ERM principle depend on the capacity of the set of functions that are implemented by the learning machine. The most preeminent results of the new theory that particularly relate to the VC dimension are the following:

1. For distribution-independent consistency of the ERM principle, the set of functions implemented by the learning machine must have a finite VC dimension.
2. Distribution-free bounds on the rate of uniform convergence depend on the VC dimension, the number of errors, and the number of observations.

The bounds for the rate of uniform convergence not only provide the main theoretical basis for the ERM inference, but also motivate a new method of inductive inference. For any level of confidence, an equivalent form of the bounds define bounds on the probability of the test error *simultaneously for all functions of the learning machine* as a function of the training errors, of the VC dimension of the set of functions implemented by the learning machine, and of the number of observations. This form of the bounds led to a new idea for controlling the generalization ability of learning machines:

In order to achieve the smallest bound on the test error by minimizing the number of training errors, the machine (set of functions) with the smallest VC dimension should be used.

These two requirements define a pair of contradictory goals involving the simultaneous minimization of the number of training errors and the utilization of a learning machine (set of functions) with a small VC dimension. In order to minimize the number of training errors, one needs to choose a function from a wide set of functions, rather than from a narrow set, with small VC dimension. Therefore, to find the best guaranteed solution, one has to compromise between the accuracy of approximation of the training data and the capacity (VC dimension) of the machine that is used for the minimization of errors. The idea of minimizing the test error by controlling two contradictory factors was formalized within the context of a new induction principle, the so-called Structural Risk Minimization (SRM) principle.

The fundamental philosophy behind the SRM principle is the so-called Occam's razor which was originally proposed by William of Occam in the fourteenth century, stating that *entities should not be multiplied beyond necessity*. In particular, the most common interpretation of Occam's razor is that *the simplest explanation is the best*. The assertion coming from SRM theory, however, is different and suggests that one should choose the explanation provided by the machine with the smallest capacity (VC dimension).

The SRM principle constitutes an integral part of the model prediction paradigm which was established by the pioneering work of Vapnik and Chervonenkis. Specifically, one of the most important achievements of the new theory concerns the discovery that the generalization ability of a learning machine depends on the capacity of the set of functions which are implemented by the learning machine which is different from the number of free parameters. Moreover, the notion of capacity determines the necessary and sufficient conditions ensuring the consistency of the learning process and the rate of convergence. In other words, it reflects intrinsic properties of inductive inference.

In order to extend the model prediction paradigm, Vapnik introduced the *Transductive Inference* paradigm in the 1980s. The goal of transductive inference is to estimate the values of an unknown predictive function at a given point of interest, but not in the whole domain of its definition. The rationale behind this approach is that it is possible to achieve more accurate solutions by solving less demanding problems. The more general philosophical underpinning behind the transductive paradigm can be summarized by the following imperative:

If you possess a restricted amount of information for solving some general problem, try to solve the problem directly and never solve a more general problem as an intermediate step. It is possible that the available information is sufficient for a direct solution but is insufficient for solving a more general intermediate problem.

In many real-life problems, the goal is to find the values of an unknown function only at points of interest, namely the testing data points. In order to solve this problem the model prediction approach uses a two-stage procedure which is particularly illustrated in Fig. 2.1.

At the first stage (inductive step) a function is estimated from a given set of functions, while at the second stage (deductive step) this function is used in order to evaluate the values of the unknown function at the points of interest. It is obvious that at the first stage of this two-stage scheme one addresses a problem that is more general than the one that needs to be solved. This is true since estimating an unknown

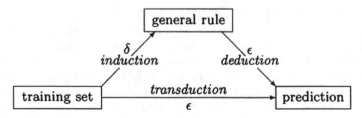

Fig. 2.1 Inference models

function involves estimating its values at all points in the function domain when only a few are of practical importance. In situations when there is only a restricted amount of information, it is possible to be able to estimate the values of the unknown function reasonably well at the given points of interest but cannot estimate the values of the function well at any point within the function domain. The direct estimation of function values only at points of interest using a given set of functions forms the transductive type of inference. As clearly depicted in Fig. 2.1, the transductive solution derives results in one step, directly from particular to particular (transductive step).

2.3 Machine Learning Categorization According to the Amount of Inference

Although machine learning paradigms can be categorized according to the type of inference that is performed by the corresponding machines, a common choice is to classify learning systems based on the amount of inference. Specifically, this categorization concerns the amount of inference that is performed by the learner which is one of the two primary entities in machine learning, the other being the supervisor (teacher). The supervisor is the entity that has the required knowledge to perform a given task, while the learner is the entity that has to learn the knowledge in order to perform a given task. In this context, the various learning strategies can be distinguished by the amount of inference the learner performs on the information given by the supervisor.

Actually, there are two extreme cases of inference, namely performing no inference and performing a remarkable amount of inference. If a computer system (the learner) is programmed directly, its knowledge increases but it performs no inference since all cognitive efforts are developed by the programmer (the supervisor). On the other hand, if a systems independently discovers new theories or invents new concepts, it must perform a very substantial amount of inference since it is deriving organized knowledge from experiments and observations. An intermediate case could be a student determining how to solve a math problem by analogy to problem solutions contained in a textbook. This process requires inference but much less than discovering a new theorem in mathematics.

Increasing the amount of inference that the learner is capable of performing, the burden on the supervisor decreases. The following taxonomy of machine learning paradigms captures the notion of trade-off in the amount of effort that is required of the learner and of the supervisor. Therefore, there are four different learning types that can be identified, namely *rote learning*, *learning from instruction*, *learning by analogy* and *learning from examples*.

2.3.1 Rote Learning

Rote learning consists in the direct implanting if knowledge into a learning system. Therefore, there is no inference or other transformation of the knowledge involved on the part of the learner. There are, of course, several variations of this method such as:

- Learning by being programmed or modified by an external entity. This variation requires no effort in the part of the learner. A typical paradigm is the usual style of computer programming.
- Learning by memorization of given facts and data with no inference drawn from incoming information. For instance, the primitive database systems.

2.3.2 Learning from Instruction

Learning from instruction (or *learning by being told*) consists in acquiring knowledge from a supervisor or other organized source, such as a textbook, requiring that the learner transforms the knowledge from the input language to an internal representation. The new information is integrated with the prior knowledge for effective use. The learner is required to perform some inference, but a large fraction of the cognitive burden remains with the supervisor, who must present and organize knowledge in a way that incrementally increases the learner's actual knowledge. In other words, learning from instruction mimics education methods. In this context, the machine learning task involves building a system that can accept and store instructions in order to efficiently cope with a future situation.

2.3.3 Learning by Analogy

Learning by analogy consists in acquiring new facts or skills by transforming and increasing existing knowledge that bears strong similarity to the desired new concept or skill into a form effectively useful in the new situation. A learning-by-analogy system could be applied in order to convert an existing computer program into one that performs a closely related function for which it was not originally designed. Learning by analogy requires more inference on the part of the learner that does rote learning or learning from instruction. A fact or skill analogous in relevant parameters must be retrieved from memory which will be subsequently transformed in order to be applied to the new situation.

2.4 Learning from Examples

Learning from examples is a model addressing the problem of functional dependency estimation within the general setting of machine learning. The fundamental
components of this model, as they are illustrated in Fig. 2.2, are the following:

1. The generator of the data G.
2. The target operator or *supervisor's operator* S.
3. The learning machine LM.

The generator G serves as the main environmental factor generating the *independently and identically distributed* (i.i.d) random vectors $\mathbf{x} \in \mathbf{X}$ according to some
unknown (but fixed) probability distribution function $F(\mathbf{x})$. In other words, the generator G determines the common framework in which the supervisor and the learning
machine act. The random vectors $\mathbf{x} \in \mathbf{X}$ are subsequently fed as inputs to the target
operator (supervisor S) which finally returns the output values y. It is important to
note that although there is no information concerning the transformation of input
vector to output values, it is known that the corresponding target operator exists and
does not change. The learning machine observes l pairs

$$(\mathbf{x}_1, y_1), \ldots, (\mathbf{x}_l, y_l) \tag{2.18}$$

(the training set) which contains input vectors $\mathbf{x}$ and the supervisor's response y.
During this period the learning machine constructs some operator which will be
used for prediction of the supervisor's answer y_i on an particular observation vector
$\mathbf{x}$ generated by the generator G. Therefore, the goal of the learning machine is to
construct an appropriate approximation. In order to be a mathematical statement, this
general scheme of learning from examples needs some clarification. First of all, it is
important to describe the kind of functions that are utilized by the supervisor. In this
monograph, it is assumed that the supervisor returns the output value y on the input
vector $\mathbf{x}$ according to a conditional distribution function $F(y|\mathbf{x})$ including the case
when the supervisor uses a function of the form $y = f(\mathbf{x})$. Thus, the learning machine
observes the training set, which is drawn randomly and independently according to
a joint distribution function $F(\mathbf{x}, y) = F(\mathbf{x})F(y|\mathbf{x})$ and by utilizing this training
set it constructs an approximation to the unknown operator. From a formal point

Fig. 2.2 Learning from
examples

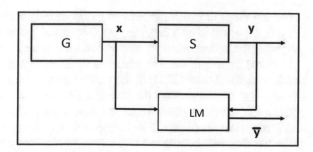

of view, the process of constructing an operator consists of developing a learning machine having the ability to implement some fixed set of functions given by the construction of the machine. Therefore, *the learning process is a process of choosing an appropriate function from a given set of functions.*

2.4.1 The Problem of Minimizing the Risk Functional from Empirical Data

Each time the problem of selecting a function with desired qualities arises, one should look in the set of possible functions for the one that satisfies the given quality criterion in the best possible way. Formally, this means that on a subset Z of the vector space $\mathbb{R}^n$, a set of admissible functions $\{g(\mathbf{z})\}$, $\mathbf{z} \in Z$, is given, and a functional

$$R = R(g(\mathbf{z})) \tag{2.19}$$

is defined as the criterion of quality for the evaluation of any given function. It is then required to find the function $g'(\mathbf{z})$ minimizing the functional (2.19) assuming that the minimum of the functional corresponds to the best quality and that the minimum of (2.19) exists in $\{g(\mathbf{z})\}$. In the case when the set of functions $\{g(\mathbf{z})\}$ and the functional $R(g(\mathbf{z}))$ were explicitly given, finding the function $g'(\mathbf{z})$ which minimizes (2.19) would a problem of the calculus of variations. In real-life problems, however, this is merely the case since the most common situation is that the risk functional is defined on the basis of a given probability distribution $F(\mathbf{z})$ defined on Z. Formally, the risk functional is defined as the mathematical expectation given by the following equation

$$R(g(\mathbf{z})) = \int L(\mathbf{z}, g(\mathbf{z})) dF(\mathbf{z}) \tag{2.20}$$

where the function $L(\mathbf{z}, g(\mathbf{z}))$ is integrable for any $g(\mathbf{z}) \in \{g(\mathbf{z})\}$. Therefore, the problem is to minimize the risk functional (2.20) in the case when the probability distribution $F(\mathbf{z})$ is unknown but the sample

$$\mathbf{z}_1, \ldots, \mathbf{z}_l \tag{2.21}$$

of observations drawn randomly and independently according to $F(\mathbf{z})$ is available.

It is important to note that there is a substantial difference between problems arising when the optimization process involves the direct minimization of the functional (2.19) and those encountered when the functional (2.20) is minimized on the basis of the empirical data (2.21). In the case of minimizing (2.19) the problem reduces to organizing the search for the function $g'(\mathbf{z})$ from the set $\{g(\mathbf{z})\}$ which minimizes (2.19). On the other hand, when the functional (2.20) is to be minimized on the basis of the empirical data (2.21), the problem reduces to formulating a constructive criterion that will be utilized in order to choose the optimal function rather than organizing the search of the functions in $\{g(\mathbf{z})\}$. Therefore, the question in the first case is: *How to obtain the minimum of the functional in the given set of functions?* On the other hand, in the second case the question is: *What should be minimized in order to select from the set $\{g(z)\}$ a function which will guarantee that the functional (2.20) is small?*

Strictly speaking, the direct minimization of the risk functional (2.20) based on the empirical data (2.21) is impossible based on the utilization of methods that are developed in optimization theory. This problem, however, lies within the core of mathematical statistics.

When formulating the minimization problem for the functional (2.20), the set of functions $g(\mathbf{z})$ will be given in a parametric form $\{g(\mathbf{z}, \alpha), \alpha \in \Lambda\}$. Here α is parameter from the set Λ such that the value $\alpha = \alpha^*$ defines the specific function $g(z, \alpha^*)$ in the set $g(z, \alpha)$. Therefore, identifying the required function is equivalent to determining the corresponding parameter $\alpha \in \Lambda$. The exclusive utilization of parametric sets of functions does not imply a restriction on the problem, since the set Λ, to which the parameter α belongs, is arbitrary. In other words, Λ can be a set of scalar quantities, a set of vectors, or a set of abstract elements. Thus, in the context of the new notation the functional (2.20) can be rewritten as

$$R(a) = \int Q(\mathbf{z}, \alpha) dF(\mathbf{z}), \alpha \in \Lambda, \tag{2.22}$$

where

$$Q(\mathbf{z}, \alpha) = L(\mathbf{z}, g(\mathbf{z}, \alpha)). \tag{2.23}$$

The function $Q(\mathbf{z}, \alpha)$ represents a *loss function* depending on the variables $\mathbf{z}$ and α.

The problem of minimizing the functional (2.22) may be interpreted in the following simple way: It is assumed that each function $Q(\mathbf{z}, \alpha)$, $\alpha \in \Lambda$ (e.g. each function of $\mathbf{z}$ for a fixed $\alpha = \alpha^*$), determines the amount of loss resulting from the realization of the vector $\mathbf{z}$. Thus, the *expected loss* (with respect to $\mathbf{z}$) for the function $Q(\mathbf{z}, \alpha^*)$ will be determined by the integral

$$R(\alpha^*) = \int Q(\mathbf{z}, \alpha^*) dF(\mathbf{z}). \tag{2.24}$$

This functional is the so-called *risk functional* or *risk*. The problem, then, is to choose in the set $Q(\mathbf{z}, \alpha)$, $\alpha \in \Lambda$, a function $Q(\mathbf{z}, \alpha_0$ which minimizes the risk when the probability distribution function is unknown but independent random observations $\mathbf{z}_1, \ldots, \mathbf{z}_l$ are given.

Let $\mathcal{P}_0$ be the set of all possible probability distribution functions on Z and $\mathcal{P}$ some subset of probability distribution functions from $\mathcal{P}_0$. In this context, the term "unknown probability distribution function", means that the only available information concerning $F(\mathbf{z})$ is the trivial statement that $F(\mathbf{z}) \in \mathcal{P}_0$.

2.4.2   Induction Principles for Minimizing the Risk Functional on Empirical Data

The natural problem that arises at this point concerns the minimization of the risk functional defined in Eq. (2.24) which is impossible to perform directly on the basis of an unknown probability distribution function $F(\mathbf{x})$ (which defines the risk). In order to address this problem Vapnik and Chervonenkis introduced a new induction principle, namely the principle of *Empirical Risk Minimization*. The principle of empirical risk minimization suggests that instead of minimizing the risk functional (2.22) one could alternatively minimize the functional

$$R_{emp}(\alpha) = \frac{1}{l} \sum_{i=1}^{l} Q(\mathbf{z}_i, \alpha), \qquad (2.25)$$

which is called the *empirical risk functional*. The empirical risk functional is constructed on the basis of the data $\mathbf{z}_1, \ldots, \mathbf{z}_l$ which are obtained according to the distribution $F(\mathbf{z})$. This functional is defined in explicit form and may be subject to direct minimization. Letting the minimum of the risk functional (2.22) be attained at $Q(\mathbf{z}, \alpha_0)$ and the minimum of the empirical risk functional (2.25) be attained at $Q(\mathbf{z}, \alpha_l)$, then the latter may be considered as an approximation to the function $Q(\mathbf{z}, \alpha_0)$. This principle of solving the empirical risk minimization problem is called the empirical risk minimization (induction) principle.

2.4.3   Supervised Learning

In supervised learning (or learning *with a teacher*), the available data are given in the form of input-output pairs. In particular, each data sample consists of a particular input vector and the related output value. The primary purpose of this learning paradigm is to obtain a concise description of the data by finding a function which yields the correct output value for a given input pattern. The term supervised learning stems from the fact that the objects under consideration are already associated with target values which can be either integer class identifiers or real values. Specifically, the type of the output values distinguishes the two branches of the supervised learning paradigm corresponding to the learning problems of *classification* and *regression*.

The Problem of Pattern Recognition

The problem of pattern recognition was formulated in the late 1950s. In essence, it can be formulated as follows: A supervisor observes occurring situations and determines to which of k classes each one of them belongs. The main requirement of the problem is to construct a machine which, after observing the supervisor's classification, realizes an approximate classification in the same manner as the supervisor. A formal definition of the pattern recognition learning problem could be obtained by considering the following statement: In a certain environment characterized by a probability distribution function $F(\mathbf{x})$, situation $\mathbf{x}$ appears randomly and independently. The supervisor classifies each one of the occurred situations into one of k classes. It is assumed that the supervisor carries out this classification by utilizing the conditional probability distribution function $F(\omega|\mathbf{x})$, where $\omega \in \{0, 1, \ldots, k-1\}$. Therefore, $\omega = p$ indicates that the supervisor assigns situation $\mathbf{x}$ the class p. The fundamental assumptions concerning the learning problem of pattern recognition is that neither the environment $F(\mathbf{x})$ nor the decision rule of the supervisor $F(\omega|\mathbf{x})$ are known. However, it is known that both functions exist meaning yielding the existence of the joint distribution $F(\omega, x) = F(\mathbf{x})F(\omega|\mathbf{x})$.

Let $\phi(\mathbf{x}, \alpha)$, $\alpha \in \Lambda$ be a set of functions which can take only k discrete values contained within the $\{0, 1, \ldots, k-1\}$ set. In this setting, by considering the simplest loss function

$$L(\omega, \phi) = \begin{cases} 0, & \text{if } \omega = \phi; \\ 1, & \text{if } \omega \neq \phi. \end{cases} \tag{2.26}$$

the problem of pattern recognition may be formulated as the minimization of the risk functional

$$R(\alpha) = \int L(\omega, \phi(\mathbf{x}, \alpha))dF(\omega, \mathbf{x}) \tag{2.27}$$

on the set of functions $\phi(\mathbf{x}, \alpha)$, $\alpha \in \Lambda$. The unknown distribution function $F(\omega, \mathbf{x})$ is implicitly described through a random independent sample of pairs

$$(\omega_1, \mathbf{x}_1), \ldots, (\omega_1, \mathbf{x}_l) \tag{2.28}$$

For the loss function (2.26), the functional defined in Eq. (2.27) determines the average probability of misclassification for any given decision rule $\phi(\mathbf{x}, \alpha)$. Therefore, the problem of pattern recognition reduces to the minimization of the average probability of misclassification when the probability distribution function $F(\omega, \mathbf{x})$ is unknown but the sample data (2.28) are given.

In this way, the problem of pattern recognition is reduced to the problem of minimizing the risk functional on the basis of empirical data. Specifically, the empirical risk functional for the pattern recognition problem has the following form:

$$R_{emp}(\alpha) = \frac{1}{l} \sum_{i=1}^{l} L(\omega_i, \phi(\mathbf{x}_i, \alpha)), \quad \alpha \in \Lambda. \tag{2.29}$$

The special feature of this problem, however, is that the set of loss functions $Q(\mathbf{z}, \alpha)$, $\alpha \in \Lambda$ is not as arbitrary as in the general case defined by Eq. (2.23). The following restrictions are imposed:

- The vector $\mathbf{z}$ consists of $n + 1$ coordinates: coordinate ω, which takes on only a finite number of values and n coordinates $(x^1, \ldots, x^n)$ which form the vector $\mathbf{x}$.

- The set of functions $Q(\mathbf{z}, \alpha)$, $\alpha \in \Lambda$, is given by $Q(\mathbf{z}, \alpha) = L(\omega, \phi(\mathbf{x}, \alpha))$, $\alpha \in \Lambda$ taking only a finite number of values.

This specific feature of the risk minimization problem characterizes the pattern recognition problem. In particular, the pattern recognition problem forms the simplest learning problem because it deals with the simplest loss function. The loss function in the pattern recognition problem describes a set of *indicator* functions, that is functions that take only binary values.

The Problem of Regression Estimation

The problem of regression estimation involves two sets of elements X and Y which are connected by a functional dependence. In other words, for each element $\mathbf{x} \in X$ there is a unique corresponding element $y \in Y$. This relationship constitutes a function when X is a set of vectors and Y is a set of scalars. However, there exist relationships (stochastic dependencies) where each vector $\mathbf{x}$ can be mapped to a number of different y's which are obtained as a result of random trials. This is mathematically described by considering the conditional distribution function $F(y|\mathbf{x})$, defined on Y, according to which the selection of the y values is realized. Thus, the function of the conditional probability expresses the stochastic relationship between y and $\mathbf{x}$.

Let the vectors $\mathbf{x}$ appear randomly and independently in accordance with a distribution function $F(\mathbf{x})$. Then, it is reasonable to consider that the y values are likewise randomly sampled from the *conditional distribution function $F(y|\mathbf{x})$*. In this case, the sample data points may be considered to be generated according to a *joint probability distribution function $F(\mathbf{x}, y)$*. The most intriguing aspect of the regression estimation problem is that the distribution functions $F(\mathbf{x})$ and $F(y|\mathbf{x})$ defining the joint distribution function $F(y, \mathbf{x}) = F(\mathbf{x})F(y|\mathbf{x})$ are *unknown*. Once again, the problem of regression estimation reduces to the approximation of the true joint distribution function $F(y|\mathbf{x})$ through a series of randomly and independently sampled data points of the following form

$$(y_1, \mathbf{x}_1), \dots, (y_l, \mathbf{x}_l). \tag{2.30}$$

However, the knowledge of the function $F(y, \mathbf{x})$ is often not required as in many cases it is sufficient to determine one of its characteristics, for example the function of the conditional mathematical expectation:

$$r(\mathbf{x}) = \int y F(y|\mathbf{x}) \tag{2.31}$$

This function is called the *regression* and the problem of its estimation in the set of functions $f(\mathbf{x}, \alpha)$, $\alpha \in \Lambda$, is referred to as the problem of regression estimation. Specifically, it was proved that the problem of regression estimation can be reduced to the model of minimizing risk based on empirical data under the following conditions:

$$\int y^2 dF(y, \mathbf{x}) < \infty \text{ and } \int r^2(\mathbf{x}) dF(y, \mathbf{x}) < \infty \tag{2.32}$$

Indeed, on the set $f(\mathbf{x}, \alpha)$ the minimum of the functional

$$R(\alpha) = \int (y - f(\mathbf{x}, \alpha))^2 dF(y, \mathbf{x}) \tag{2.33}$$

(provided that it exists) is attained at the regression function if the regression $r(\mathbf{x})$ belongs to $f(\mathbf{x}, \alpha)$, $\alpha \in \Lambda$. On the other hand, the minimum of this functional is attained at the function $f(\mathbf{x}, a^*)$, which is the closest to the regression $r(\mathbf{x})$ in the metric $L_2(P)$, defined as

$$L_2(f_1, f_2) = \sqrt{\int (f_1(\mathbf{x}) - f_2(\mathbf{x}))^2 dF(\mathbf{x})} \tag{2.34}$$

if the regression $r(\mathbf{x})$ does not belong to the set $f(\mathbf{x}, \alpha)$, $\alpha \in \Lambda$.

Thus, the problem of estimating the regression may be also reduced to the scheme of minimizing a risk functional on the basis of a given set of sample data. Specifically, the empirical risk functional for the regression estimation problem has the following form:

$$R_{emp}(\alpha) = \frac{1}{l} \sum_{i=1}^{l} (y_i - f(\mathbf{x}_i, \alpha))^2, \quad \alpha \in \Lambda \tag{2.35}$$

The specific feature of this problem is that the set of functions $Q(\mathbf{z}, \alpha)$, $\alpha \in \Lambda$, is subject to the following restrictions:

- The vector $\mathbf{z}$ consists of $n + 1$ coordinates: the coordinate y and the n coordinates $(x^1, \ldots, x^n)$ which form the vector $\mathbf{x}$. However, in contrast to the pattern recognition problem, the coordinate y as well as the function $f(\mathbf{x}, a)$ may take any value in the interval $(-\infty, \infty)$
- The set of loss functions $Q(\mathbf{z}, \alpha)$, $\alpha \in \Lambda$, is of the form $Q(\mathbf{z}, a) = (y - f(\mathbf{x}, \alpha))^2$.

2.4.4 Unsupervised Learning

If the data is only a sample of objects without associated target values, the problem is known as *unsupervised learning*. In unsupervised learning, there is no teacher. Hence, a concise description of the data can be a set of clusters or a probability density stating how likely it is to observe a certain object in the future. The primary objective of unsupervised learning is to extract some structure from a given sample of training objects.

The Problem of Density Estimation

Let $p(\mathbf{x}, \alpha)$, $\alpha \in \Lambda$, be a set of probability densities containing the required density

$$p(\mathbf{x}, \alpha_0) = \frac{dF(\mathbf{x})}{d\mathbf{x}}. \tag{2.36}$$

It was shown that the minimum of the risk functional

$$R(\alpha) = \int \ln p(\mathbf{x}, \alpha) dF(\mathbf{x}) \tag{2.37}$$

(if it exists) is attained at the functions $p(\mathbf{x}, \alpha^*)$ which may differ from $p(\mathbf{x}, \alpha_0)$ only on a set of zero measure. Specifically, Bretagnolle and Huber [1] proved the following inequality

$$\int |p(\mathbf{x}, \alpha)) - p(\mathbf{x}, \alpha_0))| d\mathbf{x} \le 2\sqrt{R(\alpha) - R(\alpha_0)} \tag{2.38}$$

according to which the problem of estimating the density in L_1 is reduced to the minimization of the functional (2.37) on the basis of empirical data. The general form of the L_p metric in a k-dimensional metric space is given by the following equation

$$\|x\|_p = \left(\sum_{i=1}^{k} |x_i|^p \right)^{\frac{1}{p}} \tag{2.39}$$

for $1 \le p < \infty$ and $x \in \mathbb{R}^k$. In particular, the corresponding empirical risk functional has the following form

$$R_{emp}(\alpha) = - \sum_{i=1}^{l} \ln p(\mathbf{x}_i, \alpha) \tag{2.40}$$

The special feature of the density estimation problem is that the set of functions $Q(\mathbf{z}, \alpha)$ is subject to the following restrictions:

- The vector $\mathbf{z}$ coincides with the vector $\mathbf{x}$,
- The set of functions $Q(\mathbf{z}, \alpha)$, $\alpha \in \Lambda$, is of the form $Q(\mathbf{z}, \alpha) = -\log p(\mathbf{x}, \alpha)$, where $p(\mathbf{x}, \alpha)$ is a set of density functions. The loss function $Q(\mathbf{z}, \alpha)$ takes on arbitrary values on the interval $(-\infty, \infty)$.

Clustering

A general way to represent data is to specify a similarity between any pair of objects. If two objects share much structure, is should be possible to reproduce the data from the same *prototype*. This is the primary idea underlying *clustering methods* which form a rich subclass of the unsupervised learning paradigm. Clustering is one of the most primitive mental activities of humans, which is used in order to handle the huge amount of information they receive every day. Processing every piece of information as a single entity would be impossible. Thus, humans tend to categorize entities (i.e. objects, persons, events) into clusters. Each cluster is then characterized by the common attributes of the entities it contains.

The definition of clustering leads directly to the definition of a single "cluster". Many definitions have been proposed over the years, but most of them are based on loosely defined terms such as "similar" and "alike" or are oriented to a specific kind of clusters. Therefore, the majority of the proposed definitions for clustering are of vague or of circular nature. This fact reveals that it is not possible to provide a universally accepted formal definition of clustering. Instead, one can only provide an intuitive definition stating that given a fixed number of clusters, the clustering procedure aims at finding a grouping of objects (*clustering*) such that similar objects will be assigned to same group (*cluster*). Specifically, if there exists a partitioning of the original data set such that the similarities of the objects in one cluster are much greater than the similarities among objects from different clusters, then it is possible to extract structure from the given data. Thus, it is possible to represent a whole cluster by one representative data point. More formally, by letting

$$X = \{\mathbf{x}_1, \ldots, \mathbf{x}_l\} \tag{2.41}$$

be the original set of available data the m - *clustering* $\mathcal{R}$ of X may be defined as the partitioning of X into m sets (*clusters*) $C_1, \ldots, C_m$ such that the following three conditions are met:

- $C_i \neq \emptyset,\ i = 1, \ldots, m$
- $\cup_{i=1}^{m} C_i = X$
- $C_i \cap C_j = \emptyset,\ i \neq j,\ i, j = 1, \ldots, m$

It must be noted that the data points contained in a cluster C_i are more "similar" to each other and less similar to the data points of the other clusters. The quantification, however, of the terms "similar" and "dissimilar" is highly dependent on the type of the clusters involved. The type of the clusters is determinately affected by the shape of the clusters which in tern depends on the particular measure of *dissimilarity* or *proximity* between clusters.

2.4.5 Reinforcement Learning

Reinforcement learning is learning how to map situations to actions in order to maximize a numerical reward signal. The learner is not explicitly told which actions to take, as in most forms of machine learning, but instead must discover which actions yield the most reward by trying them. In most interesting and challenging cases, actions may affect not only the immediate reward, but also the next situation and, through that, all subsequent rewards. These two characteristics (trial and error, and delayed reward) are the most important distinguishing features of reinforcement learning. Reinforcement learning does not define a subclass of learning algorithms, but rather a category of learning problems which focuses on designing learning agents which cope with real-life problems. The primary features of such problems involves the interaction of the learning agents with their environments in order to achieve a particular goal. Clearly, this kind of agents must have the ability to sense the state of the environment to some extent and must be able to take actions affecting that state. The agent must also have a goal or goals relating to the state of the environment.

Reinforcement learning is different from the classical supervised learning paradigm where the learner is explicitly instructed by a knowledgable external supervisor through a series of examples that indicate the desired behavior. This is an important kind of learning, but it is not adequate on its own to address the problem of learning from interaction. In interactive problems, it is often impractical to obtain correct and representative examples of all the possible situations in which the agent has to act. In uncharted territory, where one would expect learning to be more beneficial, an agent must be able to learn from its own experience.

One of the challenges that arises in reinforcement learning and not in other kinds of learning is the tradeoff between exploration and exploitation. To obtain a lot of reward, a reinforcement learning agent must prefer actions that it has tried in the past and found effective in producing reward. However, the discovery of such actions requires that the agent has to try actions that he has not selected before. In other words, the agent has to *exploit* what is already known in order to obtain reward, but it is also important to *explore* new situations in order to make better action selections in the future. The dilemma is that neither exploitation nor exploration can be pursued exclusively without failing at the task. The agent must try a variety of actions and progressively favor those that appear to be best. Moreover, when the learning problems involves a stochastic task, each action must be tried many times in order to reliably estimate the expected reward.

Another key feature of reinforcement learning is that it explicitly considers the *whole* problem of a goal-directed agent interacting with an uncertain environment. This is in contrast with many learning approaches that address subproblems without investigating how they fit into a larger picture. Reinforcement learning, on the other hand, starts with a complete, interactive goal-seeking agent. All reinforcement learning agents have explicit goals, can sense aspects of their environments, and can choose actions to influence their environments. Moreover, it is usually assumed from the beginning that the agent has to operate despite significant uncertainty about the environment it faces. For learning research to make progress, important subproblems have to be isolated and studied, but they should be incorporated in the larger picture as subproblems that are motivated by clear roles in complete, interactive, goal-seeking agents, even if all the details of the complete agent cannot yet be filled in.

2.5   Theoretical Justifications of Statistical Learning Theory

Statistical learning theory provides the theoretical basis for many of today's machine learning algorithms and is arguably one of the most beautifully developed branches of artificial intelligence in general. Providing the basis of new learning algorithms, however, was not the only motivation for the development of statistical learning theory. It was just as much a philosophical one, attempting to identify the fundamental element which underpins the process of drawing valid conclusions from empirical data.

The best-studied problem in machine learning is the problem of classification. Therefore, the theoretical justifications concerning Statistical Learning Theory will be analyzed within the general context of supervised learning and specifically pattern classification. The pattern recognition problem, in general, deals with two kind of spaces: the input space $\mathbf{X}$, which is also called the space of *instances*, and the output space $\mathbf{Y}$, which is also called the *label* space. For example, if the learning task is to classify certain objects into a given, finite set of categories, then $\mathbf{X}$ consists of the space of all possible objects (instances) in a certain, fixed representation, while $\mathbf{Y}$ corresponds to the discrete space of all available categories such that $\mathbf{Y} = \{0, \ldots, k-1\}$. This discussion, however, will be limited to the case of binary classification for simplicity reasons which yields that the set of available categories will be restricted to $\mathbf{Y} = \{-1, +1\}$. Therefore, the problem of classification may be formalized as the procedure of estimating a functional dependence of the form $\phi : \mathbf{X} \rightarrow \mathbf{Y}$, that is a relationship between input and output spaces $\mathbf{X}$ and $\mathbf{Y}$ respectively. Moreover, this procedure is realized on the basis of a given set of *training examples* $(\mathbf{x}_1, y_1), \ldots, (\mathbf{x}_l, y_l)$, that is pairs of objects with the associated category label. The primary goal when addressing the pattern classification problem is to find such a mapping that yields the smallest possible number of classification errors. In other words, the problem of pattern recognition is to find that mapping for which the number of objects in $\mathbf{X}$ that are assigned to the wrong category is as small as possible. Such a mapping is referred to as a *classifier*. The procedure for determining such a mapping on the basis of a given set of training examples is referred to as a *classification algorithm* or *classification rule*. A very important issue concerning the definition of the pattern recognition problem is that no particular assumptions are made on the spaces $\mathbf{X}$ and $\mathbf{Y}$. Specifically, it is assumed that there exists a *joint distribution function F* on $\mathbf{Z} = \mathbf{X} \times \mathbf{Y}$ and that the training examples $(\mathbf{x}_i, y_i)$ are sampled independently from this distribution F. This type of sampling is often denoted as *iid* (independently and identically distributed) sampling.

It must be noted that any particular discrimination function $\phi(\mathbf{x})$ is parameterized by a unique parameter $\alpha \in \Lambda_{all}$ which can be anything from a single parameter value to a multidimensional vector. In other words, Λ_{all} denotes the set of all measurable functions from $\mathbf{X}$ to $\mathbf{Y}$ corresponding to the set of all possible classifiers for a given pattern recognition problem. Of particular importance is the so-called Bayes Classifier $\phi_{Bayes}(\mathbf{x})$, identified by the parameter α_{Bayes}, whose discrimination function has the following form

$$\phi_{Bayes}(\mathbf{x}, \alpha_{Bayes}) = \arg \min_{\omega \in \mathbf{Y}} P(Y = \omega | X = \mathbf{x}). \qquad (2.42)$$

The Bayes classifier operates by assigning any given pattern to the class with the maximum a posteriori probability. The direct computation of the Bayes classifier, however, is impossible in practice since the underlying probability distribution is completely unknown to the learner. Therefore, the problem of pattern recognition may be formulated as the procedure of constructing a function $\phi(\mathbf{x}, \alpha) : \mathbf{X} \to \mathbf{Y}$, uniquely determined by the parameter α, through a series of training points $(\mathbf{x}_1, y_1), \ldots, (\mathbf{x}_l, y_l)$ which has risk $R(\alpha)$ as close as possible to the risk $R(\alpha_{Bayes})$ of the Bayes classifier.

2.5.1 Generalization and Consistency

Let $(\mathbf{x}_1, y_1), \ldots, (\mathbf{x}_l, y_l)$ be a sequence of training patterns and α_l be the function parameter corresponding to the classifier obtained by the utilization of some learning algorithm on the given training set. Even though it is impossible to compute the true underlying risk $R(\alpha_l)$ for this classifier according to Eq. (2.27), it is possible to estimate the empirical risk $R_{emp}(\alpha_l)$ according to Eq. (2.29) accounting for the number of errors on the training points.

Usually, for a classifier α_n trained on a particular training set, the empirical risk $R_{emp}(\alpha_l)$ is relatively small since otherwise the learning algorithm will not even seem to be able to explain the training data. A natural question arising at this point is whether a function α_l which makes a restricted number of errors on the training set will perform likewise on the rest of the $\mathbf{X}$ space. This question is intimately related to the notion of *generalization*. Specifically, a classifier α_l is said to generalize well if the difference $|R(\alpha_l) - R_{emp}(\alpha_l)|$ is small. This definition, however, does not imply that the classifier α_l will have a small overall error R_{emp}, but it just means that the empirical error $R_{emp}(\alpha_l)$ is a good estimate of the true error $R(\alpha_l)$. Particularly bad in practice is the situation where $R_{emp}(\alpha_l)$ is much smaller than $R(\alpha_l)$ misleading to the assumption of being overly optimistic concerning the quality of the classifier.

The problem concerning the generalization ability of a given machine learning algorithm may be better understood by considering the following regression example. One is given a set of observations $(\mathbf{x}_1, y_1), \ldots, (\mathbf{x}_l, y_l) \in \mathbf{X} \times \mathbf{Y}$, where for simplicity it is assumed that $\mathbf{X} = \mathbf{Y} = \mathbb{R}$. Figure 2.3 shows a plot of such a dataset, indicated by the round points, along with two possible functional dependencies that could underlly the data.

The dashed line α_{dashed} represents a fairly complex model that fits the data perfectly resulting into a zero training error. The straight line, on the other hand, does not completely explain the training data, in the sense that there are some residual errors, leading to a small training error. The problem regarding this example concerns the inability to compute the true underlying risks $R(\alpha_{dashed})$ and $R(\alpha_{straight})$ since the two possible functional dependencies have very different behavior. For example, if the straight line classifier $\alpha_{straight}$ was the true underlying risk, then the dashed line classifier α_{dashed} would have a high true risk, as the L_2 distance between the true and the estimated function is very large. The same also holds when the true functional dependence between the spaces $\mathbf{X}$ and $\mathbf{Y}$ is represented by the dashed line while the straight line corresponds to the estimated functional dependency. In both cases, the true risk would be much higher than the empirical risk.

This example emphasizes the need to make the correct choice between a relatively complex function model, leading to a very small training error, and a simpler function

Fig. 2.3 Regression
example

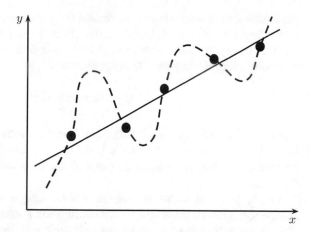

model at the cost of a slightly higher training error. In one form or another, this issue
was extensively studied within the context of classical statistics as the *bias-variance*
dilemma. The bias-variance dilemma involves the following dichotomy. If a linear
fit is computed for any given data set, then every functional dependence discovered
would be linear but as a consequence of the *bias* imposed from the choice of the
linear model which does not necessarily comes from the data. On the hand, if a
polynomial model of sufficiently high degree is fit for any given data set, then the
approximation ability of the model would fit the data perfectly but it would suffer
from a large variance depending on the initial accuracy of the measurements. In
other words, within the context of applied machine learning, complex explanations
show *overfitting*, while overly simple explanations imposed by the learning machine
design lead to *underfitting*. Therefore, the concept of generalization can be utilized
in order to determine the amount of increase in the training error in order to tolerate
for a fitting a simpler model and quantify the way in which a given model is simpler
than another one.

Another concept, closely related to generalization, is the one of consistency. However, as opposed to the notion of generalization discussed above, consistency is not
a property of an individual function, but a property of a set of functions. The notion
of consistency, as it is described in classical statistics, aims at making a statement
about what happens in the limit of infinitely many sample points. Intuitively, it seems
reasonable to request that a learning algorithm, when presented with more and more
training points, should eventually converge to an optimal solution.

Given any particular classification algorithm and a set of l training points, α_l
denotes the parameter identifying the obtained classifier where the exact procedure
for its determination is not of particular importance. Note that any classification
algorithm chooses its functions from some particular function space identified by
the complete parameter space Λ such that $\mathcal{F} = \{\phi(\mathbf{x}, \alpha) : \alpha \in \Lambda\}$. For some
algorithms this space is given explicitly, while for others it only exists implicitly via
the mechanism of the algorithm. No matter how the parameter space Λ is defined,

the learning algorithm attempts to choose the parameter $\alpha_l \in \Lambda$ which it considers as the best classifier in Λ, based on the given set of training points. On the other hand, in theory the best classifier in Λ is the one that has the smallest risk which is uniquely determined by the following equation:

$$\alpha_\Lambda = \arg \min_{\alpha \in \Lambda} R(\alpha). \tag{2.43}$$

The third classifier of particular importance is the Bayes classifier α_{Bayes} introduced in Eq. (2.42). Bayes classifier, while being the best existing classifier, it may be not be included within the parameter space Λ under consideration, so that $R(\alpha_\Lambda) > R(\alpha_{Bayes})$.

Let $(\mathbf{x}_i, y_i)_{i \in \mathbb{N}}$ be an infinite sequence of training points which have been drawn independently from some probability distribution P and, for each $l \in \mathbb{N}$, let α_l be a classifier constructed by some learning algorithm on the basis of the first l training points. The following types of consistency may be defined:

1. The learning algorithm is called *consistent with respect to Λ and P* if the risk $R(\alpha_l)$ converges in probability to the risk $R(\alpha_\Lambda)$ of the best classifier Λ, that is for all $\epsilon > 0$,

$$P(R(\alpha_l) - R(\alpha_\Lambda) > \epsilon) \to 0 \text{ as } n \to \infty \tag{2.44}$$

2. The learning algorithm is called *Bayes-consistent with respect to P* if the risk $R(R(\alpha_l))$ converges to the risk $R(\alpha_{Bayes})$ of the Bayes classifier, that is for all $\epsilon > 0$,

$$P(R(\alpha_l) - R(\alpha_{Bayes}) > \epsilon) \to 0 \text{ as } n \to \infty \tag{2.45}$$

3. The learning algorithm is called universally consistent with respect to Λ (*resp universally Bayes-consistent*) if it is consistent with respect to Λ (*reps Bayes-consistent*) for all probability distributions P.

It must be noted that none of the above definitions involves the empirical risk $R_{emp}(\alpha_l)$ of a classifier. On the contrary, they exclusively utilize the true risk $R(\alpha_l)$ as a quality measure reflecting the need to obtain a classifier which is as good as possible. The empirical risk constitutes the most important estimator of the true risk of a classifier so that the requirement involving the convergence of the true risk $(R(\alpha_l) \to R(\alpha_{Bayes}))$ should be extended to the convergence of the empirical risk $(R_{emp}(\alpha_l) \to R(\alpha_{Bayes}))$.

2.5.2 Bias-Variance and Estimation-Approximation Trade-Off

The goal of classification is to get a risk as close as possible to the risk of the Bayes classifier. A natural question that arises concerns the possibility of choosing the complete parameter space Λ_{all} as the parameter space Λ utilized by a particular classifier. This question raises the subject of whether the selection of the overall best classifier, obtained in the sense of the minimum empirical risk,

$$\alpha_l = \arg \min_{\alpha \in \Lambda_{all}} R_{emp}(\alpha) \tag{2.46}$$

implies consistency. The answer for this question is unfortunately negative since the optimization of a classifier over too large parameter (function) spaces, containing all the Bayes classifiers for all probability distributions P, will lead to inconsistency. Therefore, in order to learn successfully it is necessary to work with a smaller parameter (function) space Λ.

Bayes consistency deals with the convergence of the term $R(\alpha_l) - R(\alpha_{Bayes})$ which can be decomposed in the following form:

$$R(\alpha_l) - R(\alpha_{Bayes}) = \underbrace{(R(\alpha_l) - R(\alpha_\Lambda))}_{estimation \quad error} + \underbrace{(R(\alpha_\Lambda) - R(\alpha_{Bayes}))}_{approximation \quad error} \tag{2.47}$$

The first term on the right hand side is called the *estimation error* while the second term is called the *approximation error*. The first term deals with the uncertainty introduced by the random sampling process. That is, given a finite sample, it is necessary to estimate the best parameter (function) in Λ. Of course, in this process there will be a hopefully small number of errors which is identified by the term estimation error. The second term, on the other hand, is not influenced by random qualities. It particularly deals with the error made by looking for the best parameter (function) in a small parameter (function) space Λ, rather than looking for the best parameter (function) in the entire space Λ_{all}. Therefore, the fundamental question in this context is how well parameters (functions) in Λ can be used to approximate parameters (functions) in Λ_{all}.

In statistics, the estimation error is also called the *variance*, and the approximation error is called the *bias* of an estimator. The first term measures the variation of the risk of the function corresponding to the parameter α_l estimated on the sample, while the second one measures the bias introduced in the model by choosing a relatively small function class.

In this context the parameter space Λ may be considered as the means to balance the trade-off between estimation and approximation error. This is particularly illustrated in Fig. 2.4 which demonstrates that the selection of a very large parameter space Λ yields a very small approximation error term since there is high probability that the Bayes classifier will be contained in Λ or at least it can be closely approximated by some element in Λ. The estimation error, however, will be rather large in this case since the space Λ will contain more complex functions which will lead to overfitting. The opposite effect will happen if the function class corresponding to the parameter space Λ is very small.

The trade-off between estimation and approximation error is explicitly depicted in Fig. 2.5. According to the graph, when the parameter space Λ corresponds to a small complexity function space utilized by the classification algorithm, then the

Fig. 2.4 Illustration of
estimation and
approximation error

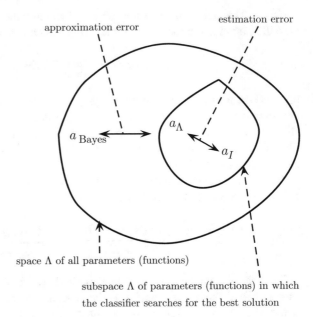

space Λ of all parameters (functions)

subspace Λ of parameters (functions) in which
the classifier searches for the best solution

Fig. 2.5 Trade-off between
estimation and
approximation error

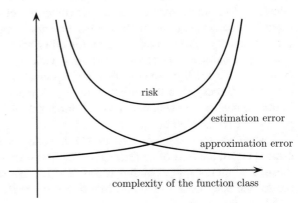

estimation error will be small but the approximation error will be large (underfitting).
On the other hand, if the complexity of Λ is large, then the estimation error will also
be large, while the approximation error will be small (overfitting). The best overall
risk is achieved for "moderate" complexity.

2.5.3 Consistency of Empirical Minimization Process

As discussed in Sect. 2.4.2, the ERM principle provides a more powerful way of
classifying data since it is impossible to directly minimize the true risk functional

given by Eq. (2.27). In particular, the ERM principle addresses the problem related with the unknown probability distribution function $F(\omega, \mathbf{x})$ which underlies the data generation process by trying to infer a function $f(\mathbf{x}, \alpha)$ from the set of identically and independently sampled training data points. The process of determining this function is based on the minimization of the so-called empirical risk functional which for the problem of pattern classification is given by Eq. (2.29).

The fundamental underpinning behind the principle of Empirical Risk Minimization is the law of large numbers which constitutes one of the most important theorems in statistics. In its simplest form it states that under mild conditions, the mean of independent, identically-distributed random variables ξ_i, which have been drawn from some probability distribution function P of finite variance, converges to the mean of the underlying distribution itself when the sample size goes to infinity:

$$\frac{1}{l} \sum_{i=1}^{l} \xi_i \rightarrow E(\xi) \text{ for } l \rightarrow \infty. \tag{2.48}$$

A very important extension to the law of large numbers was originally provided by the Chernoff inequality (Chernoff 1952) which was subsequently generalized by Hoeffding (Hoeffding 1963). This inequality characterizes how well the empirical mean approximates the expected value. Namely, if ξ_i, are random variables which only take values in the [0, 1] interval, then

$$P\left(\left|\frac{1}{l} \sum_{i=1}^{l} \xi_i - E(\xi)\right| \geq \epsilon\right) \leq \exp(-2l\epsilon^2). \tag{2.49}$$

This theorem can be applied to the case of the empirical and the true risk providing a bound which states how likely it is that for a given function, identified by the parameter α, the empirical risk is close to the actual risk:

$$P(|R_{emp}(\alpha) - R(\alpha)| \geq \epsilon) \leq \exp(-2l\epsilon^2) \tag{2.50}$$

The most important fact concerning the bound provided by Chernoff in Eq. (2.50) is its probabilistic nature. Specifically, it states that the probability of a large deviation between the test error and the training error of a function $f(\mathbf{x}, \alpha)$ is small when the sample size is sufficiently large. However, by not ruling out the presence of cases where the deviation is large, it just says that for a fixed function $f(\mathbf{x}, \alpha)$, this is very unlikely to happen. The reason why this has to be the case is the random process that generates the training samples. Specifically, in the unlucky cases when the training data are not representative of the true underlying phenomenon, it is impossible to infer a good classifier. However, as the sample size gets larger, such unlucky cases become very rare. Therefore, any consistency guarantee can only be of the form "the empirical risk is close to the actual risk, with high probability".

Another issue related to the ERM principle is that the Chernoff bound in (Eq. 2.50) is not enough in order to prove the consistency of the ERM process. This is true since the Chernoff inequality holds only for a fixed function $f(\mathbf{x}, \alpha)$ which does not depend on the training data. While this seems to be a subtle mathematical difference, this is where the ERM principle can go wrong as the classifier α_l does depend on the training data.

2.5.4   Uniform Convergence

It turns out the conditions required to render the ERM principle consistent involve *restricting the set of admissible functions*. The main insight provided by the VC theory is that the consistency of the ERM principle is determined by the *worst case* behavior over all functions $f(\mathbf{x}, \alpha)$, where $\alpha \in \Lambda$, that the learning machine could use. This worst case corresponds to a version of the law of large numbers which is *uniform* over all functions parameterized by Λ.

A simplified description of the uniform law of large numbers which specifically relates to the consistency of the learning process is given in Fig. 2.6. Both the empirical and the actual risk are plotted as functions of the α parameter and the set of all possible functions, parameterized by the set Λ, is represented by a single axis of the plot for simplicity reasons. In this context, the ERM process consists of picking the parameter α that yields the minimal value of R_{emp}. This process is consistent if the minimum of R_{emp} converges to that of R as the sample size increases. One way to ensure the convergence of the minimum of all functions in Λ is uniform convergence over Λ. Uniform convergence over Λ requires that for all functions $f(\mathbf{x}, \alpha)$, where $\alpha \in \Lambda$, the difference between $R(\alpha)$ and $R_{emp}(\alpha)$ should become small *simultaneously*. In other words, it is required that there exists some large l such that for sample size at least n, it is certain that for all functions $f(\mathbf{x}, \alpha)$, where $\alpha \in \Lambda$, the difference

Fig. 2.6 Convergence of the empirical risk to the actual risk

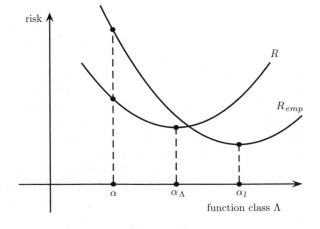

$|R(\alpha) - R_{emp}(\alpha)|$ is smaller than a given ϵ. Mathematically, this statement can be expressed using the following inequality:

$$\sup_{\alpha \in \Lambda} |R(\alpha) - R_{emp}(\alpha)| \leq \epsilon. \tag{2.51}$$

In Fig. 2.6, this means that the two plots of R and R_{emp} become so close that their distance is never larger than ϵ. This, however, does not imply that in the limit of infinite sample sizes, the *minimizer* of the empirical risk, α_l, will lead to a value of the risk that is as good as the risk of the best function, α_Λ, in the function class. The latter is true when uniform convergence is imposed over all functions that are parameterized by Λ. Intuitively it is clear that if it was known that for all functions $f(\mathbf{x}, \alpha)$, where $\alpha \in \Lambda$, the difference $|R(\alpha) - R_{emp}(\alpha)|$ is small, then this holds in particular for any function identified by the parameter α_l that might have been chosen based on the given training data. That is, for any function $f(\mathbf{x}, a)$, where $\alpha \in \Lambda$, it is true that:

$$|R(\alpha) - R_{emp}(\alpha)| \leq \sup_{\alpha \in \Lambda} |R(\alpha) - R_{emp}(\alpha)|. \tag{2.52}$$

Inequality (2.52) also holds for any particular function parameter α_l which has been chosen on the basis of a finite sample of training points. Therefore, the following conclusion can be drawn:

$$P(|R(\alpha) - R_{emp}(\alpha)| \geq \epsilon) \leq P(\sup_{\alpha \in \Lambda} |R(\alpha) - R_{emp}(\alpha)| \geq \epsilon), \tag{2.53}$$

where the quantity on the right hand side represents the very essence of the uniform law of large numbers. In particular, the law of large numbers is said to uniformly hold over a function class parameterized by Λ if for all $\epsilon > 0$,

$$P(\sup_{\alpha \in \Lambda} |R(\alpha) - R_{emp}(\alpha)| \geq \epsilon) \to 0 \text{ as } l \to \infty. \tag{2.54}$$

Inequality (2.52) can be utilized in order to show that if the uniform law of large numbers holds for some function class parameterized by Λ, then the ERM is consistent with respect to Λ. Specifically, Inequality (2.52) yields that:

$$|R(\alpha_l) - R(\alpha_\Lambda)| \leq 2 \sup_{\alpha \in \Lambda} |R(\alpha) - R_{emp}(\alpha)|, \tag{2.55}$$

which finally concludes:

$$P(|R(\alpha_l) - R(\alpha_\Lambda)| \geq \epsilon) \leq P(\sup_{\alpha \in \Lambda} |R(\alpha) - R_{emp}(\alpha)| \geq \frac{\epsilon}{2}). \tag{2.56}$$

The right hand side of Inequality (2.56) tends to 0, under the uniform law of large numbers, which then leads to consistency of the ERM process with respect to the underlying function class parameterized by Λ. Vapnik and Chervonenkis [5] proved that uniform convergence as described by Inequality (2.54) is a necessary and sufficient condition for the consistency of the ERM process with respect to Λ. It must be noted that the condition of uniform convergence crucially depends on the set of functions for which it must hold. Intuitively, it seems clear that the larger the function space parameterized by Λ, the larger the quantity $\sup_{\alpha \in \Lambda} |R(\alpha) - R_{emp}(\alpha)|$. Thus, the larger Λ, the more difficult it is to satisfy the uniform law of large numbers. That is, for larger function spaces (corresponding to larger parameter spaces Λ) consistency is harder to achieve than for smaller function spaces. This abstract characterization of consistency as a uniform convergence property, whilst theoretically intriguing, is not at all that useful in practice. This is true, since in practice it is very difficult to infer whether the uniform law of large numbers holds for a given function space parameterized by Λ. Therefore, a natural question that arises at this point is whether there are properties of function spaces which ensure uniform convergence of risks.

2.5.5   Capacity Concepts and Generalization Bounds

Uniform convergence was referred to as the fundamental property of a function space determining the consistency of the ERM process. However, a closer look at this convergence is necessary in order to make statements concerning the behavior of a learning system when it is exposed to a limited number of training samples. Therefore, attention should be focused on the probability

$$P(\sup_{\alpha \in \Lambda} |R(\alpha) - R_{emp}(\alpha)| > \epsilon) \tag{2.57}$$

which will not only provide insight into which properties of function classes determines the consistency of the ERM process, but will also provide bounds on the risk. Along this way, two notions are of primary importance, namely:

1. the *union bound* and
2. the method of *symmetrization* by a ghost sample.

The Union Bound

The union bound is a simple but convenient tool in order to transform the standard law of large numbers of individual functions into a uniform law of large numbers over a set of finitely many functions parameterized by a set $\Lambda = \{\alpha_1, \alpha_2, \ldots, \alpha_m\}$. Each of the functions $\{f(\mathbf{x}, \alpha_i) : \alpha_i \in \Lambda\}$, satisfies the standard law of large numbers in the form of a Chernoff bound provided by Inequality (2.50), that is

$$P(|R(\alpha_i) - R_{emp}(\alpha_i)| \geq \epsilon) \leq 2 \exp(-2l\epsilon^2). \tag{2.58}$$

In order to transform these statements about the individual functions $\{f(x, \alpha_i) : \alpha_i \in \Lambda\}$ into a uniform law of large numbers the following derivations are necessary

$$P(\sup_{\alpha \in \Lambda} |R(\alpha) - R_{emp}(\alpha)| > \epsilon) = P\left(\bigcup_{i=1}^{m} |R(\alpha_i) - R_{emp}(\alpha_i)| > \epsilon\right)$$

$$\leq \sum_{i=1}^{m} P(|R(\alpha_i) - R_{emp}(\alpha_i)| > \epsilon)$$

$$\leq 2m \exp(-2l\epsilon^2) \tag{2.59}$$

It is clear that the difference between the Chernoff bound given by Inequality (2.50) and the right hand side of Eq. (2.59) is just a factor of m. Specifically, if the function space $\mathcal{F} = \{f(\mathbf{x}, \alpha_i) : \alpha_i \in \Lambda\}$ is fixed, this factor can be regarded as a constant and the term $2m \exp(-2l\epsilon^2)$ still converges to 0 as $l \to \infty$. Hence, the empirical risk converges to 0 uniformly over $\mathcal{F}$ as $l \to \infty$. Therefore, it is proved that an ERM process over a finite set Λ of function parameters is consistent with respect to Λ.

Symmetrization

Symmetrization is an important technical step towards using capacity measures of function classes. Its main purpose is to replace the event $\sup_{\alpha \in \Lambda} |R(\alpha) - R_{emp}(\alpha)|$ by an alternative event which can be solely computed on a given sample size. Assume that a new *ghost sample* $\{(\mathbf{x}'_i, y'_i)\}_{i=1}^{l}$ is added to the initial training sample $\{(\mathbf{x}_i, y_i)\}_{i=1}^{l}$. The ghost sample is just another sample which is also drawn iid from the same underlying distribution and which is independent of the first sample. The ghost sample, however, is a mathematical tool that is not necessary to be physically sampled in practice. It is just an auxiliary set of training examples where the corresponding empirical risk will be denoted by $R'_{emp}(\alpha)$. In this context, Vapnik and Chervonenkis [6] proved that for $m\epsilon^2 \geq 2$

$$P(\sup_{\alpha \in \Lambda} |R(\alpha) - R_{emp}(\alpha)| > \epsilon) \leq 2P\left(\sup_{\alpha \in \Lambda} |R_{emp}(\alpha) - R'_{emp}(\alpha)| > \frac{\epsilon}{2}\right). \tag{2.60}$$

Here, the first P refers to the distribution of an iid sample l, while the second one refers to the distribution of two samples of size l, namely the original sample and the ghost one which form an iid sample of size $2l$. In the latter case, R_{emp}, measures the empirical loss on the first half of the sample, and R'_{emp} on the second half. This statement is referred to as the *symmetrization lemma* referring to the fact that the attention is focused on an event which depends on a symmetric way on a sample of size l. Its meaning is that if the empirical risks of two independent l-samples are close to each other, then they should also be close to the true risk. The main purpose of this lemma is to provide a way to bypass the need to directly estimate the quantity $R(\alpha)$ by computing the quantity $R'_{emp}(\alpha)$ on a finite sample size.

In the previous section, the uniform bound was utilized as a means to constraint the probability of uniform convergence in terms of a probability of an event referring to a finite function class. The crucial observation is now that even if Λ parameterizes an infinite function class, the different ways in which it can classify a training set of l sample points is finite. Namely, for any given training point in the training sample, a function can take only values within the set $\{-1, +1\}$ which entails that on a sample of l points $\{\mathbf{x}_1, \ldots, \mathbf{x}_l\}$, a function can act in at most 2^l different ways. Thus, even for an infinite function parameter class Λ, there are at most 2^l different ways the corresponding functions can classify the l points of finite sample. This means that when considering the term $\sup_{\alpha \in \Lambda} |R_{emp}(\alpha) - R'_{emp}(\alpha)|$, the supremum effectively runs over a finite set of function parameters. In this context, the supremum over Λ on the right hand side of Inequality (2.60) can be replaced by the supremum over a finite function parameter class with at most 2^{2l} function parameters. This number comes as a direct consequence from the fact that there is a number of $2l$ sample points for both the original and the ghost samples.

The Shattering Coefficient

For the purpose of bounding the probability in Eq. (2.57), the symmetrization lemma implies that the function parameter class Λ is effectively finite since it can be restricted to the $2l$ points appearing on the right hand side of Inequality (2.60). Therefore, the function parameter class contains a maximum number of 2^{2l} elements. This is because only the values of the functions on the sample points and the ghost sample points count. In order to formalize this, let $Z_l = \{(\mathbf{x}_1, y_1), \ldots, (\mathbf{x}_l, y_l)\}$ be a given sample of size l and let $|\Lambda_{Z_l}|$ be the cardinality of Λ when restricted to $\{\mathbf{x}_1, \ldots, \mathbf{x}_l\}$, that is, the number of function parameters from Λ that can be distinguished from their values on $\{\mathbf{x}_1, \ldots, \mathbf{x}_l\}$. Moreover, let $\mathcal{N}(\Lambda, l)$ be the maximum number of functions that can be distinguished in this way, where the maximum runs over all possible choices of samples, so that

$$\mathcal{N}(\Lambda, l) = \max \{|\Lambda_{Z_l}| : \mathbf{x}_1, \ldots, \mathbf{x}_l \in \mathbf{X}\}. \tag{2.61}$$

The quantity $\mathcal{N}(\Lambda, l)$ is referred to as the *shattering coefficient of the function class parameterized by Λ with respect to the sample size l*. It has a particularly simple interpretation: it is the number of different outputs $\{y_1, \ldots, y_l\}$ that the functions parameterized by Λ can achieve on samples of a given size l. In other words, it measures the *number of ways that the function space can separate the patterns into two classes*. Whenever, $\mathcal{N}(\Lambda, l) = 2^l$, this means that there exists a sample of size l on which all possible separations can be achieved by functions parameterized by Λ. In this case, the corresponding function space is said to *shatter l* points. It must be noted that because of the maximum in the definition of $\mathcal{N}(\Lambda, l)$, shattering means that there *exists* a sample of l patterns which can be shattered in all possible ways. This definition, however, does not imply that all possible samples of size l will be shattered by the function space parameterized by Λ. The shattering coefficient can be considered as a capacity measure for a class of functions in the sense that it measures the "size" of a function class in a particular way. This way involves counting the number of functions in relation to a given sample of finite training points.

Uniform Convergence Bounds

Given an arbitrary, possibly infinite, class of function parameters consider the right hand side of Inequality (2.60) where the sample of $2l$ points will be represented by a set Z_{2l}. Specifically, the set Z_{2l} may be interpreted as the combination of l points from the original sample and l points from the ghost sample. The main idea is to replace the supremum over Λ by the supremum over $\Lambda_{Z_{2l}}$ where the set Z_{2l} contains at most $\mathcal{N}(\Lambda, l) \leq 2^{2l}$ different functions, then apply the union bound on this finite set and then the Chernoff bound. This leads to a bound as in Inequality (2.59), with $\mathcal{N}(\Lambda, l)$ playing the role of m. Essentially, those steps can be written down as follows:

$$P(\sup_{\alpha \in \Lambda} |R(\alpha) - R_{emp}(\alpha)| > \epsilon) \leq 2P\left(\sup_{\alpha \in \Lambda} |R_{emp}(\alpha) - R'_{emp}(\alpha)| > \frac{\epsilon}{2} \right)$$

$$= 2P\left(\sup_{\alpha \in \Lambda_{Z_{2l}}} |R_{emp}(\alpha) - R'_{emp}(\alpha)| > \frac{\epsilon}{2} \right)$$

$$\leq 2\mathcal{N}(\Lambda, 2l) \exp\left(\frac{-l\epsilon^2}{4} \right), \tag{2.62}$$

yielding the following inequality

$$P(\sup_{\alpha \in \Lambda} |R(\alpha) - R_{emp}(\alpha)| > \epsilon) \leq 2\mathcal{N}(\Lambda, 2l) \exp\left(\frac{-l\epsilon^2}{4} \right). \tag{2.63}$$

The notion of uniform bound may be utilized in order to infer whether the ERM process is consistent for a given class of function parameters Λ. Specifically, the right hand side of Inequality (2.63) guarantees that the ERM process is consistent for a given class of function parameters Λ when it converges to 0 as $l \to \infty$. In this context, the most important factor controlling convergence is the quantity $\mathcal{N}(\Lambda, 2l)$. This is true since the second factor of the product $2\mathcal{N}(\Lambda, 2l) \exp(\frac{-l\epsilon^2}{4})$ is always the same for any given class of function parameters. Therefore, when the shattering coefficient is considerably smaller than 2^{2l}, say $\mathcal{N}(\Lambda, 2l) \leq (2n)^k$, it is easy to derive that the right hand side of the uniform bound takes the form

$$2\mathcal{N}(\Lambda, 2l) \exp\left(\frac{-l\epsilon^2}{4} \right) = 2\exp\left(k \log(2l) - l\frac{\epsilon^2}{4} \right), \tag{2.64}$$

which converges to 0 as $l \to \infty$. On the other hand, when the class of function parameters coincides with the complete parameter space Λ_{all} then the shattering coefficient takes its maximum value such that $\mathcal{N}(\Lambda, 2l) = 2^{2l}$. This entails that the right hand side of Inequality (2.63) takes the form

$$2\mathcal{N}(\Lambda, 2l) \exp\left(\frac{-l\epsilon^2}{4} \right) = 2\exp\left(l(2\log(2)) - \frac{\epsilon^2}{4} \right), \tag{2.65}$$

which does not converge to 0 as $l \to \infty$.

The union bound, however, cannot directly guarantee the consistency of the ERM process when utilizing the complete parameter space Λ_{all}. The reason is that Inequality (2.63) gives an upper bound on $P(\sup_{\alpha \in \Lambda} |R(\alpha) - R_{emp}(\alpha)| > \epsilon)$ which merely provides a sufficient condition for consistency but not a necessary one. According to

(Devroye et al. 1996) a necessary and sufficient condition for the consistency of the ERM process is that

$$\log \frac{\mathcal{N}(\Lambda, 2l)}{l} \to 0. \tag{2.66}$$

2.5.6 Generalization Bounds

Sometimes it is useful to reformulate the uniform convergence bound so that the procedure of initially fixing ϵ and subsequently computing the probability that the empirical risk deviates from the true risk more than ϵ is reversed. In other words, there are occasions when it would be reasonable to initially specify the probability of the desired bound and then get a statement concerning the proximity between the empirical and the true risk. This can be achieved by setting the right hand side of Inequality (2.63) equal to some $\delta > 0$ and then solving for ϵ. The resulting statement declares that with probability at least $1 - \delta$, any function in $\{f(\mathbf{x}, a) : \alpha \in \Lambda\}$ satisfies

$$R(\alpha) \le R_{emp}(\alpha) + \sqrt{\frac{4}{l}(\log(2\mathcal{N}(\Lambda, 2l) - \log(\delta)))}. \tag{2.67}$$

Consistency bounds can also be derived by utilizing Inequality (2.67). In particular, it is obvious that the ERM process is consistent for a given function class parameterized by Λ when the term $\frac{\sqrt{\log(2\mathcal{N}(\Lambda, 2l))}}{l}$ converges to 0 as $l \to \infty$. The most important aspect concerning the generalization bound provided by Inequality (2.67) is that it holds for any function in $\{f(\mathbf{x}, a) : \alpha \in \Lambda\}$. This constitutes a highly desired property since the bound holds in particular for the function which minimizes the empirical risk, identified by the function parameter α_l. On the other hand, the bound holds for learning machines that do not truly minimize the empirical risk. This is usually interpreted as a negative property since by taking into account more information about a function, one could hope to obtain more accurate bounds.

Essentially, the generalization bound states that when both $R_{emp}(\alpha)$ and the square root term are small simultaneously then it is highly probable that the error on future points (actual risk) will be small. Despite sounding like a surprising statement this claim does not involve anything impossible. It only says that the utilization of a function class $\{f(\mathbf{x}, a) : \alpha \in \Lambda\}$ with relatively small $\mathcal{N}(\Lambda, l)$, which can nevertheless explain data sampled from the problem at hand, is not likely to be a coincidence. In other words, when a relatively small function class happens to "explain" data sampled from the problem under consideration, then there is a high probability that this function class captures some deeper aspects of the problem. On the other hand, when the problem is too difficult to learn from the given amount of training data, then it is necessary to use a function class so large that can "explain" nearly everything. This results in a small empirical error but at the same time increases the magnitude of the square root term. Therefore, according to the insight provided by the generalization bound, the difficulty of a particular learning problem is entirely determined by the suitability of the selected function class and by the prior knowledge available for the problem.

The VC Dimension

So far, the various generalization bounds were expressed in terms of the shattering coefficient $\mathcal{N}(\Lambda, l)$. Their primary downside is that they utilize capacity concepts that are usually difficult to evaluate. In order to avoid this situation, Vapnik and Chervonenkis [2] introduced the so-called VC dimension which is one of the most well known capacity concepts. Its primary purpose is to characterize the growth behavior of the shattering coefficient using a single number.

A sample of size l is said to be *shattered by the function parameter class* Λ if this class parameterizes functions that can realize any labelling on the given sample, that is $|\Lambda_{Z_l}| = 2^l$. The VC dimension of Λ, is now defined as the largest number l such that there exists a sample of size l which is shattered by the functions parameterized by Λ. Formally,

$$VC(\Lambda) = \max\{l \in \mathbb{N} : |\Lambda_{Z_l}| = 2^l \text{ for some } Z_l\} \tag{2.68}$$

If the maximum does not exist, the VC dimension is defined to be infinity. For example, the VC dimension of the set of *liner indicator functions*

$$Q(\mathbf{z}, \alpha) = \theta\left\{\sum_{p=1}^{l} a_p z_p + a_0\right\} \tag{2.69}$$

in l-dimensional coordinate space $\mathbf{Z} = (\mathbf{z}_1, \dots, \mathbf{z}_l)$ is equal to $l + 1$, since using functions from this set one can shattered at most $l + 1$ vectors. Moreover, the VC dimension of the set of linear functions

$$Q(\mathbf{z}, \alpha) = \sum_{p=1}^{l} a_p z_p + a_0 \tag{2.70}$$

in l-dimensional coordinate space $\mathbf{Z} = (\mathbf{z}_1, \dots, \mathbf{z}_l)$ is equal to $l + 1$, since a linear function can shatter at most $l + 1$ points.

A combinatorial result proved simultaneously by several people [3, 4, 7] characterizes the growth behavior of the shattering coefficient and relates it to the VC dimension. *Let Λ be a function parameter class with finite VC dimension d. Then*

$$\mathcal{N}(\Lambda, l) \leq \sum_{i=0}^{d} \binom{n}{i} \tag{2.71}$$

for all $l \in \mathbb{N}$. In particular, for all $l \geq d$ the following inequality holds

$$\mathcal{N}(\Lambda, l) \leq \left(\frac{en}{d}\right)^d. \tag{2.72}$$

The importance of this statement lies in the last fact. If $l \geq d$, then the shattering coefficient behaves like a polynomial function of the sample size l. According to this result, when the VC dimension of a function parameter class is finite then the corresponding shattering coefficients will grow polynomially with l. Therefore, *the ERM process is consistent with respect to a function parameter space Λ if and only if $VC(\Lambda)$ is finite.*

A fundamental property shared by both the shattering coefficient and the VC dimension is that they do not depend on the underlying probability distribution P, since they only depend on the function parameter class Λ. On the one hand, this is an advantage, as the capacity concepts apply to all possible probability distributions. On the other hand, this can be considered as a disadvantage, as the capacity concepts do not take into account particular properties of the distribution at hand.

A particular class of distribution independent bounds is highly related with the concept of Structural Risk Minimization. Specifically, these bounds concern the subset of totally bounded functions

$$0 \leq Q(\mathbf{z}, \alpha) \leq B, \ \alpha \in \Lambda \tag{2.73}$$

with finite VC dimension such as the set of indicator functions. The main result for this set of functions is the following theorem: *With probability at least $1 - \delta$, the inequality*

$$R(\alpha) \leq R_{emp}(\alpha) + \frac{B\epsilon}{2}\left(1 + \sqrt{1 + \frac{4R_{emp}\alpha}{B\epsilon}}\right) \tag{2.74}$$

holds true simultaneously for all functions of the set (2.73) where

$$\epsilon = 4\frac{d(\ln\frac{2l}{d} + 1) - \ln\delta}{l} \tag{2.75}$$

and $B = 1$.

The Structural Risk Minimization Principle

The ERM process constitutes a fundamental learning principle which efficiently deals with problems involving training samples of large size. This fact is specifically justified by considering Inequality (2.74) which formulates the conditions that guarantee the consistency of the ERM process. In other words, when the ratio l/d is large, the second summand on the right hand side of Inequality (2.74) will be small. The actual risk is then close to the value of the empirical risk. In this case, a small value of the empirical risk ensures a small value of the actual risk. On the other hand, when the ratio l/d is small, then even a small value for the empirical risk will not guarantee a small value for the actual risk. The latter case indicates the necessity for a new learning principle which will focus on acquiring a sufficiently small value for the actual risk $R(\alpha)$ by simultaneously minimizing both terms on the right hand side of Inequality (2.74). This is the basic underpinning behind the principle of Structural Risk Minimization (SRM). In particular, SRM is intended to

Fig. 2.7 Admissible
structure of function sets

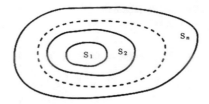

minimize the risk functional $R(\alpha)$ with respect to both the empirical risk and the VC
dimension of the utilized set of function parameters Λ.

The SRM principle is based on a nested structural organization of the function set
$S = Q(\mathbf{z}, \alpha), \ \alpha \in \Lambda$ such that

$$S_1 \subset S_2 \cdots \subset S_n \cdots , \tag{2.76}$$

where $S_k = \{Q(\mathbf{z}, \alpha) : \alpha \in \Lambda_k\}$ are subsets of the original function space such that
$S^* = \cup_k S_k$ as it is illustrated in Fig. 2.7

Moreover, the set of utilized functions must form an *admissible structure* which
satisfies the following three properties:

1. The VC dimension d_k of each set S_k of functions is finite,
2. Any element S_k of the structure contains totally bounded functions

$$0 \leq Q(\mathbf{z}, \alpha) \leq B_k, \ \alpha \in \Lambda_k,$$

3. The set S^* is everywhere dense in S in the $L_1(F)$ metric where $F = F(\mathbf{z})$ is the
   distribution function from which examples are drawn.

Note that in view of (2.76) the following assertions are true:

1. The sequence of values of VC dimensions d_k for the elements S_k of the structure
   S in nondecreasing with increasing k

$$d_1 \leq d_2 \leq \cdots \leq d_n \leq \cdots ,$$

2. The sequence of values of the bounds B_k for the elements S_k of the structure S in
   nondecreasing with increasing k

$$B_1 \leq B_2 \leq \cdots \leq B_n \leq \cdots ,$$

Denote by $Q(z, \alpha_l^k)$ the function that minimizes the empirical risk in the set of
functions S_k. Then with probability $1 - \delta$ one can assert that the actual risk for this
function is bounded by the following inequality

$$R(\alpha_l^k) \leq R_{emp}(\alpha_l^k) + B_k \epsilon_k(l) \left(1 + \sqrt{1 + \frac{4 R_{emp} \alpha_l^k}{B \epsilon_k(l)}} \right), \tag{2.77}$$

Fig. 2.8 Admissible
structure of function sets

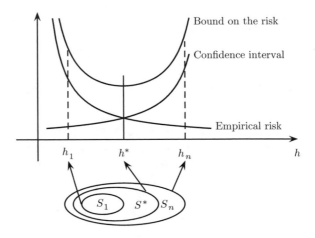

where

$$\epsilon_k(l) = 4\frac{d_k(\ln\frac{2l}{d_k} + 1) - \ln\frac{\delta}{4}}{l}. \tag{2.78}$$

For a given set of observations $\mathbf{z}_1, \ldots, \mathbf{z}_l$, the SRM method actually suggests that one should choose the element S_k of the structure for which the smallest bound on the risk is achieved. In other words, the SRM principle introduces the notion of a *trade-off between the quality of approximation and the complexity of the approximating function* as it is particularly illustrated in Fig. 2.8.

Therefore, the SRM principle is based upon the following idea: *To provide the given set of functions with an admissible structure and then to find the function that minimizes risk (2.77) over given elements of the structure.* This principle is called the principle of structural risk minimization in order to stress the importance of choosing the element of the structure that possesses an appropriate capacity.

References

1. Bousquet, O., Boucheron, S., Lugosi, G.: Introduction to statistical learning theory. In: Advanced Lectures on Machine Learning, pp. 169–207. Springer, Heidelberg (2004)
2. Pednault, E.P.: Statistical Learning Theory. Citeseer (1997)
3. Sauer, N.: On the density of families of sets. J. Comb. Theory Ser. A **13**(1), 145–147 (1972)
4. Shelah, S.: A combinatorial problem; stability and order for models and theories in infinitary languages. Pac. J. Math. **41**(1), 247–261 (1972)
5. Vapnik, V.: The Nature of Statistical Learning Theory. Springer Science & Business Media, New York (2013)
6. Vapnik, V.N.: An overview of statistical learning theory. IEEE Trans. Neural Netw. **10**(5), 988–999 (1999)
7. Vapnik, V.N., Chervonenkis, A.J.: Theory of Pattern Recognition (1974)

Chapter 3
The Class Imbalance Problem

Abstract We focus on a special category of pattern recognition problems that arise in cases when the set of training patterns is significantly biased towards a particular class of patterns. This is the so-called Class Imbalance Problem which hinders the performance of many standard classifiers. Specifically, the very essence of the class imbalance problem is unravelled by referring to a relevant literature review. It is shown that there exist a wide range of real-world applications involving extremely skewed (imbalanced) data sets and the class imbalance problem stems from the fact that the class of interest occupies only a negligible volume within the complete pattern space.

3.1 Nature of the Class Imbalance Problem

As the field of machine learning makes a rapid transition from the status of "academic discipline" to that of "applied science", a myriad of new issues, not previously considered by the machine learning community are coming into light. One such issue is the class imbalance problem. The class imbalance problem corresponds to the problem encountered by inductive learning systems when the assumption concerning the similarity of the prior probabilities of target classes to be learnt, is grossly violated. In other words, the class imbalance problem arises on domains for which one class is represented by a large number of examples while the other is represented only by a few. This entails that the ratios of the prior probabilities between classes are extremely skewed. The same situation arises when the target classes are assigned with extremely imbalanced misclassification costs.

The class imbalance problem is of crucial importance as it is encountered in a wide range of machine learning applications, which include:

- environmental
- cultural
- vital
- medical
- commercial
- astronomical

© Springer International Publishing AG 2017
D.N. Sotiropoulos and G.A. Tsihrintzis, *Machine Learning Paradigms*,
Intelligent Systems Reference Library 118, DOI 10.1007/978-3-319-47194-5_3

- military
- person identification - face recognition - authorship verification
- information retrieval
- computer security - intrusion detection
- user modelling - recommendation

Although practitioners might have already encountered the class imbalance problem early, it made its appearance in the machine learning / data mining literature about 15 years ago. Its importance grew as more and more researchers realized that their data sets were imbalanced, resulting in suboptimal classification performance. For example, the problem occurs and hinders classification in applications such as the detection of fraudulent telephone calls or credit card transactions in Fawcett and Provost [10] and Maes et al. [5, 23] respectively. A similar classification problem is addressed in Ezawa et al. [9] where the authors try to identify unreliable telecommunications customers. The highly imbalanced nature of these classification problems comes from the fact that the class of interest occupies only a negligible fraction of the pattern space so that corresponding instances are extremely infrequent to occur. Problems of kindred nature are also addressed in applications as diverse as the detection of oil spills in radar images [16], in direct marketing [22] where it is common to have a small response rate (about 1 %) for most marketing campaigns, in prediction of failures in manufacturing processes [28] or in diagnosing rare diseases [19]. Extremely imbalanced classes also arise in information retrieval applications [21], in learning word pronunciations [34] or in predicting telecommunication equipment failures [39].

In Japkowicz and Stephen [13], Japkowicz [12], the authors attempt to unify the relative research by focusing on the nature of the class imbalance problem. Their primary motivation is to investigate the reasons why standard learners are often biased towards the majority class. According to the authors, this is the primary effect of the class imbalance problem attributed to the fact that such classifiers attempt to reduce global quantities such as the error rate without taking the data distribution into consideration. As a result, examples from the overwhelming class are well-classified, whereas examples from the minority class tend to be misclassified. Moreover, the authors are interested in identifying the particular domain characteristics that increase the severity of the problem. Specifically, their research is focused on determining the exact way in which the degree of concept complexity, the size of the training set and the degree of class imbalance influence the classification performance of standard classifiers. Their findings indicate that the problem is aggravated by increasing the degree of class imbalance or the degree of concept complexity. However, the problem is simultaneously mitigated when the size of the training set is sufficiently large. On the contrary, class imbalances do not hinder the classification of simple problems such as the ones that are linearly separable.

The author in Estabrooks et al. [8] considers the general setting of a binary classification problem where the two classes of interest are namely the positive and negative classes of patterns. In this setting, the class imbalance problem is identified as a major factor influencing the generalization ability of standard classifiers. Specifically, the author's inquiry is focused on revealing the factors that cause learners to deviate from their typical behavior according to which they should be able to generalize over unseen instances of any class with equal accuracy. This of course is the ideal situation which is violated in many real-world applications where learners are faced with imbalanced data sets causing them to be biased towards one class. This bias is the result of one class being heavily under-represented in the training data compared to the other classes. According to the author, this bias can be attributed to two factors that specifically relate to the way in which learners are designed.

Firstly, inductive learners are typically designed to minimize errors over the training examples. Classes containing few examples can be largely ignored by learning algorithms because the cost of performing well on the over-represented class outweighs the cost of doing poorly on the smaller class. Another factor contributing to the bias is over-fitting. Over-fitting occurs when a learning algorithm creates a hypothesis that performs well over the training data but does not generalize well over unseen data. This can occur on an under-represented class.

The nature of the class imbalance problem was also studied in Weiss and Provost [37, 40]. According to the authors, many classifier induction algorithms assume that the training and test data are drawn from the same fixed underlying distribution D. In particular, these algorithms assume that r_{train} and r_{test}, the fractions of positive examples in the training and test sets, respectively, approximate the true prior probability p of encountering a positive example. These induction algorithms use, either implicitly or explicitly, the estimated class priors based on r_{train}, to construct a model and to assign classifications. If the estimated value of the class priors is not accurate, then the posterior probabilities of the model will be improperly biased. Specifically, increasing the prior probability of a class increases the posterior probability of the class, moving the classification boundary for that class so that more cases are classified into the class.

Weiss in [38] identifies the direct link between the class imbalance problem and rarity which may be encountered in two different situations. The first type of rarity is called "rare classes" concerning the poor presence of instances of a particular class. The second type of rarity concerns "rare cases", which correspond to a meaningful but relatively small subset of the data representing a sub-concept or sub-class that occurs infrequently. Thus, rare cases are particularly related to "small disjuncts" that cover a negligible volume of training examples. According to the author small disjuncts play a fundamental role in the rarity problem since the lack of data makes it difficult to detect regularities within the rare classes or cases. Specifically, it is documented that in some cases small disjuncts may not represent rare or exceptional cases, but rather noisy data. Thus, it is very important to keep only the meaningful small disjuncts that may be identified by the utilization of statistical significance tests. The problem is that the significance of small disjuncts cannot be reliably estimated and consequently significant small disjuncts may be eliminated along with the insignificant ones. Moreover, eliminating all small disjuncts results in increasing the overall error rate and therefore is not a good strategy. The lack of data for the rare classes or cases is also responsible for the inability of data mining systems to recognize existing correlations between highly infrequent items. Specifically, in order to recognize the correlation between two highly infrequent items, a data mining system has to lower the corresponding threshold value. This, however, will cause a combinatorial explosion since frequently occurring items will be associated with one another in an enormous number of ways. Another issue investigated by the author concerns the divide-and-conquer approach adapted by many data mining algorithms. Decision tree algorithms are a good example of this approach functioning by decomposing the original problem into smaller and smaller problems. This process, however, results in data fragmentation which constitutes a very significant problem since regularities can then be found within each individual partition which contains less data.

The authors in [14, 33] are also interested in examining the contribution of rare or exceptional cases to the class imbalance problem since they cause small disjuncts to occur. In more detail, learning systems usually create concept definitions that consist of several disjuncts where each disjunct constitutes a conjunctive definition of a sub-concept of the original concept. The coverage of a disjunct corresponds to the number of training examples it correctly classifies. A disjunct is considered to be small if that coverage is low. Thus, the formations of such sub-clusters are very difficult to be recognized by standard classifiers, justifying the fact that small disjuncts are more prone to errors than large ones. However, the authors elaborate that small disjuncts are not inherently more error prone than large disjuncts. On the contrary, this is a result attributed to the bias of the classifiers, the class noise and the training set size on the rare cases. Their experiments involved several artificially generated domains of varying concept complexity, training set size and degree of imbalance. In particular, they provide evidence that when all the sub-clusters are of size 50, even at the highest degree of concept complexity, the error is below 1 % which entails that the erroneous classification effect is practically negligible. This suggests that it is not the class imbalance per se that causes a performance decrease, but that it is rather the small disjunct problem created by the class imbalance (in highly complex and small-sized domains) that causes that loss of performance. Additionally, they examined the relevant classification performance of several standard classifiers such as the C5.0 decision tree classifier, Multi-Layer Perceptrons(MLPs) and Support Vector Machines (SVMs). Their findings demonstrate that C5.0 is the most sensitive to class imbalances and that is because C5.0 works globally ignoring specific data points. MLPs were found to be less prone to errors due to the class imbalance problem than C5.0. This is because of their flexibility: their solution gets adjusted by each data point in a bottom-up manner as well as by the overall data set in a top-down manner. SVMs are even less prone to errors due to the class imbalance problem than MLPs because they are only concerned with a few support vectors, namely the data points located close to the boundaries. Similar results on extremely imbalanced classification problems have been reported by Sotiropoulos and Tsihrintzis in [31, 32].

The problem of small disjuncts has also been reported in the literature by Lampropoulos et al. [18] in the context of recommender systems. According to the authors, users' preferences tend to concentrate into a small fraction of disjoint clusters relative to the complete set of items available to the user. This fact significantly affects the recommendation ability of any system attempting to model the particular preferences of a given individual.

The authors in [24] develop a systematic study aiming to question whether class imbalances are truly to blame for the loss of performance of learning systems or whether the class imbalance problem is not a problem per se. Specifically, they are interested in investigating whether a learning system will present low performance with a highly imbalanced dataset even when the classes are far apart. To this end, the authors conducted a series of experiments on artificially generated datasets revealing that the distance between the class clusters is an important factor contributing to the poor performance of learning systems independently of the presence of class imbalance. In other words, their findings indicate that it is not the class probabilities the main cause of the hinder in the classification performance, but instead the degree of overlapping between the classes. Thus, dealing with class imbalances will not always improve the performance of classifiers.

In Daskalaki et al. [6], it is stated that classification problems with uneven class distributions present several difficulties during the training as well as during the evaluation process of the classifiers. The context in which the authors conducted their experiments was the customer insolvency problem which is characterized by (a) very uneven distributions for the two classes of interest, namely the solvent and insolvent class of customers (b) small number of instances within the insol-vent class (minority class) and (c) different misclassification costs for the two classes. In order to assess the effect of imbalances in class distributions on the accuracy of classification performance, several classifiers were employed such as Neural Networks (MLPs), Multinomial Logistic Regression, Bayesian Networks (hill climbing search), Decision Tree (pruned C4.5), SVMs and Linear Logistic Regression. The classification results based on the True Positive Ratio which represents the ability of the classifiers in recognizing the minority (positive) class ($TP = Pr\{predicted Minority | actually Minority\}$), demonstrate poor efficiency.

3.2 The Effect of Class Imbalance on Standard Classifiers

3.2.1 Cost Insensitive Bayes Classifier

In order to acquire a clearer understanding of the nature of the class imbalance problem it is imperative to formulate it within the more general setting of a binary classification problem. Specifically, binary classification problems will be addressed within the broader framework of Bayesian Decision Theory. For this class of problems one is given a pair of complementary hypotheses:

- H_0: the null hypothesis and
- H_1: the alternative hypothesis

Given a multidimensional vector space X and a particular instance represented by the random observation vector $\mathbf{x}$ in X, the two hypotheses may be defined as follows:

- H_0: $\mathbf{x}$ belongs to Class Ω^+ and
- H_1: $\mathbf{x}$ belongs to Class Ω^-

where Class $\Omega^- = $ Class $\neg\Omega^+$, i.e. Class Ω^- is the complement of Class Ω^+ in the vector space X. Thus, the ultimate purpose of Bayesian Decision Theory consists of finding out which one of the given hypotheses, H_0 or H_1, is true. A natural solution to this binary hypothesis testing problem (link this with some theoretical result form Chap. 1) involves calculating the a posteriori probabilities, $P(H_0|\mathbf{x})$ and $P(H_1|\mathbf{x})$, that $\mathbf{x}$ satisfies hypotheses H_0 and H_1 respectively. While the a posteriori probabilities are unknown, it is assumed that the conditional probability density functions of the observation vector given a specific hypothesis, $p(\mathbf{x}|H_0)$ and $p(\mathbf{x}|H_1)$, can be estimated. This is an underlying assumption of Bayesian Decision Theory stating that the probability density function corresponding to the observation vector $\mathbf{x}$ depends on its class. Moreover, the prior probabilities for each hypothesis, $P(H_0)$ and $P(H_1)$, as well as the prior probability of the observation vector $\mathbf{x}$, can also be estimated. Then, the a posteriori probabilities $P(H_0|\mathbf{x})$ and $P(H_1|\mathbf{x})$ can be worked out by utilizing Bayes Theorem:

$$P(H_j|\mathbf{x}) = \frac{p(\mathbf{x}|H_j)P(H_j)}{p(\mathbf{x})}, \text{ where } j \in \{0, 1\}. \tag{3.1}$$

In Eq. (3.1), $p(\mathbf{x})$ corresponds to the probability density function of the observation vector $\mathbf{x}$ for which we have:

$$p(\mathbf{x}) = \sum_j p(\mathbf{x}|H_j)P(H_j), \text{ where } j \in \{0, 1\}. \tag{3.2}$$

The estimation of the data probability density function constitutes an essential prerequisite in order to solve the binary classification problem, but it is not sufficient to solve the problem per se. In addition, a decision rule must be defined to choose one hypothesis over the other. The milestone of Bayesian Decision Theory is the so-called "Bayes Decision Rule" which takes as input the class-conditional probability density functions and the a priori probabilities. The Bayes Decision Rule may be formally defined as:

$$\begin{aligned} &\text{If } P(H_0|\mathbf{x}) > P(H_1|\mathbf{x}), \text{ decide } H_0 \\ &\text{If } P(H_1|\mathbf{x}) > P(H_0|\mathbf{x}), \text{ decide } H_1 \end{aligned} \tag{3.3}$$

The case of equality is detrimental and the observation vector can be assigned to either of the two classes without affecting the overall probability of erroneous classification. Thus, either of the two hypotheses may be considered to be true. Using Eq. (3.3) the decision can equivalently be based on the inequalities:

$$p(\mathbf{x}|H_0)P(H_0) \gtrless p(\mathbf{x}|H_1)P(H_1) \tag{3.4}$$

A intrinsic property of Bayes Decision Rule is that it provides the optimal decision criterion by minimizing the probability of error, given by the following equation:

$$P(error) = \int_X P(error, \mathbf{x})d\mathbf{x} = \int_X P(error|\mathbf{x})p(\mathbf{x})dx, \qquad (3.5)$$

where the conditional probability of error may be computed as follows:

$$P(error|\mathbf{x}) = \begin{cases} P(H_0|\mathbf{x}), & \text{if we decide } H_1, \\ P(H_1|\mathbf{x}), & \text{if we decide } H_0. \end{cases} \qquad (3.6)$$

Having in mind that decisions are made on the basis of the Bayes rule, Eq. (3.6) may be written in the form

$$P(error|x) = \min [P(H_0|\mathbf{x}), P(H_1|\mathbf{x})], \qquad (3.7)$$

where $P(H_0|\mathbf{x}) + P(H_1|\mathbf{x}) = 1$. Equation (3.7), then, entails that the conditional probability of error $P(error|\mathbf{x})$ is the smallest possible for any given observation vector $\mathbf{x}$. As a consequence the probability of error given by the integral in Eq. (3.5) is minimized over all vector space X.

The previous results demonstrate the optimality of the Bayesian classifier for any given binary classification problem parameterized by the class-conditional and prior probability density functions. Thus, it is of vital importance to unfold the very nature of the class imbalance problem within the framework of binary classification problems and particularly to investigate its influence on the classification behavior of the optimal Bayesian classifier. The class imbalance problem arises in situations when the class priors are extremely skewed entailing that the volumes of positive X^+ and negative X^- subspaces are proportionately asymmetric in magnitude. It is essential to understand the exact way in which the class imbalance problem determinately affects the decision making process which is formalized by Inequalities (3.4) of the Bayes Decision Rule.

Letting $q = p(\mathbf{x}|H_0)$ and consequently $1 - q = p(\mathbf{x}|H_1)$, Inequalities (3.4) may be written in the following form:

$$qP(H_0) \gtrless (1-q)P(H_1). \qquad (3.8)$$

If q^* corresponds to the class-conditional probability for which both sides of Inequalities (3.8) are equal, then q^* may be computed by the following equation:

$$q^*P(H_0) = (1-q^*)P(H_1), \qquad (3.9)$$

which finally results in the following threshold value:

$$q^* = p(H_1). \qquad (3.10)$$

Equation (3.10) then entails that the original form of the Bayes Decision Rule may
be reformulated as:

$$\text{If } q > q^*, \text{ decide } H_0$$
$$\text{If } q < q^*, \text{ decide } H_1.$$
(3.11)

Since q^* provides the class-conditional probability threshold for making optimal
decisions as a function of the negative class prior, it is reasonable to investigate
how the prior class imbalance influences the threshold and consequently the decision making process. Letting $P_0 = P(H_0)$ and $P_1 = P(H_1)$ be the class priors for
the positive and negative classes of patterns respectively, so that $P_0 + P_1 = 1$, the
decision threshold can be given by the following equation:

$$q^*(P_0) = 1 - P_0, \ 0 \leq P_0 \leq 1.$$
(3.12)

Equation (3.12) can be utilized in order to infer the behavior of the Bayesian
classifier over the complete range of class priors. However, the situations of primary
concern are those corresponding to extreme prior class probabilities when the positive
(or negative) class prior approaches the upper and lower limits of the [0, 1] interval.
Formally, in order to identify the behavior of q^* with respect to the imbalance factor
P_0, it would be reasonable to compute the following quantities:

- $\lim_{P_0 \to \frac{1}{2}} q^*(P_0)$: which denotes the behavior of the class conditional probability
  threshold when the class priors are completely balanced,
- $\lim_{P_0 \to 1} q^*(P_0)$: which denotes the behavior of the class conditional probability threshold when the class priors are completely imbalanced towards the positive Ω^+ class
  of patterns,
- $\lim_{P_0 \to 0} q^*(P_0)$: which denotes the behavior of the class conditional probability
  threshold when the class priors are completely imbalanced towards the negative
  Ω^- class of patterns.

The first limit is given by the following equation:

$$\lim_{P_0 \to \frac{1}{2}} q^*(P_0) = \frac{1}{2},$$
(3.13)

corresponding to the situation when class prior probabilities are completely balanced.
It is clear that such a situation does not affect the decision making process. The second
limit is given by the equation:

$$\lim_{P_0 \to 0} q^*(P_0) = 1,$$
(3.14)

corresponding to the situation when the volume occupied by the subspace of positive
patterns is negligible in comparison with the volume occupied by the subspace of
negative patterns. Thus, according to Eq. (3.14) when $P_0 \to 0$ then $q^* \to 1$, which

yields that the Bayes Decision Rule may be re-expressed in the following form:

$$\text{If } q > 1, \text{ decide } H_0$$
$$\text{If } q < 1, \text{ decide } H_1.$$
(3.15)

Equation (3.15) then entails that it is impossible to choose H_0 over H_1 since the probability threshold approaches its maximum possible value which cannot be exceeded. As a consequence all decisions will be made in favor of the majority class of negative patterns.

Finally, the third limit is given by the equation:

$$\lim_{P_0 \to 1} q^*(P_0) = 0$$
(3.16)

corresponding to the situation when the volume occupied by the subspace of negative patterns is negligible in comparison with the volume occupied by the subspace of positive patterns. Equation (3.16) states that when $P_0 \to 1$ then $q^* \to 0$, which yields that the Bayes Decision Rule may be re-expressed in the following form:

$$\text{If } q > 0, \text{ decide } H_0$$
$$\text{If } q < 0, \text{ decide } H_1$$
(3.17)

Equation (3.17) then entails that it is impossible to choose H_1 over H_0 since the probability threshold approaches its minimum possible value which is always exceeded. As a consequence all decisions will be made in favor of the majority class of positive patterns.

3.2.2 Bayes Classifier Versus Majority Classifier

The previous discussion reveals the presumable behavior of the Bayes classifier when the class priors are extremely skewed, within the general setting of a binary classification problem. The main conclusion so far is that the decision making process is biased towards the majority class of patterns, since the Bayes classifier will never predict that a given pattern belongs to the minority class, when either one of the corresponding prior class probabilities approaches the lower limit of the $[0, 1]$ interval. Even so, the Bayes classifier demonstrates the optimal classification performance in terms of the error probability. Specifically, the probability of error becomes infinitesimally small as the degree of class imbalance increases. Let X_0 and X_1 be subspaces of X defined as follows:

$$X_0 = \{\mathbf{x} \in X : a(\mathbf{x}) = H_0\}$$
$$X_1 = \{\mathbf{x} \in X : a(\mathbf{x}) = H_1\},$$
(3.18)

where $a(\mathbf{x}) \in \{H_0, H_1\}$ denotes the decision concerning the true hypothesis. Equations (3.18) then, may be utilized in order to re-express the error probability in the following form:

$$P(error) = P(\mathbf{x} \in X_1, H_0) + P(\mathbf{x} \in X_0, H_1), \tag{3.19}$$

where $P(.\,,\,.)$ denotes the joint probability of two events. Subsequently, by taking advantage of the Bayes Theorem, Eq. (3.19) can be re-written as:

$$P(error) = P(\mathbf{x} \in X_1 | H_0) P(H_0) + P(\mathbf{x} \in X_0 | H_1) P(H_1), \tag{3.20}$$

which finally yields that the probability of error can be estimated by the following equation:

$$P(error) = P_0 \int_{X_1} p(\mathbf{x}|H_0) d\mathbf{x} + P_1 \int_{X_0} p(\mathbf{x}|H_1) d\mathbf{x}. \tag{3.21}$$

Thus, when $P_0 \to 0$ then $|X_0| \to 0$ which entails that $X_1 \approx X$ and as a consequence Eq. 3.21 may be rewritten as:

$$P(error) = P_0 \int_X p(\mathbf{x}|H_0) d\mathbf{x}, \tag{3.22}$$

which finally yields that $P(error) \to 0$. On the other hand when $P_1 \to 0$ then $|X_1| \to 0$ which entails that $X_0 \approx X$ and as a consequence Eq. (3.21) may be re-written as:

$$P(error) = P_1 \int_X p(\mathbf{x}|H_1) d\mathbf{x}. \tag{3.23}$$

This finally yields that $P(error) \to 0$. It must be mentioned that the probability of error for Bayes classifier for any given pair of class priors (P_0, P_1) can be estimated by utilizing Eq. (3.21) where the subspaces X_0 and X_1 may be determined from Eq. (3.9) by solving with respect to $\mathbf{x}$.

Since the probability of error for the Bayesian classifier is infinitesimally small when the class priors are extremely skewed, in order to reveal the exact nature of the class imbalance problem it is desirable to estimate the error probability for any class prior in the [0, 1] interval. In other words, in order to understand the exact way in which highly asymmetric class priors affect the performance of the Bayes classifier, it is important to extend the monitoring of its classification performance within the full range of class priors. Moreover, it is vital to identify the degree of overlapping between the class-conditional probability density functions as a supplementary factor which significantly affects the severity of the class imbalance problem.

Having in mind that the predictions of the Bayesian classifier are biased towards the class with the dominant prior probability, it is natural to compare its performance against the majority classifier. The special feature of the majority classifier is that its decisions are always made in favor of the majority class of patterns for any given pair of class priors pertaining to the binary classification setting. Thus, the classification accuracy of any binary classifier can be measured in terms of the relative error probability reduction with respect to the probability of error achieved by the majority classifier. In other words, the majority classifier provides an upper bound for the error probability as a function of the class priors.

In order to compare the classification performance of any binary classifier against the majority one, it is necessary to express the error probability as a function of the class priors P_0 and P_1. According to Eq. (3.20), it is possible to deduce for the error probability the following equation:

$$P(error) = P(H_1|H_0)P(H_0) + P(H_0|H_1)P(H_1) \qquad (3.24)$$

where the quantity $P(H_i|H_j)$, for $\{i, j\} \in \{0, 1\}$, corresponds to the probability of deciding in favor of H_i when H_j is true. Thus, the error probability of a binary classifier may be formulated as a convex combination of the likelihood functions $P(H_1|H_0)$ and $P(H_0|H_1)$ weighted by the class priors P_0 and P_1 respectively.

The quantities False Positive Rate(FPR) and False Negative Rate(FNR) may serve as good approximations of the likelihood functions under consideration. Thus, the error probability of any binary classifier may be estimated by the following equation:

$$P(error) = P_0 FNR + P_1 FPR, \qquad (3.25)$$

where the first term corresponds to the Type I Error and the second term corresponds to the Type II Error. Type I errors occur when patterns originating from the positive class are misclassified as negative while Type II errors occur when patterns originating from the negative class are misclassified as positive. By substituting $P_1 = 1 - P_0$ in Eq. (3.25) the latter may be rewritten in the following form:

$$P(error) = P_0(FNR - FPR) + FPR \qquad (3.26)$$

Equation (3.26) describes the probability of error for any binary classifier as a linear function of the positive class prior.

Generally, the graphical representation of any error probability function with respect to some class prior defines an error curve summarizing the classification accuracy of any binary classifier. Specifically, the error curve in Eq. (3.26) corresponds to a straight line which lies within the plane defined by the horizontal axis of positive class priors and the vertical axis of error probabilities. In other words, the error curve in Eq. (3.26) describes the exact way in which the positive class prior probability influences the variation of the error probability in the $[min(FPR, FNR), max(FPR, FNR)]$ interval. Thus, the upper and lower bounds for the error probability will be given by the following equation

$$P(error) = \begin{cases} FPR, & P_0 = 0 \\ FNR, & P_0 = 1, \end{cases} \qquad (3.27)$$

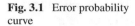

Fig. 3.1 Error probability curve

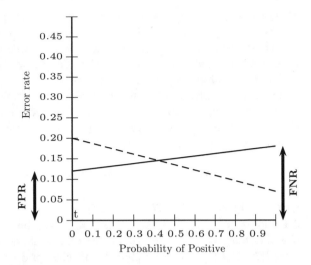

corresponding to the points of intersection of the error curve with the vertical lines at $P_0 = 0$ and $P_0 = 1$.

This is explicitly illustrated in Fig. 3.1 which depicts the error probability curves for two different binary classifiers over the full range of the $[0, 1]$ interval. This illustration can be utilized in order to directly determine the classifier that achieves the minimum probability of error for any prior probability. Therefore, the relative location of the two error curves indicates the relative classification performance of the corresponding classifiers. Particularly, an error curve which is strictly below any other curve within the previously defined plane, signifies a classifier with superior classification performance. However, the existence of an intersection point between the error curves suggests that there is no classifier outperforming the others for the complete range of prior probabilities. On the contrary, there is a subrange of dominance for each binary classifier.

Of significant importance are the two trivial classifiers, namely the one that always predicts that instances are negative and the one that always predicts that instances are positive. Specifically, the trivial classifier that assigns patterns exclusively to the negative class is dominated by the Type I Error, since there are no instances that are classified as positive. This entails that the second term of Eq. (3.25) can be ignored so that the error probability for this particular classifier degenerates to the following equation:

$$P(error) = P_0 FNR. \tag{3.28}$$

Moreover, the quantity FNR appearing in Eq. (3.28) corresponds to the fraction of truly positive instances that are misclassified as negative. It is easy to deduce then that this quantity will be equal to 1 when the positive class of patterns is completely ignored. In other words, $FNR = 1$, when the complete set of positive instances will be assigned to the negative class. Thus, the probability of error given by Eq. (3.28) can be re-expressed in the following form:

$$P(error) = P_0. \tag{3.29}$$

On the other hand, the trivial classifier that exclusively decides that patterns originate from the positive class is dominated by the Type II Error since there are no instances that are classified as negative. This yields that the second term of Eq. (3.25) cancels out so that the error probability for this particular classifier may be simplified to:

$$P(error) = (1 - P_0)FPR. \tag{3.30}$$

Here, the quantity FPR denotes the fraction of truly negative instances that are misclassified as positive. This entails that the quantity FPR will be equal to 1 since the entire set of negative instances will be misclassified as positive. Thus, the probability of error given by Eq. (3.30) may be reformulated as:

$$P(error) = 1 - P_0. \tag{3.31}$$

The combination of these two trivial classifiers forms the majority classifier which always predicts the most common class. Therefore, the probability of error for this classifier will be given by unifying Eqs. (3.29) and (3.31) within a single equation as:

$$P(error) = \begin{cases} P_0, & 0 \leq P_0 \leq \frac{1}{2}, \\ (1 - P_0), & \frac{1}{2} \leq P_0 \leq 1. \end{cases} \tag{3.32}$$

Figure 3.2 depicts the error curve summarizing the classification performance of the majority classifier compared against the error curve of a given binary classifier.

Fig. 3.2 Majority classifier

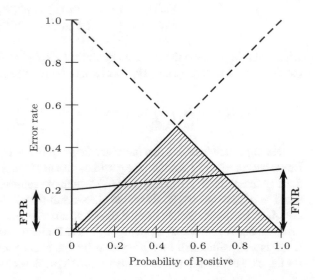

This graphical representation suggests that any single classifier with a non-zero error rate will always be outperformed by the majority classifier if the prior probabilities are sufficiently skewed and therefore of little use. This entails that a useful classifier must appreciably outperform the trivial solution of choosing the majority class. More importantly, it seems intuitive that a practical classifier must do much better on the minority class, often the one of greater interest, even if this means sacrificing performance on the majority class.

In order to complete the discussion concerning the effects of the class imbalance problem on the classification performance of the Bayes classifier, it is necessary to measure how much it reduces the error probability of the majority classifier. Specifically, the amount of relative reduction on the error probability of the majority classifier may serve as an ideal measure of accuracy since the classification behavior of the Bayes classifier was proven to be biased towards the majority class of patterns when class priors were severely skewed. Thus, an important issue that remains to be addressed concerns the evaluation of this quantity within the full range of class priors and under different degrees of overlapping between the class-conditional probability density functions.

Therefore, the graphical comparison of the error curve corresponding to the Bayes classifier against the error curve of the majority one will reveal the relative classification performance of the two classifiers. Having defined the error probability for the majority classifier as a function of the positive class prior it is necessary to act likewise for the error probability of the Bayes classifier. A function of this form is provided by Eq. (3.21) which explicitly represents the strong dependence between the error probability of the Bayesian classifier and the prior probabilities of the positive and negative classes of patterns. Specifically, the prior class probabilities affect the decision subspaces X_0 and X_1 which may be expressed by refining Eqs. (3.18) in the following form:

$$X_0(P_0) = \{\mathbf{x} \in X : q(\mathbf{x}) > q^*(P_0)\}$$
$$X_1(P_0) = \{\mathbf{x} \in X : q(\mathbf{x}) < q^*(P_0)\}, \tag{3.33}$$

where $q(\mathbf{x})$ represents the decision quantity $p(\mathbf{x}|H_0)$ and $q^*(P_0) = 1 - P_0$ denotes the decision threshold. Thus, Eq. (3.21) may be rewritten as:

$$P_{error}(P_0) = P_0 \int_{X_1(P_0)} q(\mathbf{x})d\mathbf{x} + (1 - P_0) \int_{X_0(P_0)} (1 - q(\mathbf{x}))d\mathbf{x}. \tag{3.34}$$

It is important to notice that the exact shape of the error curve corresponding to the Bayesian classifier depends on the particular form of the class-conditional probability density functions $p(\mathbf{x}|H_0)$ and $p(\mathbf{x}|H_1)$ which are represented by the quantities $q(\mathbf{x})$ and $1 - q(\mathbf{x})$ respectively. When the class-conditional probability density functions are given by the univariate normal distributions $N_1(\mu_1, \sigma_1)$ and $N_2(\mu_2, \sigma_2)$ and the distance between the means, $|\mu_1 - \mu_2|$, is sufficiently large, the error curve for the Bayesian classifier will be represented by the solid curve in Fig. 3.3. Specifically, for the case of univariate normal distributions the degree of overlapping between the

Fig. 3.3 Majority classifier

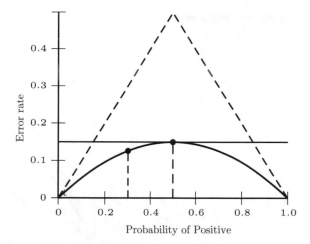

class-conditional probability density functions can be measured by the quantity:

$$D(\mu_1, \mu_2, \sigma_1, \sigma_2) = |\mu_1 - \mu_2| - |(\mu_{max} - 3\sigma_{max}) - (\mu_{min} + 3\sigma_{min})|, \quad (3.35)$$

where:

$$\begin{aligned}
\mu_{min} &= \min(\mu_1, \mu_2) \\
\mu_{max} &= \max(\mu_1, \mu_2) \\
\sigma_{min} &= \sigma_{argmin(\mu_1, \mu_2)} \\
\sigma_{max} &= \sigma_{argmax(\mu_1, \mu_2)}.
\end{aligned} \quad (3.36)$$

Figure 3.3 illustrates the relative locations of the error curves corresponding to the Bayes and Majority classifiers for the full range of positive class prior probabilities when $D \gg 0$. Therefore, it is easy to observe that the classification performance of both classifiers coincides when the degree of overlapping is sufficiently small and the class priors are severely skewed.

However, the remaining issue that needs to be addressed concerns the classification behavior of the Bayesian classifier when the degree of overlapping increases so that $D \ll 0$. This issue was particularly investigated in Drummond and Holte [7], Weiss [38] where the authors conducted a sequence of experiments in order to reveal the classification behavior of the Bayes classifier under increasing degrees of overlapping between the class-conditional probability density functions. Their results are summarized in Fig. 3.4 which shows the error curve for the Bayesian classifier (from top to bottom) for three different distances between the means of the normal distributions. Specifically, the distances were chosen in order to reduce the relative error probability by a factor of 20, 50 and 80 % when the classes are balanced. The series of progressively smaller triangles in Fig. 3.4, made of dotted lines, were called "error probability reduction contours". Each error probability reduction contour indicates a specific percentage gain on the error probability achieved by the majority classifier.

Fig. 3.4 Different distances

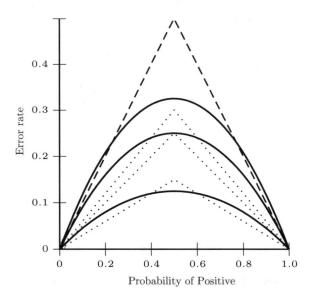

The continuous curves cross multiple contours indicating a decreasing relative error probability reduction as imbalance increases. Specifically, focusing on the lower left hand corner of Fig. 3.4, where the negative instances are much more common than the positives, it is easy to observe that the upper two curves have become nearly indistinguishable from the majority classifier for ratios about 20 : 1. The lowest cost curve has crossed the 0.5 cost reduction contour at an imbalance of about 10 : 1 and crossed the 0.25 cost reduction contour at about 50 : 1. Thus, even a Bayes-optimal classifier with good performance, say an error probability of 0.1 with no imbalance, fares a lot worse under severe imbalance. With imbalances as low as 10 : 1, and certainly for imbalances of 100:1 and greater, the performance gain over the majority classifier is minimal.

Therefore, even the Bayes-optimal classifier performs only marginally better than a trivial classifier with severe imbalance. This is a very important conclusion since real classifiers will do worse than Bayes-optimal ones and often even worse than the trivial classifier.

3.2.3 Cost Sensitive Bayes Classifier

While straightforward, however, the Bayes Decision Rule does not take into consideration the possibility of different misclassification costs. Under certain circumstances, the cost of misclassifying a pattern originating from Class Ω^+ (Type I Error) is vital to be assigned with a different value than that corresponding to the cost of misclassifying a pattern originating from Class Ω^- (Type II Error). A necessity of this kind naturally arises within the framework of medical machine learning applications, where Class Ω^+ could correspond to "Suffering from Cancer" while Class Ω^- could correspond to "Healthy". In this case, Type I errors are much costlier than Type II errors since it is far more dangerous for a patient to be misdiagnosed as healthy, when in fact suffering from some dangerous disease, than it is to be misdiagnosed as having the disease when in fact healthy.

This issue is taken into consideration by the Bayes Criterion for Minimum Cost. The Bayes Criterion for Minimum Cost predicts for a particular observation vector $\mathbf{x}$ the hypothesis that leads to the lowest expected cost given a specification of costs for correct and incorrect predictions. Formally, letting $C(H_i|H_j)$ be the cost of predicting hypothesis H_i when H_j is true, where $\{i, j\} \in \{0, 1\}$, this criterion states that the optimal decision for a specific observation vector is in favor of hypothesis H_i that minimizes the risk functional given by the following equation:

$$R(H_i|\mathbf{x}) = \sum_{j \in \{0,1\}} P(H_j|\mathbf{x})C(H_i|H_j). \tag{3.37}$$

The risk functional in Eq. (3.37) corresponds to the risk taken when predicting that H_i is the true hypothesis given the observation vector $\mathbf{x}$. A basic assumption concerning the cost specification is the "reasonableness" condition stating that the cost for labelling an example incorrectly should always be greater than the cost of labelling it correctly. Mathematically, it should always be the case that $C(H_1|H_0) > C(H_0|H_0)$ and $C(H_0|H_1) > C(H_1|H_1)$. Then Bayes' Criterion for Minimum Cost may be summarized as:

$$\begin{aligned} &\text{If } R(H_0|\mathbf{x}) < R(H_1|\mathbf{x}), \text{ decide } H_0 \\ &\text{If } R(H_1|\mathbf{x}) < R(H_0|\mathbf{x}), \text{ decide } H_1. \end{aligned} \tag{3.38}$$

The case of equality is also detrimental which entails that either one of the two hypotheses may be considered to be true. Using Eq. (3.37), the decision can equivalently be based on the inequalities:

$$P(H_0|\mathbf{x})C(H_0|H_0) + P(H_1|\mathbf{x})C(H_0|H_1) \gtrless P(H_0|\mathbf{x})C(H_1|H_0) + P(H_1|\mathbf{x})C(H_1|H_1) \tag{3.39}$$

The Bayes Criterion for Minimum Cost minimizes the average misclassification cost in the same way the Bayes Decision Rule minimizes the probability of error. Specifically, the average misclassification cost will be given by the following equation:

$$R = \int_X R(a(\mathbf{x})|\mathbf{x})p(\mathbf{x})d\mathbf{x}, \tag{3.40}$$

where $R(a(\mathbf{x})|\mathbf{x})$ corresponds to the risk incurred when the decision $a(\mathbf{x})$ is made given a particular observation vector $\mathbf{x}$. The risk incurred by making a particular decision on a given example is mathematically described as:

$$R(a(\mathbf{x})|\mathbf{x}) = \begin{cases} R(H_0|\mathbf{x}), & \text{if we decide } H_0 \\ R(H_1|\mathbf{x}), & \text{if we decide } H_1. \end{cases} \tag{3.41}$$

Having in mind that the decisions are made on the basis of the Bayes Criterion for Minimum Cost, Eq. (3.41) may be written in the following form:

$$R(a(\mathbf{x})|\mathbf{x}) = \min\left[R(H_0|\mathbf{x}), R(H_1|\mathbf{x})\right] \tag{3.42}$$

Equation (3.42), then, entails that the conditional cost is as small as possible for any given observation vector $\mathbf{x}$. An immediate consequence of this fact is that the average cost given by the integral in Eq. (3.40) is minimized.

Given that $p = P(H_0|\mathbf{x})$ and consequently $1 - p = P(H_1|\mathbf{x})$, Inequalities (3.39) may be written in the following form:

$$p\, C(H_0|H_0) + (1 - p)\, C(H_0|H_1) \gtrless p\, C(H_1|H_0) + (1 - p)\, C(H_1|H_1) \tag{3.43}$$

If p^* corresponds to the probability for which both sides of Eq. (3.43) are equal, then p^* may be computed via the equation:

$$p^*\, C(H_0|H_0) + (1 - p^*)\, C(H_0|H_1) = p^*\, C(H_1|H_0) + (1 - p^*)\, C(H_1|H_1). \tag{3.44}$$

Thus, p^* corresponds to the a posteriori probability threshold for making optimal decisions. Its value can be directly computed by rearranging Eq. (3.44) and solving for p^* which leads to the following threshold value:

$$p^* = \frac{C(H_0|H_1) - C(H_1|H_1)}{C(H_1|H_0) - C(H_0|H_0) + C(H_0|H_1) - C(H_1|H_1)}. \tag{3.45}$$

According to this interpretation the Bayes Criterion for Minimum Cost may be reformulated as follows:

$$\begin{array}{l} \text{If } p > p^*, \text{ decide } H_0 \\ \text{If } p < p^*, \text{ decide } H_1, \end{array} \tag{3.46}$$

where the prediction concerning the true hypothesis is exclusively based on the probability threshold provided by Eq. (3.45).

The misclassification cost imbalance problem arises in situations when Type I and Type II Errors are assigned values which are extremely skewed in magnitude. Specifically, these situations occur when the cost of misclassifying a pattern from the positive (negative) class Ω^+ (Ω^-) is significantly greater than the cost of misclassifying a pattern from the negative (positive) class Ω^- (Ω^+). Mathematically, this problem is associated with cost assignments for which $C(H_1|H_0) >> C(H_0|H_1)$ or $C(H_0|H_1) >> C(H_1|H_0)$. A crucial point in obtaining a deeper understanding of the misclassification-cost imbalance problem involves the identification of the primary factor affecting the decision making process. This entails that one should focus on the probability threshold p^* which constitutes the fundamental component of the decision making process. Specifically, it is essential to understand that assigning extremely imbalanced misclassification costs results in extreme modifications of the probability threshold. Having in mind the "reasonableness" condition, the probability threshold value in Eq. (3.45) may be written in the following form:

$$p^* \approx \frac{C(H_0|H_1)}{C(H_1|H_0) + C(H_0|H_1)} \tag{3.47}$$

since the contribution of the terms $C(H_0|H_0)$ and $C(H_1|H_1)$ is practically negligible.

When the case is that Type I Errors are significantly greater than Type II Errors, it is reasonable to assume that $C(H_1|H_0) + C(H_0|H_1) \approx C(H_1|H_0)$ which yields that the probability threshold will be given by the following equation:

$$p^* \approx \frac{C(H_0|H_1)}{C(H_1|H_0)} = 0 \tag{3.48}$$

Thus, the Bayes Criterion for Minimum Cost degenerates to the following form:

$$\begin{aligned} &\text{If } p > 0, \text{ decide } H_0 \\ &\text{If } p < 0, \text{ decide } H_1, \end{aligned} \tag{3.49}$$

where it is obvious that the decision making process is biased towards the hypothesis with the minimum misclassification cost, which is H_0 for this particular occasion. On the other hand, when the case is that Type II Errors are significantly greater than Type I Errors, it is reasonable to assume that $C(H_1|H_0) + C(H_0|H_1) \approx C(H_0|H_1)$ which yields that the probability threshold will be given by the following equation:

$$p^* \approx \frac{C(H_0|H_1)}{C(H_0|H_1)} \approx 1. \tag{3.50}$$

Thus, the Bayes Criterion for Minimum Cost degenerates to the following form:

$$\begin{aligned} &\text{If } p > 1, \text{ decide } H_0 \\ &\text{If } p < 1, \text{ decide } H_1, \end{aligned} \tag{3.51}$$

where it is obvious that the decision making process is biased towards the hypothesis with the minimum misclassification cost, which is H_1 for this particular occasion. When the misclassification costs are completely balanced it is easy to deduce for the probability threshold that $p^* \approx \frac{1}{2}$ and the misclassification cost imbalance problem vanishes.

In order to incorporate class conditional and prior probabilities within the decision making process, the a posteriori probability density functions, $P(H_0|\mathbf{x})$ and $P(H_1|\mathbf{x})$, are written as follows:

$$\begin{aligned} P(H_0|\mathbf{x}) &= \frac{P(\mathbf{x}|H_0)P(H_0)}{p(\mathbf{x})} \\ P(H_1|\mathbf{x}) &= \frac{P(\mathbf{x}|H_1)P(H_1)}{p(\mathbf{x})}. \end{aligned} \tag{3.52}$$

Letting, $p(\mathbf{x}|H_0) = q$ and consequently $p(\mathbf{x}|H_1) = 1 - q$, then Inequalities (3.43) may be written in the following form:

$$q\, C(H_0|H_0)P_0 + (1 - q)\, C(H_0|H_1)P_1 \geqslant q\, C(H_1|H_0)P_0 + (1 - q)\, C(H_1|H_1)P_1 \tag{3.53}$$

If q^* corresponds to the probability for which both sides of Inequalities (3.53) are equal, then q^* may be computed by the equation:

$$q^*\, C(H_0|H_0)P_0 + (1 - q^*)\, C(H_0|H_1)P_1 = q^*\, C(H_1|H_0)P_0 + (1 - q^*)\, C(H_1|H_1)P_1 \tag{3.54}$$

Thus, q^* corresponds to the class conditional probability threshold for making optimal decisions. Its value can be directly computed by rearranging Eq. (3.54) and solving for q^* which yields that:

$$q^* = \frac{P_1(C(H_0|H_1) - C(H_1|H_1))}{P_0(C(H_1|H_0) - C(H_0|H_0)) + P_1(C(H_0|H_1) - C(H_1|H_1))}. \tag{3.55}$$

According to this interpretation, the Bayes Criterion for Minimum Cost may be reformulated as follows:

$$\begin{aligned} &\text{If } q > q^*, \text{ decide } H_0 \\ &\text{If } q < q^*, \text{ decide } H_1, \end{aligned} \tag{3.56}$$

where the prediction concerning the true hypothesis is exclusively based on the class conditional probability threshold provided by Eq. (3.55).

Equation (3.55) may serve as an ideal starting point in order to conduct a further investigation concerning the intrinsic nature of the class imbalance problem. Specifically, it can be used to unfold the reason why severe imbalance on the class priors determinately affects the decision making process. In other words, Eq. (3.55) plays an essential role in order to identify the direct link between the degree of class imbalance and the class conditional probability threshold q^*. Suppose that the prior probabilities are given by the following equations:

$$\begin{aligned} P_0 &= P(\Omega^+) = A + z \\ P_1 &= P(\Omega^-) = B - z, \end{aligned} \tag{3.57}$$

where $A, B > 0$ and $A + B = 1$ so that $P_0 + P_1 = 1$. Both class prior probabilities vary within the $[0, 1]$ interval which indicates that z will vary within the $[-A, B]$ interval. In other words z is a factor accounting for the degree of imbalance between the class prior probabilities. Thus, Eq. (3.55) may be rewritten in the following form:

$$q^*(z) = \frac{(B - z)(C(H_0|H_1) - C(H_1|H_1))}{(A + z)(C(H_1|H_0) - C(H_0|H_0)) + (B - z)(C(H_0|H_1) - C(H_1|H_1))} \tag{3.58}$$

where $-A \leq z \leq B$. Equation 3.58 describes the class conditional probability threshold as a function of the parameter z which quantifies the degree of prior class imbalance. Thus, in order to identify the behavior of q^* with respect to the imbalance factor z it would be reasonable to compute the following quantities:

- $\lim\limits_{z \to 0} q^*(z)$ where $A = B = \frac{1}{2}$: which denotes the behavior of the class conditional probability threshold when the class priors are completely balanced.
- $\lim\limits_{z \to -A} q^*(z)$: which denotes the behavior of the class conditional probability threshold when the class priors are completely imbalanced towards the negative Ω^- class of patterns.
- $\lim\limits_{z \to B} q^*(z)$: which denotes the behavior of the class conditional probability threshold when the class priors are completely imbalanced towards the positive Ω^+ class of patterns.

The first limit, in which $A = B = \frac{1}{2}$, corresponds to the situation when the class priors are completely balanced which yields that:

$$\lim_{z \to 0} q^*(z) = \frac{B(C(H_0|H_1) - C(H_1|H_1))}{A(C(H_1|H_0) - C(H_0|H_0)) + B(C(H_0|H_1) - C(H_1|H_1))}. \quad (3.59)$$

When both classes are also assigned equal costs for correct and incorrect classifications it is clear from Eq. (3.59) that when $z \to 0$, then $q^* \to \frac{1}{2}$ and the class imbalance problem vanishes.

The second limit corresponds to the situation when the class priors are completely imbalanced towards the negative class Ω^- of patterns which yields that:

$$\lim_{z \to -A} q^*(z) = \frac{(B + A)(C(H_0|H_1) - C(H_1|H_1))}{(B + A)(C(H_0|H_1) - C(H_1|H_1))} = 1. \quad (3.60)$$

In other words, when $z \to -A$ then $q^* \to 1$, which according to the Bayes Criterion for Minimum Cost in Eq. (3.56) yields that the decision making process will be biased in favor of the negative class of patterns.

Finally, the third limit corresponds to the situation when the class priors are completely imbalanced towards the positive class Ω^+ of patterns which yields that

$$\lim_{z \to B} q^*(z) = 0. \quad (3.61)$$

Thus, when $z \to B$ then $q^* \to 0$, which according to the Bayes Criterion for Minimum Cost in Eq. (3.56) yields that the decision making process will be biased in favor of the positive class of patterns.

The same kind of behavior will be observed when the class priors and misclassification costs are simultaneously taken into consideration. For this purpose it is convenient to express the class conditional probability threshold in the following form:

$$q^* = \frac{P(H_1)C(H_0|H_1) - P(H_1)C(H_1|H_1)}{P(H_0)C(H_1|H_0) - P(H_0)C(H_0|H_0) + P(H_1)C(H_0|H_1) - P(H_1)C(H_1|H_1)}.$$

$$(3.62)$$

Then, according to the "reasonableness" conditions the quantities $P(H_1)C(H_1|H_1)$ and $P(H_0)C(H_0|H_0)$ may be ignored since practically they contribute nothing. Thus, the class conditional probability threshold will be approximately equal to:

$$q^* \approx \frac{P(H_1)C(H_0|H_1)}{P(H_0)C(H_1|H_0) + P(H_1)C(H_0|H_1)}. \tag{3.63}$$

When $P(H_1)C(H_0|H_1) >> P(H_0)C(H_1|H_0)$ then the class conditional probability threshold will be written in the following form:

$$q^* \approx \frac{P(H_1)C(H_0|H_1)}{P(H_1)C(H_0|H_1)} = 1 \tag{3.64}$$

and as a consequence all decisions will be made in favor of hypothesis H_1. On the other hand, when $P(H_1)C(H_0|H_1) << P(H_0)C(H_1|H_0)$ the class conditional probability threshold will be written in the following form:

$$q^* \approx \frac{P(H_1)C(H_0|H_1)}{P(H_0)C(H_1|H_0)} \approx 0 \tag{3.65}$$

and as a consequence all decisions will be made in favor hypothesis H_0.

3.2.4   Nearest Neighbor Classifier

The class imbalance effect on the Nearest Neighbor classifier was the subject of several works such as Robert et al. [29], Kubat and Matwin [17], Batista et al. [3] where the authors investigate the reasons why abundant negative instances hinder the performance of standard classifiers. These approaches are focused on designing binary classifiers addressing classification problems which involve two highly imbalanced and overlapping classes. Specifically, the authors are interested in investigating the problem related to the designing of a binary classifier that should be able to discriminate the more infrequent (positive) class against the majority (negative) one. It must be mentioned that the positive class of patterns, despite being under-represented in the training set, is equally important to the negative class.

The authors are primarily interested in explaining the performance of the Nearest Neighbor (1-NN) classifier in such a setting. Specifically, they document that as the number of negative examples in a noisy domain grows (with the number of positives being held constant) so does the likelihood that the nearest neighbor of any example will be negative. Thus, many positive examples will be misclassified. According to the authors, this effect will be unavoidably magnified as the number of negatives is infinitely increased so that the positive class of patterns will be totally ignored. In other words, the (1-NN) classifier will fail to detect any of the sparse positive examples in a domain thriving on negative instances. Therefore, the k-nearest neighbor rule will correctly recognize most examples of the majority class.

3.2.5   Decision Trees

The class imbalance effect on decision trees classifiers was studied in several works such as in Weiss and Provost [37, 40], Batista et al. [3]. Specifically, the authors are interested in providing a qualitative description revealing the effect of class distribution on designing learning algorithms which are based on a limited-size training set due to cost concerns. Their research is focused on investigating the reasons for which classifiers perform worse on the minority class within the general context of decision trees learning algorithms. Specifically, they elaborate on two major observations that became clear throughout their experimentation which involved 26 different data sets and a decision tree classifier trained on these data sets.

The first observation is that the classification rules predicting the minority class tend to have a much higher error rate than those predicting the majority class. Specifically, it is documented that minority-biased classification rules perform worse than their majority-biased counterparts mainly because there is an uneven distribution of positive (minority) and negative (majority) class patterns within the testing set. The authors elaborate that when decision trees classifiers are evaluated on a test set with a 9 : 1 proportion in favor of the negative class patterns the leaves predicting the minority class demonstrate an expected error rate of 90 % while the leaves predicting the majority class have an expected error rate of only 10 %! Moreover, they argue that minority-biased rules perform so poorly because they are generally formed from fewer training examples than the majority-biased rules.

The second observation is that test examples belonging to the minority class are misclassified more often than test examples belonging to the majority class. According to the authors, decision tree classifiers are supposed to handle all feature values, even if they are not observed in the training data, so that it is possible for a leaf in the decision tree to cover no training examples. In this case, if a learner adopts the strategy of labelling the leaf with the majority class, the performance of the classifier on the majority-class examples will improve, but at the expense of the minority-class examples. A second reason that classifiers perform worse on the minority-class test examples is that a classifier is less likely to fully flesh out the boundaries of the minority concept in the concept space because there are fewer examples of that class to learn from. Thus, some of the minority-class test examples may be classified as belonging to the majority class. However, if additional examples of the minority class were available for training, one would expect that the minority concept to grow to include additional regions of the concept space, such as regions that previously were not sampled at all.

The effects of the class imbalance problem on decision tree classifiers were also investigated in Robert et al. [29], Kubat and Matwin [17]. Specifically, they provide evidence that the induction of decision tree classifiers will also be affected. Decision trees are known to be universal classifiers having the ability to realize any dichotomy of points in general position in a n-dimensional continuous space by appropriately adjusting their size. Such classifiers attempt to partition the instance space into regions labelled with the class that has the majority in the region. This entails that when imbalanced classes are involved, the regions with mixed positives and negatives will tend to be labelled with the preponderant class. According to the authors, each positive example may be viewed as being separated from other positives by a "wall" of negatives which forces the tree generator either to stop splitting or, in which case negatives are a majority, or to keep splitting until it forms a tiny region.

Finally, their experimentation involved the definition of an artificial test-bed where random examples are generated for both the positive and negative classes of patterns. Specifically, positive examples were drawn from a two-dimensional normal distribution with $\mu_+ = [0, 0]$ and $\sigma_+ = [1, 1]$ while negative examples were drawn from a two-dimensional normal distribution with $\mu_- = [2, 0]$ and $\sigma_- = [2, 2]$. In all runs, the same 50 training positives were used, while the number of negatives grew from 50 to 800 in increments of 50. Performance was measured on an independent testing set with the same proportion of positive and negative examples as the training set. The classification results obtained by utilizing the C4.5 classifier [25] and 1-NN demonstrate that as the set of negative instances outnumbers the corresponding set of positive instances, the performance on the majority class exceeds 90 % while the performance on the minority class collapses.

Further investigation is conducted by the authors in [3], where they focus on decision tree classifiers in the presence of class overlapping. The authors argue that they may need to create many tests to distinguish the minority class cases from majority class cases. Moreover, they document that pruning the decision tree might not necessarily alleviate the problem. This is due to the fact that pruning removes some branches considered too specialized which results in labelling new leaf nodes with the dominant class of this node. Thus, there is a high probability that the majority class will also be the dominant class of those leaf nodes.

3.2.6   Neural Networks

It has been theoretically shown [11, 15, 27, 30, 36] that Neural Network-based classifiers approximate Bayesian a posteriori probabilities when the desired network outputs are 1 of M and squared error or cross-entropy cost functions are used during the training phase. However, this theoretical result relies on a number of assumptions that guarantee the accurate estimation of these classifiers. Specifically, it is assumed that

- the network size is large enough
- infinite training data are available so that the training procedure finds a global minimum
- the a priori class probabilities of the test set are correctly represented in the training set.

Notwithstanding, a commonly encountered problem in neural network-based classification is related to the case when the class frequencies in the training set are highly skewed. This corresponds to the particular occasion when the number of training examples from each class varies significantly between classes so that the resulting classifier is biased towards the majority class. This is also the problem addressed by the authors in [2, 4, 20, 26] where they show that the class imbalance problem generates unequal contributions to the mean square error (MSE) in the training phase. Thereby, the major contribution to the MSE is produced by the majority class.

In order to investigate the exact way in which the class imbalance problem affects the Back Propagation algorithm, it is convenient to consider the general framework of a binary classification problem where the two classes of interest are namely Ω^+ and Ω^- and the number of samples from each class is assumed to be in proportion to the a priori probability of class membership. Therefore, the network output $F(\mathbf{x}; \Theta)$ will provide approximations to the a posteriori probabilities $P(\Omega^+|\mathbf{x})$ and $P(\Omega^-|\mathbf{x})$ given the observation vector $\mathbf{x}$ and the vector Θ of internal network parameters. The optimal network parameter vector $\widehat{\Theta}$ will be determined by the utilization of the Back Propagation algorithm which minimizes the MSE cost functional given by the equation:

$$E(\Theta) = E^+(\Theta) + E^-(\Theta) = \frac{1}{n^+} \sum_{i=1}^{n^+} (y_i - F(\mathbf{x}_i; \Theta))^2 + \frac{1}{n^-} \sum_{j=1}^{n^-} (y_j - F(\mathbf{x}_j; \Theta))^2,$$

(3.66)

where $y_i = +1, \forall i \in [n^+]$ denotes the target network output when presented with a positive observation vector $\mathbf{x}_i$ and $y_j = -1, \forall j \in [n^-]$ denotes the target network output when presented with a negative observation vector $\mathbf{x}_j$. Moreover, n^+, n^- correspond to the numbers of training samples from the positive and negative classes respectively. Nonetheless, the class imbalance problem refers to the situation when the positive class of patterns (Ω^+) corresponds to the minority class while the negative class of patterns (Ω^-) corresponds to the majority class so that $n^+ << n^-$. If this is the case, then $E^+(\Theta) << E^-(\Theta)$ and $\|\nabla E^+(\Theta)\| << \|\nabla E^-(\Theta)\|$ which finally yields that $\|\nabla E(\Theta)\| = \|\nabla E^-(\Theta)\|$. Thus, $-\|\nabla E(\Theta)\|$ is not always the best direction to minimize the MSE in both classes. In addition, the angle between the two vectors ∇E^+ and ∇E^- is larger than 90° which entails that during the gradient descent, E^- tends to decrease and E^+ tends to increase rapidly. As a consequence, when the value of E^+ approaches its upper limit convergence becomes slow.

3.2.7   Support Vector Machines

Support Vector Machines (SVMs) are extensively discussed in Sect. 5.1 where the reasons for their remarkable success in many applications are thoroughly explained. However, the classification performance of SVMs is significantly hindered when learning from imbalanced datasets in which negative instances heavily outnumber the positive ones. This particular problem is studied in Akbani et al. [1], Veropoulos et al. [35], Wu and Chang [41] where the authors are interested in revealing the effects of class imbalance on SVMs. According to the authors, the source of the boundary skewness may be attributed to the fact that the positive points lie further from the ideal boundary. In particular, they mention that the imbalance in the training data ratio means that the positive instances may lie further away from the "ideal" boundary than the negative instances.

This may be illustrated by considering the situation when one is to draw n randomly chosen numbers between 1–100 from a uniform distribution. In such a case the chances of drawing a number close to 100 would improve with increasing values of n, even though the expected mean of the draws is invariant of n. As a result of this phenomenon, SVMs learn a boundary that is too close to and skewed towards the positive instances. Moreover, SVMs are trained to minimize an objective function which simultaneously accounts for the margin between the positive and the negative subspaces of patterns, and the associated misclassification error. For this purpose there is a tradeoff parameter tolerating between the maximization of the margin and the minimization of the error. If this regularization parameter is not very large, SVMs simply learn to classify everything as negative because that makes the "margin" the largest, with zero cumulative error on the abundant negative examples. The only tradeoff is the small amount of cumulative error on the few positive examples, which does not count for much. This explains why SVMs fail completely in situations with a high degree of imbalance. According to the authors the proportion of positive and negative support vectors for the trained classifier may be identified as a supplementary source of boundary skew. Specifically, it is argued that the degree of class imbalance within the training data set determinately affects the degree of imbalance related to the ratio between the positive and negative support vectors. Thus, a data set thriving in negative example will result in a set of support vectors highly skewed towards the negative class. This entails that the neighborhood for any data point close to the boundary will be dominated by negative support vectors and hence the decision function would be more likely to assign a boundary point to the negative class of patterns.

References

1. Akbani, R., Kwek, S., Japkowicz, N.: Applying support vector machines to imbalanced datasets. In: Proceedings of the 15th European Conference on Machine Learning (ECML), pp. 39–50 (2004)

2. Alejo, R., Sotoca, J.M., Casa n, G.A.: An empirical study for the multi-class imbalance problem with neural networks. In: CIARP '08: Proceedings of the 13th Iberoamerican Congress on Pattern Recognition, pp. 479–486. Springer, Berlin, Heidelberg (2008)

3. Batista, G.E.A.P.A., Prati, R.C., Monard, M.C.: A study of the behavior of several methods for balancing machine learning training data. SIGKDD Explor. Newsl. **6**(1), 20–29 (2004)

4. Bruzzone, L., Serpico, S.B.: Classification of imbalanced remote-sensing data by neural networks. Pattern Recognit. Lett. **18**(11–13), 1323–1328 (1997)

5. Chan, P., Stolfo, S.J.: Toward scalable learning with non-uniform class and cost distributions: A case study in credit card fraud detection. In: Proceedings of the Fourth International Conference on Knowledge Discovery and Data Mining, pp. 164–168. AAAI Press (1998)

6. Daskalaki, S., Kopanas, I., Avouris, N.: Evaluation of classifiers for an uneven class distribution problem. Appl. Artif. Intell. **20**, 1–37 (2006)

7. Drummond, C., Holte, R.: Learning to live with false alarms. In: Proceedings of the KDD Data Mining Methods for Anomaly Detection Workshop, pp. 21–24 (2005)

8. Estabrooks, A., Jo, T., Japkowicz, N.: A multiple resampling method for learning from imbalanced data sets. Comput. Intell. **20**(1), 18–36 (2004)

9. Ezawa, K., Singh, M., Norton, S.W.: Learning goal oriented bayesian networks for telecommunications risk management. In: Proceedings of the 13th International Conference on Machine Learning, pp. 139–147. Morgan Kaufmann (1996)

10. Fawcett, T., Provost, F.: Combining data mining and machine learning for effective user profiling, pp. 8–13. AAAI Press (1996)

11. Hampshire, J.B., II, Pearlmutter, B.A.: Equivalence proofs for multi-layer perceptron classifiers and the bayesian discriminant function (1990)

12. Japkowicz, N.: . Learning from Imbalanced Data Sets: A Comparison of Various Strategies, pp. 10–15. AAAI Press (2000)

13. Japkowicz, N., Stephen, S.: The class imbalance problem: a systematic study. Intell. Data Anal. **6**(5), 429–449 (2002)

14. Jo, T., Japkowicz, N.: Class imbalances versus small disjuncts. SIGKDD Explor. **6**(1), 40–49 (2004)

15. Kanaya, F., Miyake, S.: Bayes statistical behavior and valid generalization of pattern classifying neural networks. In: COGNITIVA 90: Proceedings of the third COGNITIVA symposium on At the Crossroads of Artificial Intelligence, Cognitive Science, and Neuroscience, pp. 35–44. Amsterdam, The Netherlands, The Netherlands. North-Holland Publishing Co (1991)

16. Kubat, M., Holte, R.C., Matwin, S., Kohavi, R., Provost, F.: Machine learning for the detection of oil spills in satellite radar images. In: Machine Learning, pp. 195–215 (1998)

17. Kubat, M., Matwin, S.: Addressing the curse of imbalanced training sets: one-sided selection. In: Proceedings of the Fourteenth International Conference on Machine Learning, pp. 179–186. Morgan Kaufmann (1997)

18. Lampropoulos, A.S., Sotiropoulos, D.N., Tsihrintzis, G.A.: Cascade hybrid recommendation as a combination of one-class classification and collaborative filtering. Int. J. Artif. Intell. Tools **23**(4), 1 (2014)

19. Laurikkala, J.: Improving identification of difficult small classes by balancing class distribution. In: AIME '01: Proceedings of the 8th Conference on AI in Medicine in Europe, pp. 63–66. Springer-Verlag, London, UK (2001)

20. Lawrence, S., Burns, I., Back, A., Tsoi, A.C., Giles, C.L.: Neural Network Classification and Prior Class Probabilities (1998)

21. Lewis, D.D., Catlett, J.: Heterogeneous uncertainty sampling for supervised learning. In: Cohen, W.W., Hirsh, H. (eds.) Proceedings of ICML-94, 11th International Conference on Machine Learning, pp. 148–156. Morgan Kaufmann Publishers, New Brunswick, US, San Francisco, US (1994)

22. Ling, C.X., Li, C.: Data mining for direct marketing: problems and solutions. In: KDD, pp. 73–79 (1998)

23. Maes, S., Tuyls, K., Vanschoenwinkel, B., Manderick, B.: Credit card fraud detection using bayesian and neural networks. In: Maciunas R.J. (ed.) Interactive image-guided neurosurgery. American Association Neurological Surgeons, pp. 261–270 (1993)

24. Prati, R.C., Batista, G.E.A. P.A., Monard, M.C.: Class Imbalances Versus Class Overlapping: An Analysis of a Learning System Behavior (2004)
25. Quinlan, J.R.: C4.5: Programs for Machine Learning. Morgan Kaufmann Publishers Inc., San Francisco, CA, USA (1993)
26. Anand, R., Mehrota, K.G., Mohan, C.K., Ranka, S.: An imnprove algorithm for neural netowork classification of imbalanced training sets. IEEE Trans. Neural Netw. **4**, 962–969 (1993)
27. Richard, M., Lippmann, R.: Neural network classifiers estimate bayessian a posteriori probabilities. Neural Comput. **3**(4), 461–483 (1991)
28. Riddle, P., Segal, R., Etzioni, O.: Representation design and brute-force induction in a boeing manufacturing domain. Appl. Artif. Intell. **8**, 125–147 (1994)
29. Robert, M.K., Holte, R., Matwin, S.: Learning when negative examples abound. In: ECML-97, Lecture Notes in Artificial Intelligence, pp. 146–153. Springer Verlag (1997)
30. Rojas, R.: A Short Proof of the Posterior Probability Property of Classifier Neural Networks (1996)
31. Sotiropoulos, D.N., Tsihrintzis, G.A.: Artificial immune system-based classification in class-imbalanced problems. In: 2011 IEEE Workshop on Evolving and Adaptive Intelligent Systems, EAIS 2011, pp. 131–138. Paris, France, Apr 14–15 (2011)
32. Sotiropoulos, D.N., Tsihrintzis, G.A.: Artificial immune system-based classification in class-imbalanced image classification problems. In: Eighth International Conference on Intelligent Information Hiding and Multimedia Signal Processing, IIH-MSP 2012, pp. 138–141. Piraeus-Athens, Greece, July 18–20 (2012)
33. Stroulia, E. and Matwin, S., editors (2001). *Advances in Artificial Intelligence, 14th Biennial Conference of the Canadian Society for Computational Studies of Intelligence, AI 2001, Ottawa, Canada, June 7-9, 2001, Proceedings*, volume 2056 of *Lecture Notes in Computer Science*. Springer
34. van den Bosch, A., Weijters, T., Herik, H.J. V.D., Daelemans, W.: When Small Disjuncts Abound, Try Lazy Learning: A Case Study (1997)
35. Veropoulos, K., Campbell, C., Cristianini, N.: Controlling the sensitivity of support vector machines. In: Proceedings of the International Joint Conference on AI, pp. 55–60 (1999)
36. Wan, E.: Neural networks classification: a bayessian interpretation. IEEE Trans. Neural Netw. **1**(4), 303–305 (1990)
37. Weiss, G., Provost, F.: The effect of class distribution on classifier learning: An empirical study. Technical report (2001)
38. Weiss, G.M.: Mining With Rarity: A Unifying Framework. Springer, Heidelberg (2004)
39. Weiss, G.M., Hirsh, H.: Learning to predict rare events in event sequences. In: Proceedings of the Fourth International Conference on Knowledge Discovery and Data Mining, pp. 359–363. AAAI Press (1998)
40. Weiss, G.M., Provost, F.: Learning When Training Data are Costly: The Effect of Class Distribution on Tree Induction (2002)
41. Wu, G., Chang, E.Y.: Class-boundary alignment for imbalanced dataset learning. In: ICML 2003 Workshop on Learning from Imbalanced Data Sets, pp. 49–56 (2003)

Chapter 4
Addressing the Class Imbalance Problem

Abstract In this chapter, we investigate the particular effects of the class imbalance problem on standard classifier methodologies and present the various methodologies that have been proposed as a remedy. The most interesting approach within the context of Artificial Immune Systems is the one related to the machine learning paradigm of one-class classification. One-Class Classification problems may be thought of as degenerated binary classification problems in which the available training instances originate exclusively from the under-represented class of patterns.

4.1 Resampling Techniques

Resampling is probably the most-direct approach that has been proposed as a remedy for the class imbalance problem at the data level. The basic idea behind resampling is to change the class priors within the training set by either increasing the number of instances from the minority class (over-sampling) or by decreasing the number of instances from the majority class (under-sampling). As a result, the class distribution of the training data is modified so that the resulting data set becomes more balanced. In other words, resampling techniques focus on alleviating classification problems that are strongly related to the distribution of the training data within each class. As argued in Weiss and Provost [29], the original distribution of training instances may not be the optimal distribution to use for a given classifier. Specifically, when the case is that the training data set is highly skewed towards a particular class, the original distribution is almost always the most inappropriate one as evidenced by the trivial majority classifier. Thus, the utilization of better class distributions will improve the validation and testing results of the classifier. Although there is no real way to know the best distribution for a problem, resampling methods use various heuristics to modify the distribution to one that is closer to the optimal distribution. More importantly, resampling techniques can be used in combination with any given classifier. Moreover, under-sampling provides a solution for the majority of real world applications involving enormous data sets which have to be reduced in size. Finally, the use of varying sampling rates allows the fine-tuning of class imbalances and, consequently, their resulting classification models and performance.

© Springer International Publishing AG 2017   79
D.N. Sotiropoulos and G.A. Tsihrintzis, *Machine Learning Paradigms*,
Intelligent Systems Reference Library 118, DOI 10.1007/978-3-319-47194-5_4

4.1.1   Natural Resampling

Given the general setting of a binary classification problem, *Natural Resampling* refers to the process of obtaining more samples from the minority class in order to be included in the training data set. Such an approach aims at representing both classes with an equal number of training instances. Specifically, this technique may be considered as a straightforward solution to the class imbalance problem since all of the training samples remain drawn from the same natural phenomenon that originally generated them. However, this is not always the case in real-world applications where the available training instances are inherently skewed.

4.1.2   Random Over-Sampling and Random Under-Sampling

The simplest method to increase the size of the minority class corresponds to *Random Over-Sampling*, that is, a non-heuristic method that balances the class distribution through the random replication of positive examples [3, 21]. On the other hand, *Random Under-Sampling* [15, 32] aims at balancing the data set through the random removal of negative examples. Despite its simplicity, it has empirically been shown to be one of the most effective resampling methods. A common feature of both sampling techniques is that each training sample has an equal probability of being selected either to be replicated or to be eliminated. However, it must be noted that random over-sampling can increase the likelihood of occurring overfitting, since it makes exact copies of the minority class examples. In this way, a symbolic classifier, for instance, might construct rules that are apparently accurate, but actually cover one replicated example. On the other hand, the major drawback of random under-sampling is that this method can discard potentially useful data that could be important for the induction process.

4.1.3   Under-Sampling Methods

Tomek Links

Tomek links where originally defined in Tomek [28] as an under-sampling methodology providing an elimination criterion which may be utilized in order to discard majority samples from the training set. According to the author, *Tomek links* are formed between pairs of training instances $(\mathbf{x}, \mathbf{y})$ if there is no point $\mathbf{z}$ in the training data set for which it holds that $d(\mathbf{x}, \mathbf{z}) < d(\mathbf{x}, \mathbf{y})$ or $d(\mathbf{y}, \mathbf{z}) < d(\mathbf{x}, \mathbf{y})$, where $d(\mathbf{x}, \mathbf{y})$ is the distance between two given examples. Specifically, it is elaborated that if two examples form a *Tomek link*, then either one of these examples is noise or both examples are close to the class boundary. As an under-sampling methodology *Tomek links* imply that the patterns to be eliminated must be exclusively originating from the majority class. Additionally, Tomek-link under-sampling is a general methodology which can be applied in a variety of classification methods. In particular, it is very useful for noisy datasets as this algorithm potentially removes those data points that may be noise instances and could lead to classification errors. On the other hand, its major drawback is that this method can discard potentially useful data that could be important for the induction process. Finally, it must be mentioned that this method uses a time consuming algorithm of higher order complexity which will run slower than other algorithms.

Condensed Nearest Neighbor Rule

The Condensed Nearest Neighbor (CNN) [13] is an under-sampling methodology which is specific to the k-Nearest Neighbor (kNN) classification rule. Its primary purpose consists of removing minority class examples that are distant from the decision border while the remaining (safe majority or minority) examples are used for learning. The training procedure of CNN is focused on finding a consistent subset of training patterns. A subset $\widehat{E} \subseteq E$ is specifically characterized as consistent with E if, by training the 1-NN classifier on $\widehat{E}$, all patterns in E will be correctly classified. In particular, CNN starts by gathering a randomly drawn majority class example and all minority class instances in $\widehat{E}$. Afterwards, the 1-NN learning rule is utilized over $\widehat{E}$ in order to classify all examples in E. Finally, every misclassified example from E is moved to $\widehat{E}$. The basic idea behind this implementation for finding a consistent subset, which is not the smallest possible, is to eliminate training instances that are considered as less relevant for learning.

One-Sided Selection

One Sided Selection (OSS) [19] is another under-sampling method that was proposed as a data level remedy for the class imbalance problem. The fundamental underpinning behind this approach is once again a selection mechanism whose purpose is to modify the distribution of negative samples. One-sided selection operates by generating a more representative subset of negative patterns which will be more appropriate for training. In other words, it may be seen as a balancing procedure which eliminates the less informative or more error prone negative instances. Specifically, it focuses on eliminating negative patterns that suffer from class-label noise or borderline patterns that lie close to the boundary between the positive and negative examples. Redundant negative data points are also removed, even though they do not hinder the classification performance, since they can be taken over by other examples. However, the elimination procedure ignores safe negative instances and the complete set of positive patterns. It must be mentioned that the One-Sided Selection resampling technique utilizes the notion of consistency that was thoroughly discussed in Sect. 4.1.3. Thus, it is a natural consequence that an intelligent agent will try to eliminate borderline examples and examples suffering from class-label noise.

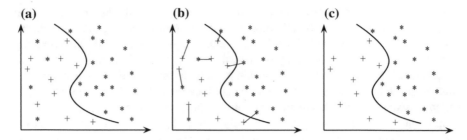

Fig. 4.1 One sided selection

Such examples may be identified by taking advantage of the notion of Tomek links that were discussed in Sect. 4.1.3. The schematic representation of the One-Sided Selection resampling technique appears in Fig. 4.1a, b, c, where Fig. 4.1a represents the original data set, Fig. 4.1b depicts the identified Tomek links and finally Fig. 4.1c represents the resulting data set after the removal data points pertaining to Tomek links.

In particular, the algorithm for the one sided selection resampling technique is the following:

1. Let S be the original training set and C be the temporary set of representative points that are not yet eliminated during One Sided Selection.
2. Initially, C contains all positive examples from S and one randomly selected negative sample.
3. Classify all elements in S with 1-NN rule using the examples in C and compare the assigned concept labels with the original ones. Move all misclassified examples into C that is now consistent with S while being smaller.
4. Remove from C all negative examples participating in Tomek links. This removes those negative examples that are believed to be borderline and/or noisy. All positive examples are retained. The resulting set of representative points is referred to as T.

Neighborhood Cleaning Rule

The Neighborhood Cleaning Rule (NCL) is an under-sampling method that was proposed in Laurikkala [20] as an extension to original OSS approach. According to the author, the major drawback of the OSS method is the utilization of the CNN rule which is extremely sensitive to noise. Specifically, the author argues that the presence of noisy examples in the final training set obtained by OSS, will hinder the classification performance of any learning algorithm. This is attributed to the fact that noisy training data will incorrectly classify several of the subsequent testing patterns. The basic reasoning behind NCL is the same as in OSS since all patterns pertaining to the class of interest will be retained in the final training set. In contrast to OSS, NCL places more emphasis on the data cleaning process than on the data reduction process. In particular, the author elaborates that the quality of classification results does not necessarily depend on the size of the class. There are small classes that identify well and large classes that are difficult to classify. Therefore, it is necessary to identify additional factors, such as noise, that may be hampering the classification performance besides the class distribution. Moreover, the author is particularly interested in developing an under-sampling methodology maintaining the original classification accuracy while the data is being reduced. This is a very important aspect since the primary purpose of any resampling technique is to improve the identification of the minority class but without sacrificing classification accuracy on the majority one. In this context the Wilson's Edited Nearest Neighbor (ENN) rule [30] is applied in order to identify the subset A_1 of noisy patterns in the original data set O. ENN operates by removing examples whose class label differs from the majority class of the three nearest neighbors. Generally, ENN retains most of the data, while maintaining a good classification accuracy. Another important feature of NCL is that it performs neighborhood cleaning for examples in O. Specifically, the three nearest neighbors in O that misclassify examples belonging to C are inserted into a temporary set A_2. However, in order to avoid excessive reduction of small classes, only examples from classes larger or equal than $0.5|C|$ are considered while forming A_2. Lastly, the union of sets A_1 and A_2 is removed from T in order to produce the reduced data set S.

Letting T be the original set of the available training patterns and C be the class of interest so that $T = C \cup O$. Then the NCL algorithm is the following:

1. Split data T into the class of interest C and the rest of data O.
2. Identify noisy data A_1 in O with the edited nearest neighbor rule.
3. For each class C_i in O if $(x \in C_i$ in the 3-nearest neighbors of misclassified $y \in C)$ and $(|C_i| \geq 0.5 |C|)$ then $A_2 = \{x\} \cup A_2$
4. Reduced data $S = T - (A_1 \cup A_2)$.

4.1.4 Over-Sampling Methods

Cluster Based Over-Sampling

The Cluster Based Over-sampling approach was proposed in Jo and Japkowicz [16] as a resampling strategy which is based on clustering. The primary purpose of this approach is to simultaneously address the problems related to between-class imbalance (the imbalance occurring between the two classes of interest) and within-class imbalance (the imbalance occurring between the subclusters of each class). This is achieved by separately clustering the training data from each class and subsequently performing random over-sampling cluster by cluster.

Specifically, the Cluster Based Over-sampling approach operates by initially utilizing the k-means clustering algorithm on the training examples pertaining to both the minority and the majority classes. In brief, k-means works by firstly selecting k training examples at random as representative data points from each cluster. The input vector of these representative examples represents the mean of each cluster. The other training examples are processed one by one. For each of these examples, the distance between it and the k-cluster centers is calculated. The example is assigned to the cluster closest to it. The cluster receiving the example, has its mean vector updated by averaging the input vectors of all its corresponding examples.

Once the training examples of each class have been clustered, the next step is the over-sampling procedure for each one of the identified clusters. In the majority class, all the clusters, except for the largest one, are randomly over-sampled so as to get the same number of training examples as the largest cluster. Let *maxclasssize* be the overall size of the largest class and *Nmajorityclass* be the number of clusters in the majority class. In the minority class, each cluster is randomly over-sampled until each cluster contains *maxclasssize* ∗ *Nmajorityclass*/*Nminorityclass* patterns, where *Nminorityclass* represents the number of subclusters in the minority class.

Synthetic Minority Over-Sampling Technique

The Synthetic Minority Over-sampling Technique (SMOTE) was proposed in Chawla et al. [6] as a methodology addressing the class imbalance problem. The main characteristic of SMOTE is that the minority class is over-sampled by creating "synthetic" examples rather than by over-sampling with replacement. Specifically, the synthetic examples are generated in a less application specific manner, by operating in "feature space" rather than in "data space".

The minority class is over-sampled by taking each minority class sample and introducing synthetic examples along the line segments joining any/all of the k minority class nearest neighbors. Depending upon the amount of over-sampling required, neighbors from the k nearest neighbors are randomly chosen.

The original implementation of SMOTE considers only the five nearest neighbors. For instance, if the amount of over-sampling needed is 200 %, only two neighbors from the five nearest neighbors are chosen and one sample is generated in the direction of each. Synthetic samples are generated in the following way: Take the difference between the feature vector (sample) under consideration and its nearest neighbor. Multiply this difference by a random number between 0 and 1 and add it to the feature vector under consideration. This causes the selection of a random point along the line segment between two specific features. This approach effectively forces the decision region of the minority class to become more general. SMOTE algorithm is given by the following pseudo code listing in Algorithm 1.

Algorithm 1 SMOTE(T, N, k)

INPUT: Number of minority class samples T; Amount of SMOTE N%; Number
of nearest neighbors k;
Output: $(N/100) * T$ synthetic minority class samples;
{If N is less than 100 %, randomize the minority class samples as only a random
percent of them will be SMOTEd}
if $N < 100$ **then**
    Randomize the T minority class samples;
    $T \leftarrow (N/100) * T$;
    $N \leftarrow 100$;
end if
$N \leftarrow (int)(N/100)$; {The amount of SMOTE is assumed to be in integral multiples of 100}
k $\leftarrow$ Number of nearest neighbors;
numattrs $\leftarrow$ Number of attributes;
Initialize Sample[ ][ ] ; {Array for original minority class samples}
newindex $\leftarrow$ 0; {counts the number of the generated synthetic class samples}
Synthetic[ ][ ] $\leftarrow$ null; {Array of synthetic samples}
{Compute the k nearest neighbors for each minority class sample only}
for $i \leftarrow 1$ to T **do**
    Compute k nearest neighbors for i, and save the indices in *nnarray*;
    Populate(N,i,*nnarray*);
end for
Populate(N, i, *nnarray*)
{Function to generate the synthetic samples}
while $N \neq 0$ **do**
    $nn \leftarrow$ rand(1,k); {This step chooses one of the k nearest neighbors of i}
    for $attr \leftarrow 1$ to *mumattrs* **do**
        diff $\leftarrow$ Sample[*nnarray*[*nn*]][*atr*] - Sample[i][*atr*];
        gap $\leftarrow$ rand(0,1);
        Synthetic[*newindex*][*atr*] $\leftarrow$ Sample[i][*atr*] + *gap* * *dif*;
    end for
    newindex $\leftarrow$ *newindex* + 1;
    $N \leftarrow N - 1$;
end while
return

Borderline-SMOTE

Borderline SMOTE was proposed in Han et al. [12] as an extension to the original
over-sampling algorithm in an attempt to achieve a better classification performance.
This is particularly realized by the utilization of a training procedure which is focused
on identifying the borderline of each class as exactly as possible. Learning this region
in the pattern space is a very important issue for any classifier since the samples in
this neighborhood are more apt to be misclassified. On the other hand, samples that
lie far enough from the borderline are in general easier to be correctly classified. In
other words, this resampling technique is explicitly interested in strengthening/over-
sampling the more error prone borderline minority samples.

Borderline-SMOTE operates initially by determining the borderline minority
examples and subsequently by generating synthetic examples from them, in order to
be included within the original training set. Let T be the complete set of available

training patterns so that P corresponds to the minority class of positive patterns and N corresponds to the majority class of negative patterns. Specifically, the subsets of positive and negative patterns may be denoted by the sets $P = \{p_1, p_2, \ldots, p_{pnum}\}$ and $N = \{n_1, n_2, \ldots, n_{nnum}\}$ respectively, so that $pnum$ is the number of positive samples and $nnum$ is the number of negative samples. In this context, the detailed description of Borderline-SMOTE is the following:

1. **Step 1**: For each pattern p_i in the minority class P (p_i $i \in [pnum]$) calculate its m nearest neighbors within the complete training set T. The number of majority samples among the m nearest neighbors is denoted by m' so that $0 \leq m' \leq m$.
2. **Step 2**: If $m' = m$, entailing that all the m nearest neighbors are majority samples, p_i is considered to be noise and it is excluded from the following steps. If $m/2 \leq m' < m$, namely the number of p_i's majority nearest neighbors is larger than the number of minority ones, p_i is considered to be easily misclassified and assigned to the *DANGER* set. If $0 \leq m' < m/2$, p_i is safe and needs not to be included in the following steps.
3. **Step 3**: The examples in *DANGER* are the borderline data of the minority class P so that $DANGER \subset P$. Setting $DANSGER = \{p'_1, p'_2, \cdots, p'_{dnum}\}$ so that $0 \leq dnum \leq pnum$ the k nearest neighbors from P are calculated for every sample in *DANGER*.
4. **Step 4**: Generate $s * dnum$ synthetic positive examples from the data in *DANGER*, where s is an integer in the $[1, k]$ interval. For each p'_i, s of the k nearest neighbors in P are randomly selected. Initially, the differences $dif_j, j \in [s]$ between p'_i and its s nearest neighbors from P are calculated and subsequently each $diff_j$ is multiplied by a random number $r_j, j \in [s]$ in the $[0, 1]$ interval so that finally s new synthetic minority examples are generated between p'_i and its nearest neighbors by utilizing the following formula: $synthetic_j = p'_j + r_j * dif_j, j \in [s]$. The above procedure is repeated for each p'_i in *DANGER* and can attain a total number of $s * dnum$ synthetic samples.

Generative Over-Sampling

The Generative Over-sampling algorithm was introduced in Liu et al. [22] as a remedy for the class imbalance problem. The characteristic feature of this approach is the utilization of a probability distribution with resampling, which is argued to be natural and well motivated. Specifically, the authors consider the general setting of a binary classification problem where the original data set X is partitioned into two disjoint sets X_{train} and X_{test} of training and testing patterns respectively. In particular, the original set of available patterns may be expressed as the union of points from sets P and Q where:

- P is a set of points from one class whose probability distribution is unknown,
- Q is a set of points from a second class whose probability distribution is also unknown
- and $\lambda = \frac{|P|}{|X|}$ corresponds to the prior probability that a given point originated from P.

According to the authors, the distributions that generate P and Q can be arbitrarily complicated yielding that any binary classifier should have the ability to learn how to distinguish points from P and points from Q. In this context, the special case of a binary classification problem where the class priors are highly skewed may be formulated in terms of P and Q by letting P representing the minority class of patterns and Q representing the majority class of patterns. Class imbalance is introduced by considering that the minority class prior λ is much less than $1 - \lambda$. Thereby, the goal of over-sampling is to increase the number of points drawn from the distribution that produces P. This entails that an ideal over-sampling technique should add points to the training set X_{train} that have been drawn directly from the distribution that originally produced P. Even though this is not the case for the majority of real world applications where the original data distribution is unknown it is possible to model this distribution in order to create additional data points from the class of interest. Thus, a fundamental prerequisite of the Generative Over-sampling approach is the existence of probability distributions that accurately model the actual data distributions.

Generative Over-sampling works as follows:

1. a probability distribution is chosen in order to model the minority class
2. based on the training data, parameters for the probability distribution are learnt
3. artificial data points are added to the resampled data set by generating points from the learned probability distribution until the desired number of minority class points in the training set has been reached.

4.1.5 Combination Methods

SMOTE + Tomek Links

SMOTE + Tomek links was first used in Batista et al. [2] in order to improve the classification performance of learning algorithms addressing the problem of annotation in Bioinformatics. According to the authors, performing over-sampling on the minority class does not completely alleviate the class imbalance problem. In many cases, there exist supplementary problems that are usually present in data sets skewed class distributions. Frequently, class clusters are not well defined since some majority class examples might be invading the minority class space. The opposite can also be true, since interpolating minority class examples can expand the minority class clusters, introducing artificial minority class examples too deeply in the majority class space. Inducing a classifier under such a situation can lead to over-fitting. In this context, Tomek links were incorporated as data cleaning procedure resulting in the creation of better-defined class clusters. Thus, instead of removing only the majority class examples that form Tomek links, examples from both classes are removed.

SMOTE + ENN

The motivation behind this method is similar to Smote + Tomek links. ENN tends to remove more examples than the Tomek links does, so it is expected that it will provide a more in depth data cleaning. Differently from NCL which is an under-sampling method, ENN is used to remove examples from both classes. Thus, any example that is misclassified by its three nearest neighbors is removed from the training set.

4.2   Cost Sensitive Learning

Most of the currently available learning algorithms for classification were initially designed based on the assumption that the same misclassification costs apply for all classes. The realization that non-uniform costs are very usual in many real-world applications has led to an increased interest in algorithms for cost-sensitive learning. The direct link between the class imbalance problem and cost sensitive learning was firstly identified in Maloof [23]. Specifically, it is argued that problems occurring when dealing with highly skewed data sets can be handled in the same manner as the problems related to learning when misclassification costs are unequal and unknown.

The problem of obtaining a classifier that is useful for cost sensitive learning was particularly investigated in Elkan [11] within the context of optimal learning and decision making when different misclassification errors incur different penalties. In particular, the author provides a theorem showing how to modify the proportion of negative samples in a training set in order to make optimal cost-sensitive decisions using a classifier induced by a standard learning algorithm. The relevant theorem was formulated within the general setting of a binary classification problem focused on designing a classifier that will be able to discriminate between a pair of complementary hypotheses H_0 and H_1. Specifically, H_0 corresponds to the hypothesis indicating that a given pattern originated from the positive class of patterns, while H_1 corresponds to the hypothesis indicating that a given pattern belongs to the negative class of patterns. According to Eq. (3.46), optimal decisions are made on the basis of a probability threshold (p^*) given by Eq. (3.45) as a function of the misclassification costs $C(H_i|H_j)$, where $(i, j) \in \{0, 1\}$. In other words, binary classification problems are reduced in designing a classifier that, given an observation vector $\mathbf{x}$, will decide whether or not $P(H_0|\mathbf{x}) \geq p^*$. Thus, the significance of the theorem proved in Elkan [11] is based on the fact that it provides a way to obtain a standard classifier whose decisions will be made on the basis of a general p^*.

The most common method of achieving this objective is to rebalance the training data set by modifying the proportion of positive and negative samples. Even though rebalancing is a common idea, the general formula describing the exact way in which it will be correctly realized is formulated via the following theorem:

Theorem: To make a target probability threshold p^* correspond to a given probability threshold p_0, the original number of negative samples N_0 in the training set should be replaced by a number N^* of negative samples given by the following equation:

$$\frac{N^*}{N_0} = \frac{1 - p^*}{p^*}\frac{p_0}{1 - p_0}. \tag{4.1}$$

It is very important to understand the directionality of the theorem. Let L be a learning algorithm that yields classifiers that make predictions based on the probability threshold p_0. Then, given a training set S and a desired probability threshold p^*, the theorem indicates how to create a training set S' by changing the number of negative training samples such that L applied to S' gives the desired classifier. The above theorem, however, does not specify the exact way in which the number of negative samples should be modified. A learning algorithm can use weights on the training samples that are specified by the ratio of negative samples given in Eq. (4.1) or it is possible to apply a resampling technique.

4.2.1 The MetaCost Algorithm

The primary objective of cost sensitive learning is the minimization of the conditional risk functional:

$$R(H_i|\mathbf{x}) = \sum_{j\in\{0,1\}} P(H_j|\mathbf{x})C(H_i|H_j), \tag{4.2}$$

which was originally defined in Eq. (3.37) and represents the expected cost of predicting that hypothesis H_i is true given a particular observation vector $\mathbf{x}$. Within the context of a binary classification problem, the Bayes optimal prediction is the one that achieves the lowest possible overall cost. As it is thoroughly described in Sect. 3.2.3, this is achieved by partitioning the complete pattern space X into two disjoint (possibly non-convex) regions X_j, where $j \in \{0, 1\}$ such that hypothesis H_j is the optimal (corresponding to lowest cost) prediction in region X_j. In other words, the goal of cost-sensitive learning is to find the boundaries between these regions, either explicitly or implicitly. This, however, can be an exceptionally difficult procedure when the cost of misclassifying patterns originating from one class is more expensive than misclassifying patterns belonging to the complementary class. In particular, regions which are associated with higher misclassification costs will expand at the expense of the complementary regions even if the conditional class probabilities $P(H_j|\mathbf{x})$ remain unchanged. In fact, the optimal predictions are unknown even for the pre-classified samples in the training set since depending on the specific misclassification costs $C(H_i|H_j)$ they may or may not coincide with the classes with which the samples are labelled.

If the training samples were labelled with their optimal classes according to the particular misclassification costs, then an error-based classifier could be applied to learn the optimal boundaries. This is true since in that case the patterns would be labelled according to those boundaries. Specifically, in the large-sample limit, a consistent error-based learner would learn the optimal, cost-minimizing boundaries. On the other hand, with a finite sample, the learner should in principle approximate the optimal zero-one loss boundaries given the original training set. Zero-one loss boundaries are those corresponding to a cost specification where $C(H_i|H_j) = 0$ for $i = j$ and $C(H_i|H_j) = 1$ for $i \neq j$.

Based on this idea, the author in Domingos [9] proposed the MetaCost algorithm as a principled method for making any classifier cost-sensitive by wrapping a cost-minimizing procedure around it. Moreover, this approach treats the underlying classifier as a black box without requiring any knowledge of its functioning or applying any modification to it. Specifically, the MetaCost algorithm utilizes a variant of Breiman's bagging [5] ensemble method in order to estimate the class probabilities $P(H_j|\mathbf{x})$. It must be mentioned that estimating these probabilities is different from finding class probabilities for unseen patterns and, more importantly, the quality of these estimates is important only in so far as it influences the final boundaries that are produced. It is true that probability estimates can be very poor and still lead to optimal classification, as long as the class that minimizes the conditional risk given the estimated probabilities is the same as the one that minimizes it given the true probabilities.

In the bagging procedure, given a training set of size s, a "bootstrap" resample of it is constructed by taking s samples with replacement from the training set. Thus, a new training set of the same size is produced, where each of the original samples may appear once, more than once or not ar all. This procedure is repeated m times and the resulting m models are aggregated by uniform voting. That is, an unclassified sample is assigned to the class which is predicted by the greatest number of votes. MetaCost differs from bagging in that the number n of examples in each resample may be smaller than the training set size s which makes it more efficient. In short, MetaCost works by initially forming multiple bootstrap replicates of the training set in order to learn a different classifier on each copy. Subsequently, each the probability of each class for each example is estimated by the fraction of votes it receives from the ensemble. The next step is the utilization of Eq. (3.37) to re-label each training sample with the estimated optimal class. Finally, the classifier is re-applied on the re-labelled training set. It should be noted that the class assignment of an unclassified sample can be performed by taking all the models generated into consideration. However, MetaCost can utilize only those models that were learnt on resamples that did not include the given example. The first type of estimate is likely to have lower variance, because it is based on a larger number of samples, while the second one is likely to have lower statistical bias, because it is not influenced by the sample class in the training set. A detailed description of the MetaCost procedure is the following: (Algorithm 2).

Algorithm 2 MetaCost(S, L, C, m, n, p, q)

INPUT: S is the training set; L is a classification algorithm; C is a cost matrix; m is the number of resamples to generate; n is the number of examples in each resample; p is *True* iff L produces class probabilities; q is *True* iff all samples are to be used for each example;
OUTPUT: M is the model produced by applying L to S;
for i=1 to m **do**
   Let S_i be a resample of S with n examples;
   Let M_i be the model produced by applying L to S_i;
end for
for each example **x** in S **do**
   for each class j **do**
      Let $P(H_j|\mathbf{x}) = \frac{1}{\sum_i} \sum_i P(H_j|\mathbf{x}, M_i)$;
      if p **then**
         $P(H_j|\mathbf{x})$ is produced by M_i
      else
         $P(H_j|\mathbf{x}, M_i) = 1$ for the class predicted by M_i for **x** and 0 for all the others;
      end if
      if q **then**
         i ranges over all M_i;
      else
         i ranges over all M_i such that $\mathbf{x} \notin S_i$
      end if
      Let **x**'s class $= \arg\min_i P(H_j|\mathbf{x})C(H_i|H_j)$
   end for
end for
return M.

4.3 One Class Learning

4.3.1 One Class Classifiers

One-class classification [27], also known as *unary classification*, refers to the particular machine learning paradigm which focuses on identifying objects of a specific class amongst all objects, by learning from a training set which exclusively consists of objects of that class. Several methods have been proposed in order to solve the one-class classification problem, which can be categorized into:

1. density estimation methods,
2. boundary methods and
3. reconstruction methods.

For each of the three approaches, different concrete models can be realized. These one-class classification methods differ in their ability to cope with or exploit different characteristics of the data. The important data characteristics that can be identified are the feature scaling, the grouping of objects into clusters, the convexity of the relative distribution and their placing in subspaces.

All one-classification methods, however, share two distinct elements. The first element is a measure for the distance $d(\mathbf{z})$ or resemblance (or probability) $p(\mathbf{z})$ of an object $\mathbf{z}$ to the target class which is represented by the given training set

$$X = \{\mathbf{x}_1, \ldots, \mathbf{x}_l\} \tag{4.3}$$

The second element is a threshold θ on this distance or resemblance. New objects are accepted by the description when the distance to the target class is smaller than the threshold θ_d:

$$f(\mathbf{z}) = I(d(\mathbf{z}) < \theta_d) \tag{4.4}$$

or when the resemblance is larger than the threshold θ_p:

$$f(\mathbf{z}) = I(p(\mathbf{z}) > \theta_p). \tag{4.5}$$

Here $I(A)$ is the indicator function defined by the following equation:

$$I(A) = \begin{cases} 1, & \text{if } A \text{ is true;} \\ 0, & \text{otherwise.} \end{cases} \tag{4.6}$$

The one-class classification methods differ in their realization of $p(\mathbf{z})$ (or $d(\mathbf{z})$), in their optimization of $p(\mathbf{z})$ (or $d(\mathbf{z})$) and thresholds with respect to the training set X. Most of the one-class classification methods focus on the optimization of the resemblance model p or distance d. The optimization of the threshold is done afterwards. Only a few one-class classification methods optimize the corresponding model of $p(\mathbf{z})$ or $d(\mathbf{z})$ to an a priori defined threshold such as the Support Vector Data Description method which will be thoroughly described within the general context of boundary methods for one-class classification.

The most important feature of the one-class classifiers is the tradeoff between the fraction of the target class that is accepted, *TPR*, and the fraction of the outliers that is rejected, *TNR*. This may be otherwise stated as the tradeoff between the error of first type E_I and the error of second type E_I as depicted in Fig. 4.2. The *TPR* can be easily measured by utilizing an independent test set drawn from the same target distribution. Measuring the *TNR*, on the other hand, requires an assumption concerning the outlier density. In other words, in order to compute the true error that corresponds to the classification performance of a particular one-class classifier one should have complete knowledge concerning the joint probability density $p(\mathbf{x}, y)$. However, in the case of one-class classification only the probability density for the target class ($p(\mathbf{x}|H_0)$) is known. This entails that only the percentage of target objects which are not accepted by the description, i.e. the *FNR* (E_I type error), can be minimized. This is trivially satisfied when the complete feature space is captured by the data description. Thereby, when there are no example outlier objects and no estimate of the target distribution $p(\mathbf{x}|H_1)$, it is not feasible to estimate the percentage of outlier objects *FPR* that will be accepted by the description (E_{II} type error).

The main complication in one-class classification is that nothing is known about the quantities *TNR* and *FPR*. Thus, in order to avoid the solution of accepting all the data, an assumption concerning the outlier distribution has to be made. A reasonable approach when there are no example from the outlier class is to assume that outliers are uniformly distributed around the target class. The utilization of a uniform outlier distribution also entails that when the E_I type error is minimized, a data description with minimal volume is obtained. Therefore, instead of minimizing both E_I and E_{II}, it is reasonable to minimize a combination of the E_I type error and the volume of the description. Of course, when the true outlier distribution deviates from the uniform distribution, it is reasonable that another data description model will demonstrate better generalization performance, but this cannot be checked without outlier examples. The generalization of a method can, then, only be given on the target data.

In order to illustrate the tradeoff between the *FNR* (E_I type error) and the *FPR* (E_{II} type error) it is possible to consider the particular situation illustrated in Fig. 4.2.

The banana shaped area corresponds to the target distribution $\mathcal{X}_T$. The circular boundary corresponds to the data distribution which should describe the data. It is obvious that some errors are unavoidable since a part of the target data is rejected and some outliers are accepted. The tradeoff between the *FNR* and *FPR* cannot be avoided since increasing the volume of the data description in order to decrease the E_I type error, will automatically increase the number of accepted outliers and, therefore, the E_{II} type error will also increase.

To obtain an estimate of the volume captured by a given description, objects can be drawn from a uniform distribution over the target data. *FPR* then, corresponds to the fraction of the outlier volume that is covered by the data description. Unfortunately, the number of test objects can become prohibitively large, especially in high dimensional feature spaces. This subject may be illustrated by assuming that the target class is distributed in d-dimensional hypersphere with radius R centered around the origin such that the subspace of target corresponds to the set $X = \{\mathbf{x} : \|\mathbf{x}\|^2 \leq R^2\}$. Moreover, the outliers may be assumed to be restricted within a hypercube in d-dimensions described by the set $Z = [-R, R]^d$. The corresponding volumes can be estimated by utilizing the following equations:

Fig. 4.2 Regions in one class classification

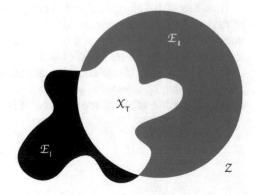

$$V_Z = (2R)^d \tag{4.7}$$

and

$$V_X = \frac{2R^d \pi^{\frac{d}{2}}}{d\Gamma(\frac{d}{2})} \tag{4.8}$$

where $\Gamma(n)$ corresponds to the gamma function defined by the following relationship:

$$\Gamma(n) = 2 \int_0^\infty e^{-r^2} r^{2n-1} dr. \tag{4.9}$$

Clearly, when d is small ($d < 6$) the relevant volumes are of the same order, but for higher dimensionalities V_X decreases to zero, while V_Z increases exponentially. As a consequence, outliers will fall with zero probability within the hypersphere. Thus, no matter how many outlier objects are drawn from the hypercube, all objects will be rejected by the hypersphere. In practice, X and Z will have more complex shapes, but when the volumes of the two regions differ, meaning that the outlier set contains the target set ($X(Z)$), their volumes will still diverge with increasing dimensionality d.

Choosing different thresholds for the distance $d(\mathbf{z})$ or resemblance $p(\mathbf{z})$ results in different tradeoffs between the *TPR* and the *TNR*. In order to conduct a comparison between methods which differ in definitions of $d(\mathbf{z})$ or resemblance $p(\mathbf{z})$, it is reasonable to fix the *TPR* and measure the *TNR* for each methodology. The distance measure is optimized for some one-class classification methods with respect to a given threshold θ so that different thresholds result in different definitions of $d(\mathbf{z})$ or resemblance $p(\mathbf{z})$. Specifically, the threshold will be adjusted in order to accept a predefined fraction of the target class. Thus, for a target acceptance rate *TPR*, the corresponding threshold θ_{TPR} will be defined as

$$\theta_{TPR} : \frac{1}{l} \sum_{i=1}^{l} I(p(\mathbf{x}_i) > \theta_{TPR}) = TPR \text{ where } \mathbf{x}_i \in X, \forall i \in [l] \tag{4.10}$$

Equation (4.10) can be utilized to compute a different threshold θ_{TPR} for a range of *TPR* values on the training set so that the *TNR* can be estimated on a set of example outliers. When for all values of *TPR* the *TNR* is measured, it is possible to obtain the Receiver Operating Characteristic (ROC) curve [24]. In order to find an error measure that fits the one-class classifiers, it is desirable to set $E_I = 1 - TPR$ and $E_{II} = 1 - TNR$ and, subsequently, integrate E_{II} over all possible thresholds corresponding to the different possible values of the E_I error. This may be formulated as:

$$\mathcal{E} = \int_0^1 E_{II}(E_I) dE_I = \int_0^1 \int_{\mathbf{z}} I(p(\mathbf{x}) > \theta) d\mathbf{z} d\theta \tag{4.11}$$

4.3.2 Density Models

The most straightforward method in order to obtain a one-class classifier is to estimate the density of the training data [26] and to set a threshold on this density. Several distributions can be obtained such as a Gaussian or a Poisson distribution, and numerous tests, namely discordancy tests, are then available to test new objects [1]. The most common density models are

1. the Normal Model,
2. the Mixture of Gaussians and
3. the Parzen Density.

The density model works quite well when it is flexible enough and the sample size is sufficiently large. This approach, however, requires a large number of training samples in order to overcome the problem related to the curse of dimensionality. Nonetheless, the curse of dimensionality problem can be addressed by restricting the dimensionality of the data and the complexity of the density model. Of course, when the model does not fit the data very well, a high degree of variance may be introduced which also results in a problematic situation. Thus, determining the right model to describe the target distribution and the given sample size constitutes a typical incarnation of the bias-variance dilemma. It must be noted that density models have an important advantage when a good probability model is assumed and the sample size is sufficiently large. Density models are based on the optimization of the threshold value, regulating the acceptance rate of target objects, which automatically results in a minimum volume probability density model. By construction, only the high density areas of the target distribution are included while superfluous areas will be discarded from the description.

Gaussian Model

The simplest model corresponds to the Normal or Gaussian density [4]. According to the Central Limit Theorem, when it is assumed that objects from one class originate from one prototype and are additively disturbed by a large number of small independent disturbances, then the model is correct. The probability distribution for a d-dimensional object is given by:

$$p_N(\mathbf{z}; \boldsymbol{\mu}, \boldsymbol{\Sigma}) = \frac{1}{(2\pi)^{\frac{d}{2}} |\boldsymbol{\Sigma}|^{\frac{1}{2}}} \exp\left\{-\frac{1}{2}(\mathbf{z} - \boldsymbol{\mu})^T \boldsymbol{\Sigma}^{-1}(\mathbf{z} - \boldsymbol{\mu})\right\}, \qquad (4.12)$$

where $\boldsymbol{\mu}$ is the mean and $\boldsymbol{\Sigma}$ is the covariance matrix. The method is very simple since it imposes a strict unimodal and convex model on the data. Thus, the number of the free parameters in the normal model is:

$$n_{freeN} = d + \frac{d(d+1)}{2} \qquad (4.13)$$

with d parameters for the mean μ and $\frac{d(d+1)}{2}$ for the covariance matrix Σ. In case of badly scaled data or data with singular directions, the inverse of the covariance matrix cannot be calculated and should be approximated by applying some regularization technique which is achieved by adding a small constant λ to the diagonal, i.e. setting $\Sigma' = \Sigma + \lambda I$. In this case, the user has to supply the regularization parameter λ which is the only magic parameter in the normal model. Moreover, this method is insensitive to data scaling since the model utilizes the complete covariance structure of the data.

Another advantage of the normal density is the possibility of obtaining an analytical computation of the optimal threshold for a given value of the *TPR*. It is known that for d independent normally distributed random variables $\mathbf{x}^{(i)}$, the new variable $\mathbf{x}' = \sum_{i=1}^{d} \frac{(\mathbf{x}^{(i)} - \mu^{(i)})^2}{\sigma_i}$ is distributed with a x_d^2 distribution where the parameter d corresponds to the degrees of freedom. By considering the (squared) Mahalanobis distance, the variable:

$$\Delta^2 = (\mathbf{x} - \mu)^T \Sigma^{-1} (\mathbf{x} - \mu) \tag{4.14}$$

for a given μ and Σ should also be distributed like x_d^2. The threshold θ_{TPR} on Δ^2 should be set at the pre-specified *TPR*:

$$\theta_{TPR} : \int_0^{TPR} x_d^2(\Delta^2)d(\Delta^2) = TPR. \tag{4.15}$$

This threshold, however, will not be used since in most cases the target data will not be normally distributed. Thus, Eq. (4.11) provides a more general reliable way to estimate the corresponding threshold given a particular value for the desired *TPR*.

Mixture of Gaussians

The Gaussian model assumes a very strong model of the data which, in particular, should be unimodal and convex. This assumption is violated for most datasets. Thus, in order to obtain a more flexible density method, the normal distribution can be extended to a mixture of Gaussians [10]. A mixture of N_{MoG} Gaussians is a linear combination of normal distributions [4]

$$p_{MoG}(\mathbf{x}) = \frac{1}{N_{MoG}} \sum_j a_j p_N(\mathbf{x}; \mu_j, \Sigma_j) \tag{4.16}$$

where the a_j's are the mixing coefficients. This approach has a smaller bias than the single Gaussian distribution, but it requires far more data for training. Thus, this method shows more variance when a limited amount of data is available. When the number of Gaussians N_{MoG} is defined beforehand by the user, the means μ_j and the covariances Σ_j of the individual components can be efficiently estimated by an expectation minimization routine [4].

The total number of free parameters in the mixture of Gaussians is given by:

$$n_{freeeMoG} = \left(d + \frac{d(d+1)}{2} + 1\right) N_{MoG}, \tag{4.17}$$

in which we count one mean μ_j, one covariance matrix Σ_j and one weight a_j for each component. In order to reduce the number of free parameters, a common approach is the utilization of diagonal covariance matrices $\Sigma_j = diag(\sigma_j)$. The number of free parameters then drops to:

$$n_{freeeMoG} = (2d + 1)N_{MoG}. \tag{4.18}$$

Parzen Density Estimation

The Parzen density estimator [25] is defined through a mixture of Gaussian Kernels which are centered on the individual training objects with often diagonal matrices $\Sigma_j = hI$ as:

$$p_p(\mathbf{x}) = \frac{1}{l} \sum_i p_l(\mathbf{x}; \mathbf{x}_i, hI) \tag{4.19}$$

The equal width h in each feature direction means that the Parzen density estimator assumes equally weighted features and it will, therefore, be sensitive to scaling of the feature values of the data, especially for lower sample sizes. In this context, the training process of a Parzen density estimator reduces to the determination of one single parameter which is the optimal width of the kernel h so that:

$$n_{freep} = 1. \tag{4.20}$$

The optimal value for the parameter h is obtained by utilizing a maximum likelihood solution [18]. The Parzen density estimation method, however, is very weak since there is a unique parameter that is user-defined and there are no supplementary magic parameters. This is true since a good description of the training set depends upon the representational capability of the density model. The computational cost for training a Parzen density estimator is almost zero, but the testing is expensive. All training objects have to be stored and, during testing, distances to all training objects have to be calculated and sorted. This is a very important drawback of the Parzen density estimator, especially when dealing with large datasets in high-dimensional feature spaces.

4.3.3 Boundary Methods

In Chap. 2, it was argued that according to Vapnik when a limited amount of data is given the best learning strategy is to avoid solving a more general problem as an intermediate step in an attempt to solve a more general problem. The solution of the more general problem might require more data than the original one. In this context, the estimation of the complete data density for a one-class classifier might be a too demanding task when only the data boundary is required. Therefore, in the boundary methods only a closed boundary around the target set is optimized. The most common boundary methods are the following:

1. k-Centers description method,
2. Nearest Neighbor description (NN-d) method and
3. Support Vector Data Description method.

Although, the volume is not always actively minimized in the boundary methods, most methods have a strong bias towards a minimal volume solution. The size of the volume depends on the fit of the method on the data. Boundary methods tend to be sensitive to feature scaling since they rely heavily on the distances between objects. On the other hand, the number of objects required is smaller than in the case of density methods. Thus, although the required sample is smaller for the boundary methods, a part of the burden is put onto well-defined distances.

A fundamental aspect of boundary methods is that they cannot be interpreted as probability. Assume that a method is trained so that a fraction f of the target set is rejected. This corresponds to obtaining a threshold θ_f for this rejection rate on its particular resemblance measure d. Decreasing the threshold θ_f results in tightening the description but does not guarantee that the high density areas will be captured. Therefore, changing the parameter f may require the retraining of the algorithm. This is exactly the case for the Support Vector Data Description method.

k - Centers

The first boundary method to be discussed is the k-center method which covers the dataset with k small balls of equal radii [31]. The centers μ_k of the balls are placed on the training objects so that the maximum distance of all minimum distances between training objects and the centers is minimized. Thus, during the training phase the method is fitted on the data through the minimization of the following error functional:

$$E_{k-center} = \max_i \min_j \|\mathbf{x}_i - \boldsymbol{\mu}_k\|^2. \tag{4.21}$$

The k-centers method uses a forward strategy which starts from a random initialization. The radius is determined by the maximum distance to the objects that the corresponding ball should capture. This, of course, entails that the k-centers method is sensitive to outliers in the training set, but works sufficiently well when clear clusters are present within the data. When the centers have been trained, the distance from a test object $\mathbf{z}$ can be calculated by utilizing the following equation:

$$d_{k-center}(\mathbf{z}) = \min_k \|\mathbf{z} - \boldsymbol{\mu}_k\|^2. \tag{4.22}$$

The training procedure incorporates several random initializations in order to avoid a sub-optimal solution. Specifically, the best solution is determined in terms of the smallest $E_{k-center}$. The number of free parameters coincides with the number

k of indices out of N that have to be selected so that:

$$n_{k-centers} = k \qquad (4.23)$$

The user should also supply the maximum number of random initializations that will be performed during the training procedure.

Nearest Neighbor

The second method to be discussed is the nearest neighbor method (NN-d), which may be derived from a local density estimation by the nearest neighbor classifier [10]. In the nearest neighbor method, a cell in d-dimensions is centered around the test object $\mathbf{z}$. Subsequently the volume of this cell is incrementally grown so that k objects from the training set are captured within the cell. The local density then can be estimated by the following equation:

$$p_{NN}(\mathbf{z}) = \frac{k/N}{V_k(\|\mathbf{z} - NN_k^{tr}(\mathbf{z})\|)}, \qquad (4.24)$$

where $NN_k^{tr}(\mathbf{z})\|)$ is the k nearest neighbor of $\mathbf{z}$ in the training set and V_k is the volume of the cell containing the object. In the one-class classifier NN-d, a test object is accepted when its local density is larger or equal to the local density of its first nearest neighbor in the training set denoted by $NN^{tr}(\mathbf{z}) = NN_1^{tr}(\mathbf{z})$. For the local density estimation, $k = 1$ is used, so that:

$$f_{NN^{tr}}(\mathbf{z}) = I\left(\frac{1/N}{V(\|\mathbf{z} - NN^{tr}(\mathbf{z})\|)} \geq \frac{1/N}{V(\|NN^{tr}(\mathbf{z}) - NN^{tr}(NN^{tr}(\mathbf{z}))\|)}\right), \qquad (4.25)$$

which is equivalent to:

$$f_{NN^{tr}}(\mathbf{z}) = I\left(\frac{V(\|\mathbf{z} - NN^{tr}(\mathbf{z})\|)}{V(\|NN^{tr}(\mathbf{z}) - NN^{tr}(NN^{tr}(\mathbf{z}))\|)} \leq 1\right). \qquad (4.26)$$

The volume of the d-dimensional cell may be computed via the following equation:

$$V(r) = V_d r^d, \qquad (4.27)$$

where V_d is the volume of a unit hypersphere in d-dimensions given by Eq. (4.8) and r corresponds to the radius of the hypersphere. Therefore, the substitution of Eq. (4.27) in Eq. (4.28) finally yields that:

$$f_{NN^{tr}}(\mathbf{z}) = I\left(\frac{\|\mathbf{z} - NN^{tr}(\mathbf{z})\|}{\|NN^{tr}(\mathbf{z}) - NN^{tr}(NN^{tr}(\mathbf{z}))\|} \leq 1\right). \qquad (4.28)$$

In other words, the distance from object $\mathbf{z}$ to its nearest neighbor in the training set $NN^{tr}(\mathbf{z})$ is compared with the distance from this nearest neighbor to its nearest neighbor.

Support Vector Data Description

The Support Vector Data Description (SVDD) is a method which directly obtains the boundary around a target dataset. In the simplest case, a hypersphere is computed so that it contains all target objects. In order to minimize the chance of accepting outliers, the volume of the hypersphere is minimized. The model can be rewritten in a form comparable to the Support Vector Classifier (SVC) [7] which justifies the name Support Vector Data Description. Specifically, it offers the ability to map the data to a new, high-dimensional feature space without much extra computational cost. Via this mapping, it is possible to obtain more flexible descriptors so that the outlier sensitivity can be flexibly controlled. A more detailed discussion of the SVDD method will be given in Chap. 5.

4.3.4   Reconstruction Methods

The reconstruction methods have not been primarily designed for one-class classification, but rather to model the data. By using prior knowledge about the data and making assumptions about the generating process, a model is chosen and fitted to the data. New objects can now be described in terms of a state of the generating model. A fundamental assumption of this approach is that it provides a more compact representation of the target data. Moreover, this representation simplifies further processing without harming the information content.

The reconstruction methods that will be discussed in this section are the following:

1. k-Means,
2. Learning Vector Quantization and
3. Self Organized Maps.

The basic functionality behind these approaches involves making assumptions concerning the data or their distribution. Moreover, they define a set of prototypes or subspaces and a reconstruction error is minimized. The simplest reconstruction methods are the k-means clustering [4] and the Learning Vector Quantization (LVQ) [17]. The basic assumption behind these approaches is that the data are clustered and can be represented by a few prototype objects or codebook vectors μ_k. Most often, the target objects are represented by the nearest prototype measured by the Euclidean distance. In other words, the prototype vectors define a Voronoi tessellation [8] of the complete feature space. In the k-means clustering, the placing of the prototypes is optimized by minimizing the following error functional:

$$E_{k-m} = \sum_i \min_k \|\mathbf{x}_i - \boldsymbol{\mu}_k\|^2. \tag{4.29}$$

The k-means clustering method resembles the k-center method, but an important difference is the choice of error to be minimized. The k-center method focuses on the worst case objects (i.e. the objects with the largest reconstruction error) and tries to optimize the centers and the radii of the balls in order to accept all the data. On the other hand, the k-means method focuses on averaging the distances to the prototypes of all objects in order to be more robust against remote outliers. Furthermore, in the k-center method the centers are placed, per definition, on some of the training objects, while in the k-means method, all center positions are free. The distance d of an object $\mathbf{z}$ to the target set is defined as the squared distance of that object to the nearest prototype:

$$d_{k-m}(\mathbf{z}) = \min_k \|\mathbf{z} - \boldsymbol{\mu}_k\|^2. \tag{4.30}$$

The Learning Vector Quantization (LVQ) algorithm is a supervised version of the k-means clustering and is primarily used for classification purposes. For each of the training objects $\mathbf{x}_i$ an extra label y_i is available indicating to which cluster it should belong. The LVQ is trained as a conventional neural network, with the exception that each hidden unit is a prototype, where for each prototype $\boldsymbol{\mu}_k$ a class label y_k is defined. The LVQ algorithm minimizes a classification error of the following form:

$$E_{0-1}(f(\mathbf{x}_i; \mathbf{w}), y_i) = \begin{cases} 0, & \text{if } f(\mathbf{x}_i; \mathbf{w}) = y_i \\ 1, & \text{otherwise,} \end{cases} \tag{4.31}$$

by updating the prototype nearest to the training object $\mathbf{x}_i$ according to the following equation:

$$\Delta\boldsymbol{\mu}_k = \begin{cases} +\eta(\boldsymbol{x}_i - \boldsymbol{\mu}_k), & \text{if } y_i = y_k; \\ -\eta(\boldsymbol{x}_i - \boldsymbol{\mu}_k), & \text{otherwise,} \end{cases} \tag{4.32}$$

where η is the learning rate. This update rule is iterated over all training objects, until convergence is reached. In both the k-means clustering and the LVQ, the k means $\boldsymbol{\mu}$ have to be estimated so that the number of free parameters is given by:

$$n_{freek-m} = n_{freeLVQ} = kd. \tag{4.33}$$

Moreover, in both the LVQ and the k-means method, the user should supply the number of clusters k while the LVQ method requires the learning rate η.

In the Self Organizing Map (SOM), the placing of the prototypes is not only optimized with respect to the data, but also constrained to form a low-dimensional manifold [17]. When the algorithm has converged, prototypes corresponding to nearby vectors on the manifold have nearby locations in the feature space. In most cases, a 2- or 3- dimensional square grid is chosen for this manifold so that data mapped on this manifold can be visualized afterwards. Higher dimensional manifolds are also possible, but they require prohibitive storage and optimization costs. Specifically, for a manifold of d_{SOM} dimensions, the number of neurons becomes $k^{d_{SOM}}$. Thus, in the optimization of the SOM $k^{d_{SOM}}$ neurons have to placed in the d-dimensional feature space. This entails that the number of free parameters in the SOM becomes:

$$n_{freeSOM} = dk^{d_{SOM}},\tag{4.34}$$

where the dimensionality d_{SOM} of the manifold the number k of prototypes per manifold dimensionality and the learning rate η are supplied by the user. The Euclidean distance is once again used in the definition of the error and the computation of the distance according to the following equation:

$$d_{SOM} = \min_k \|\mathbf{z} - \boldsymbol{\mu}_k\|^2.\tag{4.35}$$

The utilization of the Euclidean distance in the related methodologies, however, makes them extremely sensitive to scaling of the features.

4.3.5   Principal Components Analysis

Principal Components Analysis (PCA) [4] is used for data distributed in a linear subspace. In particular, the PCA mapping finds the orthonormal subspace which captures the variance in the data as the best possible in the squared-error sense. The simplest optimization procedure utilizes the eigenvalue decomposition in order to compute the eigenvectors of the target covariance matrix. The eigenvectors with the largest eigenvalues are the principal axis of the d-dimensional data pointing in the direction of the largest variance. These vectors are used as basis vectors for the mapped data. The number of basis vectors M is optimized in order to explain a certain, user-defined, fraction of the variance in the data. The basis vectors form a $d \times M$ matrix W. Since these vectors form an orthogonal basis, the number of free parameters in PCA becomes:

$$n_{freePCA} = \binom{d-1}{M}.\tag{4.36}$$

The reconstruction error of an object $\mathbf{z}$ is now defined as the squared distance from the original object and its mapped version:

$$d_{PCA}(\mathbf{z}) = \|\mathbf{z} - (WW^T)\mathbf{z}\|^2.\tag{4.37}$$

PCA works well when a clear linear subspace is present. Specifically, for very low sample sizes the data is automatically located in a subspace. When the intrinsic dimensionality of the data is smaller than the feature size, the PCA can still generalize well from the low sample size.

4.3.6 Auto-Encoders and Diabolo Networks

Auto-encoders [14] and diabolo networks focus on learning a representation of the data. Both methods are trained to reproduce the input patterns at their output layer. In other words, they perform the identity operation. Auto-encoders and diabolo networks are distinguished on the basis of the number of hidden layers and the sizes of the layers.

Figures 4.3 and 4.4 depict the basic architecture of auto-encoders and diabolo networks, where the units are represented by circles and the weights by arrows. In

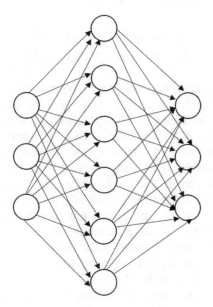

Fig. 4.3 Auto-Encoders

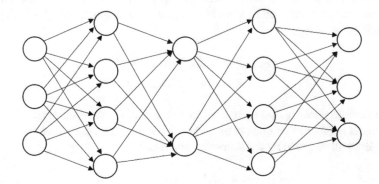

Fig. 4.4 Diabolo networks

the auto-encoder architecture just one hidden layer is present with a large number (n_{auto}) of hidden units. Each unit in an auto-encoder network utilizes conventional sigmoidal transfer functions. On the other hand, in diabolo networks, three hidden layers are used with non-linear sigmoidal transfer functions. In particular, the second layer contains a very low number of hidden units (n_{diab}) which form the bottleneck layer. The other two layers can have an arbitrary number of hidden units which is larger than the number of units in the bottleneck layer. Both types of networks are trained by minimizing the mean squared error. By doing so, it is anticipated that the target objects will be reconstructed with smaller error than outlier objects. Thus, the distance between the original object and the reconstructed one forms a measure of the distance of an object $\mathbf{z}$ to the training set:

$$d_{diab}(\mathbf{z}) = \|f_{diab}(\mathbf{z}; \mathbf{w}) - \mathbf{z}\|^2.$$
$$d_{auto}(\mathbf{z}) = \|f_{auto}(\mathbf{z}; \mathbf{w}) - \mathbf{z}\|^2. \tag{4.38}$$

Auto-encoder networks with a single hidden layer provide a linear principal component type of solution. This entails that the auto-encoder network tends to find a data description which resembles the PCA description. On the other hand, the small number of neurons in the bottleneck layer of the diabolo network act as an information compressor. Thus, in order to obtain a small reconstruction error on the target set, the network is forced to train a compact mapping of the data into the subspace coded by these hidden neurons. The number of neurons in the smallest layer gives the dimensionality of this subspace. More importantly, due to the non-linear transfer functions of the neurons in the other layers, this subspace can become non-linear. When this subspace matches the subspace in the original data, the diabolo networks can perfectly reject objects which are not in the target data space. On the other hand, when the subspace is as large as the original feature space, no distinction between target and outlier data can be expected.

The number of free parameters can be very high for both the auto-encoder as well as the diabolo network. The number of input and output neurons is given by the dimensionality d of the data. By defining n_{auto} to be the number of hidden units in the auto-encoder, the total number of weights in the auto-encoder, including bias terms will be given by:

$$n_{free_{auto}} = dn_{auto} + n_{auto}d + d \tag{4.39}$$
$$= (2d + 1)n_{auto}. \tag{4.40}$$

For the diabolo network, the number of neurons in the bottleneck layers should be added to the number of neurons in the other hidden layers so that the number of free parameters will be given by the following equation:

$$n_{free_{diab}} = n_{diab}(4d + 4n_{diab} + 5) + d. \tag{4.41}$$

References

1. Barnett, V., Barnett, V., Lewis, T.: Outliers in statistical data. Technical report (1978)
2. Batista, G.E.A.P.A., Bazzan, A.L.C., Monard, M.C.: Balancing training data for automated annotation of keywords: a case study. In: WOB, pp. 10–18 (2003)
3. Batista, G.E.A.P.A., Prati, R.C., Monard, M.C.: A study of the behavior of several methods for balancing machine learning training data. SIGKDD Explor. Newsl. 6(1), 20–29 (2004)
4. Bishop, C.M.: Neural Networks for Pattern Recognition. Oxford University Press, New York (1995)
5. Breiman, L., Breiman, L.: Bagging predictors. In: Machine Learning, pp. 123–140 (1996)
6. Chawla, N., Bowyer, K., Hall, L., Kegelmeyer, W.: Smote: synthetic minority over-sampling technique. J. Artif. Intell. Res. 16, 321–357 (2002)
7. Cortes, C., Vapnik, V.: Support-vector networks. Mach. Learn. 20(3), 273–297 (1995)
8. de Berg, M., van Kreveld, M., Overmars, M., Schwarzkopf, O.: Computational Geometry: Algorithms and Applications, 2nd edn. Springer, Heidelberg (2000)
9. Domingos, P.: Metacost: a general method for making classifiers cost-sensitive. In: Proceedings of the Fifth International Conference on Knowledge Discovery and Data Mining, pp. 155–164. ACM Press (1999)
10. Duda, R.O., Hart, P.E., et al.: Pattern Classification and Scene Analysis, vol. 3. Wiley, New York (1973)
11. Elkan, C.: The foundations of cost-sensitive learning. In: IJCAI, pp. 973–978 (2001)
12. Han, H., Wang, W., Mao, B.: Borderline-smote: a new over-sampling method in imbalanced data sets learning. In: ICIC (1), pp. 878–887 (2005)
13. Hart, P.E.: The condensed nearest neighbor rule. IEEE Trans. Inf. Theory IT–14, 515–516 (1968)
14. Japkowicz, N., Myers, C., Gluck, M., et al.: A novelty detection approach to classification. In: IJCAI, vol. 1, pp. 518–523 (1995)
15. Japkowicz, N., Stephen, S.: The class imbalance problem: a systematic study. Intell. Data Anal. 6(5), 429–449 (2002)
16. Jo, T., Japkowicz, N.: Class imbalances versus small disjuncts. SIGKDD Explor. 6(1), 40–49 (2004)
17. Kohonen, T.: Learning vector quantization. In: Self-Organizing Maps, pp. 203–217. Springer (1997)
18. Kraaijveld, M., Duin, R.: A Criterion for the Smoothing Parameter for Parzen-Estimators of Probability Density Functions. Delft University of Technology, Delft (1991)
19. Kubat, M., Matwin, S.: Addressing the curse of imbalanced training sets: one-sided selection. In: Proceedings of the Fourteenth International Conference on Machine Learning, pp. 179–186. Morgan Kaufmann (1997)
20. Laurikkala, J.: Improving identification of difficult small classes by balancing class distribution. In: AIME 2001: Proceedings of the 8th Conference on AI in Medicine in Europe, pp. 63–66. Springer, London (2001)
21. Ling, C.X., Li, C.: Data mining for direct marketing: problems and solutions. In: KDD, pp. 73–79 (1998)
22. Liu, A., Ghosh, J., Martin, C.: Generative oversampling for mining imbalanced datasets. In: DMIN, pp. 66–72 (2007)
23. Maloof, M.A.: Learning when data sets are imbalanced and when costs are unequal and unknown. In: ICML-2003 Workshop on Learning from Imbalanced Data Sets II (2003)
24. Metz, C.E.: Basic principles of roc analysis. Seminars Nuclear Med. 8, 283–298 (1978). Elsevier
25. Parzen, E.: On estimation of a probability density function and mode. Ann. Math. Stat. 33(3), 1065–1076 (1962)
26. Tarassenko, L., Hayton, P., Cerneaz, N., Brady, M.: Novelty detection for the identification of masses in mammograms. In: Fourth International Conference on Artificial Neural Networks, 1995, pp. 442–447. IET (1995)

27. Tax, D.M.J.: Concept-learning in the absence of counter-examples:an auto association-based approach to classification. Ph.D. thesis, The State University of NewJersey (1999)
28. Tomek, I.: Two modifications of cnn. IEEE Trans. Syst. Man Commun. C **SMC C–6**, 769–772 (1976)
29. Weiss, G., Provost, F.: The effect of class distribution on classifier learning: An empirical study. Technical report (2001)
30. Wilson, D.R., Martinez, T.R.: Reduction techniques for instance-based learning algorithms. Mach. Learn. **38**, 257–286 (2000)
31. Ypma, A., Duin, R.P.: Support objects for domain approximation. In: ICANN 98, pp. 719–724. Springer (1998)
32. Zhang, J., Mani, I.: knn approach to unbalanced data distributions: a case study involving information extraction, pp. 42–48 (2003)

Chapter 5
Machine Learning Paradigms

Abstract In this chapter, we discuss the machine learning paradigm of Support Vector Machines which incorporates the principles of Empirical Risk Minimization and Structural Risk Minimization. Support Vector Machines constitute a state-of-the-art classifier which is used as a benchmark algorithm to evaluate the classification accuracy of Artificial Immune System-based machine learning algorithms. In this chapter, we also present a special class of Support Vector Machines that are especially designed for the problem of One-Class Classification, namely the One-Class Support Vector Machines.

5.1 Support Vector Machines

Support Vector Machines (SVMs) are gaining much popularity in recent years as one of the most effective methods for machine learning. In pattern classification problems with two-class sets, they are generalizing linear classifiers in high-dimensional feature spaces through non-linear mappings defined implicitly by kernels in Hilbert space so that they may produce non-linear classifiers in the original space.

Linear classifiers are then optimized to give maximal margin separation between classes. This task is performed by solving some type of mathematical programming such as quadratic programming (QP) or linear programming (LP).

5.1.1 Hard Margin Support Vector Machines

Let $S = \{(\mathbf{x}_1, y_1), \ldots, (\mathbf{x}_l, y_l)\}$ be a set of training patterns such that $x_i \in \mathbb{R}^n$ and $y_i \in \{-1, 1\}$. Thus, each training input belongs to one of two disjoint classes which are associated with the labels $y_i = +1$ and $y_i = -1$, respectively.

If these data points are linearly separable, it is possible to determine a decision function of the following form:

$$g(\mathbf{x}) = \mathbf{w}^T \mathbf{x} + b = \langle \mathbf{w}, \mathbf{x} \rangle + b \tag{5.1}$$

© Springer International Publishing AG 2017
D.N. Sotiropoulos and G.A. Tsihrintzis, *Machine Learning Paradigms*,
Intelligent Systems Reference Library 118, DOI 10.1007/978-3-319-47194-5_5

which is parameterized by the n-dimensional weight vector $\mathbf{w} \in \Re^n$ and the bias term $b \in \Re$.

The decision function $g(\mathbf{x})$ defines a hyperplane in the n-dimensional vector space $\Re^n$ which has the following property:

$$\langle \mathbf{w}, \mathbf{x} \rangle + b \begin{cases} > 0, & \text{for } y_i = +1; \\ < 0, & \text{for } y_i = +1. \end{cases} \qquad (5.2)$$

Because the training data are linearly separable, there will not be any training instances satisfying $\langle \mathbf{w}, \mathbf{x} \rangle + b = 0$. Thus, in order to control separability, Inequalities (5.2), may be reformulated as:

$$\langle \mathbf{w}, \mathbf{x} \rangle + b \begin{cases} \geq +1, & \text{for } y_i = +1; \\ \leq -1, & \text{for } y_i = +1. \end{cases} \qquad (5.3)$$

Here, $+1$ and -1 on the right-hand sides of the inequalities can be replaced by a constant $k(k > 0)$ so that the values $+1$ and -1 will appear as a and $-a$. By dividing both sides of the inequalities by k the original form in (5.3) will be obtained. Inequalities (5.3) may equivalently be rewritten in the following form:

$$y_i(\langle \mathbf{w}, \mathbf{x}_i \rangle + b) \geq 1, \forall i \in [l] \qquad (5.4)$$

by incorporating the class labels.

The hyperplane $g(\mathbf{x}) = \langle \mathbf{w}, \mathbf{x} \rangle + b = c, -1 < c < +1$, forms a separating hyperplane in the n-dimensional vector space $\mathbb{R}^n$ that separates $\mathbf{x}_i, \forall i \in [l]$. When $c = 0$, the separating hyperplane lies within the middle of the two hyperplanes with $c = +1$ and $c = -1$.

The distance between the separating hyperplane and the training datum nearest to the hyperplane is called the *margin*. Assuming that the hyperplanes $g(\mathbf{x}) = +1$ and $g(\mathbf{x}) = -1$ include at least one training datum, the hyperplane $g(\mathbf{x}) = 0$ has the maximum margin for $-1 < c < +1$. The region $\{x : -1 \leq g(\mathbf{x}) \leq +1\}$ is called the generalization region of the decision function.

Figure 5.1 shows two decision functions that satisfy Inequalities (5.4), which are separating hyperplanes. It is obvious that such separating hyperplanes are not unique. However, one must choose the one that results in higher generalization ability. It must be mentioned that the generalization ability of a separating hyperplane depends exclusively on its location, which yields that the hyperplane with the maximum margin is called the optimal hyperplane.

Assuming that no outliers are included within the training data and the unknown test data will obey the same probability law as that of the training data, it is intuitively clear that the generalization ability will be maximized if the optimal hyperplane is selected as the separating hyperplane.

Now consider determining the optimal hyperplane. The Euclidean distance for a training datum $\mathbf{x}$ to the separating hyperplane parameterized by $(\mathbf{w}, b)$ is given by the following equation:

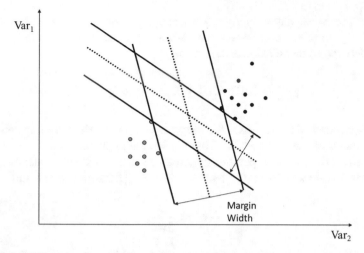

Fig. 5.1 Decision functions

$$R(\mathbf{x}; \mathbf{w}, b) = \frac{|g(\mathbf{x})|}{\|\mathbf{w}\|} = \frac{|\langle \mathbf{w}, \mathbf{x} \rangle + b|}{\|\mathbf{w}\|}. \tag{5.5}$$

This can be shown as follows. Having in mind that the vector $\mathbf{w}$ is orthogonal to the separating hyperplane, the line $l(\mathbf{x}; \mathbf{w})$ that goes through $\mathbf{x}$ and is orthogonal to the hyperplane is given by the following equation:

$$l(\mathbf{x}; \mathbf{w}) = \frac{a}{\|\mathbf{w}\|}\mathbf{w} + \mathbf{x}, \tag{5.6}$$

where $|a|$ is the Euclidean distance from $\mathbf{x}$ to the hyperplane. This line crosses the separating hyperplane at the point where $g(l(\mathbf{x}; \mathbf{w})) = 0$ which yields the following:

$$g(l(\mathbf{x}; \mathbf{w})) = 0 \Leftrightarrow \tag{5.7a}$$

$$w^T l(\mathbf{x}; \mathbf{w}) + b = 0 \Leftrightarrow \tag{5.7b}$$

$$w^T \left(\frac{a}{\|\mathbf{w}\|}\mathbf{w} + \mathbf{x} \right) + b = 0 \Leftrightarrow \tag{5.7c}$$

$$\frac{a}{\|\mathbf{w}\|}w^T w + w^T \mathbf{x} + b = 0 \Leftrightarrow \tag{5.7d}$$

$$\frac{a}{\|\mathbf{w}\|}\|w\|^2 = -w^T \mathbf{x} - b \Leftrightarrow \tag{5.7e}$$

$$a = -\frac{g(\mathbf{x})}{\|w\|} \Leftrightarrow \tag{5.7f}$$

$$|a| = \frac{g(\mathbf{x})}{\|w\|} \tag{5.7g}$$

Let $\mathbf{x}^+$, $\mathbf{x}^+$ be two data points lying on the hyperplanes $g(\mathbf{x}) = +1$ and $g(\mathbf{x}) = -1$, respectively. In order to determine the optimal hyperplane, one has to specify the parameters $(\mathbf{w}, b)$ that maximize the quantity:

$$\gamma = \frac{1}{2}\{R(\mathbf{x}^+; \mathbf{w}, b) + R(\mathbf{x}^-; \mathbf{w}, b))\} = \frac{1}{\|\mathbf{w}\|} \tag{5.8}$$

which is referred to as the *geometric margin*. The geometric margin quantifies the mean distance of the points $\mathbf{x}^+$, $\mathbf{x}^+$ from the separating hyperplane.

Therefore, the optimal separating hyperplane can be obtained by maximizing the geometric margin which is equivalent to minimizing the following quantity

$$f(\mathbf{w}) = \frac{1}{\|\mathbf{w}\|^2} \tag{5.9}$$

with respect to the hyperplane parameters $(\mathbf{w}, b)$ subject to the constraints defined by Inequalities (5.4).

The Euclidean norm $\|\mathbf{w}\|$ in Eq. (5.9) is used in order to transform the optimization problem involved into a quadratic programming one. The assumption of linear separability means that there exist $(\mathbf{w}, b)$ that satisfy the set of constraints defined by Inequalities (5.4). The solutions to Eq. (5.9) that satisfy Inequalities (5.4) are called *feasible* solutions.

Because the optimization problem has a quadratic objective function where the inequality constraints are defined by affine functions, even if the solutions are non-unique, the value of the objective function at the solutions is unique. The non-uniqueness is not a problem for support vector machines. This is one of the advantages of SVMs over neural networks which have several local optima.

Another point that needs to be mentioned is that the optimal separating hyperplane will remain the same even if it is computed by removing all the training patterns that satisfy the strict Inequalities (5.4). Figure 5.1 shows such points on both sides of the separating hyperplane which are called *support vectors*.

The primal optimization problem related to the Hard Margin Support Vector Machine may be formulated as follows:

$$\min_{\mathbf{w},b} \ \frac{1}{2}\|\mathbf{w}\|^2 \tag{5.10a}$$

$$\text{s.t.} \ \ y_i(\langle \mathbf{w}, \mathbf{x}_i \rangle + b) \geq 1, \forall i \in [l] \tag{5.10b}$$

The primal variables corresponding to the convex optimization problem defined in Eq. (5.10) are the parameters $(\mathbf{w}, b)$ defining the separating hyperplane. Thus, the number of primal variables is equal to the dimensionality of the input space augmented by 1, that is $n + 1$. When the dimensionality of the input space is small, the solution of (5.10) can be obtained by the quadratic programming technique. It must be mentioned that SVMs operate by mapping the input space into a high-dimensional feature space which in some cases may be of infinite dimensions. The solution of the optimization problem becomes then too difficult to be addressed in its primal form. A natural approach to overcome this obstacle is to re-express the optimization problem in its dual form, the number of variables of which coincides with the number of the training data.

In order to transform the original primal problem into its corresponding dual, it is necessary to compute the Lagrangian function of the primal form, given by the following equation:

$$L(\mathbf{w}, b, \mathbf{a}) = \frac{1}{2}\langle \mathbf{w}, \mathbf{w} \rangle - \sum_{i=1}^{l} a_i \{ y_i(\langle \mathbf{w}, \mathbf{x}_i \rangle + b) - 1 \}, \qquad (5.11)$$

where $\mathbf{a} = [a_1 \dots a_l]^T$ is the matrix of the non-negative ($a_i \geq 0$) Lagrange multipliers. The dual problem is subsequently formulated as follows:

$$\max_{\mathbf{a}} \min_{\mathbf{w}, b} \quad L(\mathbf{w}, b, \mathbf{a}) \qquad (5.12a)$$

$$\text{s.t} \quad a_i \geq 0, \forall i \in [l]. \qquad (5.12b)$$

Then according to the KT theorem [1] the necessary and sufficient conditions for a normal point $(\mathbf{w}^*, b^*)$ to be an optimum is the existence of $\mathbf{a}^*$ such that:

$$\frac{\partial L(\mathbf{w}^*, b^*, \mathbf{a}^*)}{\partial \mathbf{w}} = \mathbf{0} \qquad (5.13a)$$

$$\frac{\partial L(\mathbf{w}^*, b^*, \mathbf{a}^*)}{\partial b} = 0 \qquad (5.13b)$$

$$a_i^* \{ y_i(\langle \mathbf{w}^*, \mathbf{x}_i \rangle + b^*) - 1) \} = 0, \forall i \in [l] \qquad (5.13c)$$

$$y_i(\langle \mathbf{w}^*, \mathbf{x}_i \rangle + b^*) - 1) \geq 0, \forall i \in [l] \qquad (5.13d)$$

$$a_i^* \geq 0, \forall i \in [l] \qquad (5.13e)$$

Equation (5.13c) describes the KKT *complementarity conditions* [4] from which we may induce that:

- for active constraints for which $\mathbf{a}_i^* = 0$ we have that $y_i(\langle \mathbf{w}^*, \mathbf{x}_i \rangle + b^*) - 1) > 0$.
- for inactive constraints for which $\mathbf{a}_i^* > 0$ we have that $y_i(\langle \mathbf{w}^*, \mathbf{x}_i \rangle + b^*) - 1) = 0$.

The training data points $\mathbf{x}_i$ for which $\mathbf{a}_i^* > 0$ correspond to the support vectors that lie on the hyperplanes $g(\mathbf{x}) = +1$ and $g(\mathbf{x}) = -1$. Equation (5.13a) provides the optimal value for the parameter $\mathbf{w}$ which is given by the following equation:

$$\mathbf{w}^* = \sum_{i=1}^{l} a_i^* y_i \mathbf{x}_i. \qquad (5.14)$$

Similarly, Eq. (5.13b) yields that:

$$\sum_{i=1}^{l} a_i^* y_i = 0. \tag{5.15}$$

Substituting Eqs. (5.13a) and (5.13b) into Eq. (5.11), we get for the Lagrangian:

$$L(\mathbf{w}, b, \mathbf{a}) = \sum_{i=0}^{l} a_i - \frac{1}{2} \sum_{i=1}^{l} \sum_{j=1}^{l} a_i a_j y_i y_j \langle \mathbf{x}_i, \mathbf{x}_j \rangle. \tag{5.16}$$

According to Eqs. (5.13e), (5.15) and (5.16), the dual optimization problem originally defined in Eq. (5.12) may be re-written in the following form:

$$\max_{\mathbf{a}} \; \sum_{i=0}^{l} a_i - \frac{1}{2} \sum_{i=1}^{l} \sum_{j=1}^{l} a_i a_j y_i y_j \langle \mathbf{x}_i, \mathbf{x}_j \rangle \tag{5.17a}$$

$$\text{s.t} \; \sum_{i=1}^{l} a_i^* y_i = 0 \tag{5.17b}$$

$$\text{and} \; a_i \geq 0, \forall i \in [l]. \tag{5.17c}$$

Notice that the dependence on the original primal variables is removed in the dual formulation of the optimization problem where the number of variables is equal to the number of the training patterns. Moreover, having in mind that:

$$\sum_{i=1}^{l} \sum_{j=1}^{l} a_i a_j y_i y_j \langle \mathbf{x}_i, \mathbf{x}_j \rangle = \langle \sum_{i=1}^{l} a_i y_i \mathbf{x}_i, \sum_{j=1}^{l} a_j y_j \mathbf{x}_j \rangle \geq 0, \tag{5.18}$$

the dual optimization problem in Eq. (5.17) corresponds to a concave quadratic programming problem. Thus, if a solution exists, namely the classification problem is linearly separable, then there exists a global solution for $\mathbf{a}^*$. The optimal weight vector, given by Eq. (5.14), realizes the maximal margin hyperplane with geometric margin:

$$\gamma^* = \frac{1}{\|\mathbf{w}^*\|}. \tag{5.19}$$

Let SV denote the set of indices corresponding to the support vectors for which the associated Lagrange multipliers are $a_i^* \neq 0$. Then according to the KKT complementarity conditions in Eq. (5.13c), we have that:

$$y_i \{ (\langle \mathbf{w}^*, \mathbf{x}_i \rangle + b^*) - 1) \} = 0, \forall i \in [SV]. \tag{5.20}$$

Additionally, let SV^+ be the support vectors for which $y_i = +1$ such that:

$$\langle \mathbf{w}^*, \mathbf{x}_i \rangle + b^* = +1, \forall i \in SV^+ \tag{5.21}$$

and SV^- the support vectors for which $y_i = -1$ such that:

$$\langle \mathbf{w}^*, \mathbf{x}_i \rangle + b^* = -1, \forall i \in SV^-. \tag{5.22}$$

Subsequently, by summing Eqs. (5.21) and (5.22) over all the corresponding indices, we get that:

$$\sum_{i \in SV^+} \langle \mathbf{w}^*, \mathbf{x}_i \rangle + b^* + \sum_{i \in SV^-} \langle \mathbf{w}^*, \mathbf{x}_i \rangle + b^* = n^+(+1) + n^-(-1), \tag{5.23}$$

which finally yields that:

$$b^* = \frac{1}{n^+ + n^-} \left\{ (n^+ - n^-) - \sum_{i \in SV} \langle \mathbf{w}^*, \mathbf{x}_i \rangle \right\}. \tag{5.24}$$

Finally, the optimal separating hyperplane can be expressed in the dual representation by the following equation:

$$g(\mathbf{x}) = \sum_{i=1}^{l} a_i^* y_i \langle \mathbf{x}_i, \mathbf{x} \rangle + b^* = \sum_{i \in SV} a_i^* y_i \langle \mathbf{x}_i, \mathbf{x} \rangle + b^*. \tag{5.25}$$

5.1.2 Soft Margin Support Vector Machines

Within the framework of Hard Margin Support Vector Machines, it is assumed that the training data are linearly separable. However, when the data are linearly inseparable, there is no feasible solution and as a consequence the optimization problem corresponding to the Hard Margin Support Vector Machine is unsolvable. A remedy for this problem is the extension of the original Hard Margin paradigm by the so called Soft Margin Support Vector Machine. The main idea of the Soft Margin Support Vector Machine consists of allowing for some slight error which is represented by the slack variables ξ_i, ($\xi_i \geq 0$). The introduction of slack variables within the original notation yields that Inequalities (5.10) will be reformulated as:

$$y_i(\langle \mathbf{w}, \mathbf{x}_i \rangle + b) \geq 1 - \xi_i, \forall i \in [l]. \tag{5.26}$$

The primary reason for the utilization of slack variables is that there will always exist feasible solutions for the reformulated optimization problem.

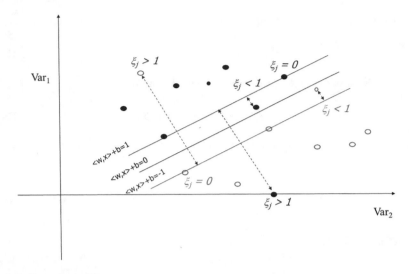

Fig. 5.2 Soft margin SVMs

Figure 5.2 shows that the optimal separating hyperplane correctly classifies all training data $\mathbf{x}_i$ for which $0 < \xi_i < 1$, even if they do not have the maximum margin. However, the optimal separating hyperplane fails to correctly classify those training patterns for which $\xi_i > 1$, as they lie beyond the location of the $g(\mathbf{x}) = 0$ hyperplane.

The Primal optimization problem corresponding to the Soft Margin Support Vector Machine is formulated by introducing a tradeoff parameter C between the margin maximization and the minimization of the sum of slack variables. The first quantity directly influences the generalization ability of the classifier while the second quantity quantifies the empirical risk of the classifier. Thus, the primal form of the optimization problem is mathematically defined as follows:

$$\min_{\mathbf{w},b,\xi} \; \frac{1}{2}\|\mathbf{w}\|^2 + C\sum_{i=1}^{l}\xi_i \tag{5.27a}$$

$$\text{s.t} \quad y_i(\langle\mathbf{w},\mathbf{x}_i\rangle + b) \geq 1 - \xi_i, \forall i \in [l] \tag{5.27b}$$

$$\text{and} \quad \xi_i \geq 0, \forall i \in [l]. \tag{5.27c}$$

In order to form the dual representation of the original optimization problem, it is necessary to estimate the corresponding Lagrangian function which may be obtained by the following equation:

$$L(\mathbf{w},b,\xi,\mathbf{a},\beta) = \frac{1}{2}\langle\mathbf{w},\mathbf{w}\rangle + C\sum_{i=1}^{l}\xi_i - \sum_{i=1}^{l}a_i\{y_i(\langle\mathbf{w},\mathbf{x}_i\rangle + b) - 1 + \xi_i\} - \sum_{i=1}^{l}\beta_i\xi_i. \tag{5.28}$$

Equation (5.28) subsequently may be re-written as:

$$L(\mathbf{w}, b, \mathbf{a}) = \frac{1}{2}\langle \mathbf{w}, \mathbf{w} \rangle - \sum_{i=1}^{l} a_i y_i (\langle \mathbf{w}, \mathbf{x}_i \rangle) - b \sum_{i=1}^{l} a_i y_i + \sum_{i=1}^{l} \{C - a_i - \beta_i\}\xi_i,$$

(5.29)

where $\mathbf{a} = [a_1 \ldots a_l]^T$ and $\boldsymbol{\beta} = [\beta_1 \ldots \beta_l]^T$ are the matrices of the non-negative $(a_i \geq 0)$ and $(\beta_i \geq 0)$ Lagrange multipliers. Moreover, the vector $\boldsymbol{\xi} = [\xi_1 \ldots \xi_l]^T$ groups the set of slack variables. The dual problem is subsequently formulated as follows:

$$\max_{\mathbf{a},\boldsymbol{\beta}} \min_{\mathbf{w},b,\boldsymbol{\xi}} \quad L(\mathbf{w}, b, \mathbf{a}) \tag{5.30a}$$

$$\text{s.t} \quad a_i \geq 0, \forall i \in [l] \tag{5.30b}$$

$$\text{and} \quad \beta_i \geq 0, \forall i \in [l]. \tag{5.30c}$$

Then, according to the KT theorem, the necessary and sufficient conditions for a normal point $(\mathbf{w}^*, b^*, \boldsymbol{\xi}^*)$ to be an optimum is the existence of $(\mathbf{a}^*, \boldsymbol{\beta}^*)$ such that:

$$\frac{\partial L(\mathbf{w}^*, b^*, \boldsymbol{\xi}^*, \mathbf{a}^*, \boldsymbol{\beta}^*)}{\partial \mathbf{w}} = \mathbf{0} \tag{5.31a}$$

$$\frac{\partial L(\mathbf{w}^*, b^*, \boldsymbol{\xi}^*, \mathbf{a}^*, \boldsymbol{\beta}^*)}{\partial \boldsymbol{\xi}} = \mathbf{0} \tag{5.31b}$$

$$\frac{\partial L(\mathbf{w}^*, b^*, \boldsymbol{\xi}^*, \mathbf{a}^*, \boldsymbol{\beta}^*)}{\partial b} = 0 \tag{5.31c}$$

$$a_i^*\{y_i(\langle \mathbf{w}^*, \mathbf{x}_i \rangle + b^*) - 1 + \xi_i)\} = 0, \forall i \in [l] \tag{5.31d}$$

$$\beta_i \xi_i = 0, \forall i \in [l] \tag{5.31e}$$

$$y_i(\langle \mathbf{w}^*, \mathbf{x}_i \rangle + b^*) - 1 + \xi_i) \geq 0, \forall i \in [l] \tag{5.31f}$$

$$a_i^* \geq 0, \forall i \in [l] \tag{5.31g}$$

$$\beta_i^* \geq 0, \forall i \in [l]. \tag{5.31h}$$

Equation (5.31a) yields that the optimal weight vector corresponding to the optimal separating hyperplane will be given by the following equation:

$$\mathbf{w}^* = \sum_{i=1}^{l} a_i^* y_i \mathbf{x}_i. \tag{5.32}$$

On the other hand Eq. (5.31b) yields that:

$$C - a_i^* - \beta_i^* = 0, \forall i \in [l]. \tag{5.33}$$

Finally, Eq. (5.31c) entails that:

$$\sum_{i=1}^{l} a_i^* y_i = 0. \tag{5.34}$$

Notice that the constraints defined by Eq. (5.33) and the set of Inequalities (5.31h) can be combined in the following way:

$$0 \le a_i^* \le C. \tag{5.35}$$

Subsequently, by substituting Eqs. (5.32)–(5.34) in (5.29), the Lagrangian may be simplified into the following form:

$$L(\mathbf{w}, b, \mathbf{a}) = \sum_{i=0}^{l} a_i - \frac{1}{2} \sum_{i=1}^{l} \sum_{j=1}^{l} a_i a_j y_i y_j \langle \mathbf{x}_i, \mathbf{x}_j \rangle. \tag{5.36}$$

By taking into consideration Eqs. (5.34) and (5.35) and the simplified version of the Lagrangian in Eq. (5.36), the dual optimization problem may once again be reformulated as:

$$\max_{\mathbf{a}} \sum_{i=0}^{l} a_i - \frac{1}{2} \sum_{i=1}^{l} \sum_{j=1}^{l} a_i a_j y_i y_j \langle \mathbf{x}_i, \mathbf{x}_j \rangle \tag{5.37a}$$

$$\text{s.t} \sum_{i=1}^{l} a_i^* y_i = 0 \tag{5.37b}$$

$$\text{and} \quad a_i \ge 0, \forall i \in [l] \tag{5.37c}$$

$$\text{and} \quad \beta_i \ge 0, \forall i \in [l]. \tag{5.37d}$$

It is clear, then, from Eqs. (5.17) and (5.37) that both Hard and Soft Margin Support Vector Machines share the same objective function with only difference being that each constituent of the vector $\mathbf{a}$ of Lagrangian multipliers cannot exceed the tradeoff parameter C.

Equations (5.31d) and (5.31e) correspond to the KKT complementarity conditions which combined with Eq. (5.34) identify three cases for the Lagrangian multipliers (a_i):

- $a_i^* = 0 \Rightarrow \beta_i = C \ne 0 \Rightarrow \xi_i = 0$. Thus the training instance $\mathbf{x}_i$ is correctly classified.
- $0 < a_i^* < C \Rightarrow \beta_i \ne 0 \Rightarrow \xi_i = 0 \Rightarrow y_i(\langle \mathbf{w}^*, \mathbf{x}_i \rangle + b^*) = 1$, which means that $\mathbf{x}_i$ is a support vector which is called *unbounded support vector*.
- $a_i^* = C \Rightarrow \beta_i = 0 \Rightarrow \xi_i \ne 0 \Rightarrow y_i(\langle \mathbf{w}^*, \mathbf{x}_i \rangle + b^*) - 1 + \xi_i = 0$, which means that $\mathbf{x}_i$ is a support vector which is called *bounded support vector*.

The decision boundary for the Soft Margin Support Vector Machine is given by the equation:

$$g(\mathbf{x}) = \sum_{i=1}^{l} a_i^* y_i \langle \mathbf{x}_i, \mathbf{x} \rangle + b^* = \sum_{i \in SV} a_i^* y_i \langle \mathbf{x}_i, \mathbf{x} \rangle + b^*, \tag{5.38}$$

which is exactly the same with the one computed for the Hard Margin Support Vector Machine.

A last issue concerning the analysis of the Soft Margin Support Vector Machine that needs to be addressed is the derivation of the optimal values for the parameters $\boldsymbol{\xi}$ and b of the primal optimization problem. Let $X_u = \{x_i \in S : 0 < a_i^* < C\}$ and $X_b = \{x_i \in S : a_i^* = C\}$ be the sets of unbounded and bounded support vectors respectively. Moreover, let SV_u and SV_b be the sets of indices corresponding to the sets X_u and X_b respectively. Thus, the following equations hold:

$$y_i(\langle \mathbf{w}, \mathbf{x}_i^* \rangle + b^*) = 1, \forall \mathbf{x}_i \in X_u. \tag{5.39}$$

Subsequently, let $X_u^+ = \{\mathbf{x}_i \in X_u : y_i = +1\}$ be the set of unbounded support vectors that lie on the hyperplane $g(\mathbf{x}) = +1$ and $X_u^- = \{\mathbf{x}_i \in X_u : y_i = -1\}$ be the set of unbounded support vectors that lie on the hyperplane $g(\mathbf{x}) = -1$. Thus, the set of unbounded support vectors X_u may be expressed as the union $X_u = X_u^+ \cup X_u^-$ of the positive X_u^+ and negative X_u^- unbounded support vectors. Assuming that $|X_u^+| = n_u^+$ and $|X_u^-| = n_u^+$ are the cardinalities of the sets of positive and negative unbounded support vectors respectively, the cardinality of the original set of unbounded support vectors may be expressed as $|X_u| = n_u^+ + n_u^-$. The same reasoning may be applied to the corresponding sets of indices SV_u^+ and SV_u^- where $|SV_u^+| = n_u^+$ and $|SV_u^-| = n_u^-$ so that $|SV_u| = n_u^+ + n_u^-$. Equation (5.39) then entails that:

$$\langle \mathbf{w}^*, \mathbf{x}_i \rangle + b^* = +1, \quad \forall i \in SV_u^+ \tag{5.40a}$$

$$\langle \mathbf{w}^*, \mathbf{x}_i \rangle + b^* = +1, \quad \forall i \in SV_u^-. \tag{5.40b}$$

Taking the sum over all possible indices for Eq. (5.40) yields that:

$$\sum_{i \in SV_u^+} \langle \mathbf{w}^*, \mathbf{x}_i \rangle + b^* + \sum_{i \in SV_u^-} \langle \mathbf{w}^*, \mathbf{x}_i \rangle + b^* = n_u^+(+1) + n_u^-(-1), \tag{5.41}$$

which finally gives the optimal solution for the parameter b of the primal optimization problem as:

$$b^* = \frac{1}{n_u^+ + n_u^-} \{(n_u^+ - n_u^-) - \sum_{i \in SV_u} \langle \mathbf{w}^*, \mathbf{x}_i \rangle\}. \tag{5.42}$$

Finally, the utilization of the following equation:

$$\xi_i^* = 1 - y_i(\langle \mathbf{w}^*, \mathbf{x}_i \rangle) + b^*, \forall i \in SV_b \tag{5.43}$$

provides the optimal values for the parameters ξ_i of the primal optimization problem. Since, $\xi_i^* \geq 0, \forall i \in S$ we may write that:

$$\xi_i^* = \max(0, 1 - y_i(\langle \mathbf{w}^*, \mathbf{x}_i \rangle) + b^*) \tag{5.44}$$

Once again the optimal weight vector given by Eq. (5.32) corresponds to the optimal separating hyperplane realizing the maximum geometric margin given by following equation:

$$\gamma^* = \frac{1}{\|\mathbf{w}^*\|}. \tag{5.45}$$

5.2 One-Class Support Vector Machines

One-class SVMs constitute a specialized type of SVMs that are particularly designed in order to address one-class classification problems where patterns originating from the majority (negative) class are completely unavailable. Therefore, the training set for the case of one-class SVMs may be defined as $X^{tr} = \{(\mathbf{x}_1, y_1), \ldots, (\mathbf{x}_l, y_l)\}$ such that $x_i \in \mathbb{R}^n, \forall i \in [l]$ and $y_i = 1, \forall i \in [l]$.

5.2.1 Spherical Data Description

This approach falls under the category of boundary methods discussed in Chap. 4 building a model which gives a closed boundary around the data. Thus, a hypersphere constitutes an ideal choice for such a model. The sphere, in general, is characterized by a center $\mathbf{a}$ and a radius R which should be appropriately determined so that the corresponding sphere contains all the training objects in X^{tr}. Demanding that all training objects in X^{tr} are contained within the sphere is equivalent to setting the empirical error to 0. Therefore, the structural error may be defined analogously to the SVM classifier according to the following equation:

$$E_{struct}(R, \mathbf{a}) = R^2, \tag{5.46}$$

which has to be minimized subject to the constraints:

$$\|\mathbf{x}_i - a\|^2 \leq R^2, \forall i \in [l] \tag{5.47}$$

In order to allow the possibility for outliers in the training set and, thus, make the method more robust, the distance from objects $\mathbf{x}_i$ to the center $\mathbf{a}$ should not be strictly smaller than R^2, but larger distances should be penalized. This entails that the empirical error should not be identically zero. Therefore, the primal optimization problem for the SVDD has the following form:

$$\min_{R,\mathbf{a},\xi} \quad R^2 + C \sum_{i=1}^{l} \xi_i \tag{5.48a}$$

$$\text{s.t} \quad \|\mathbf{x}_i - \mathbf{a}\|^2 \le R^2 + \xi_i, \ \forall i \in [l] \tag{5.48b}$$

$$\text{and} \quad \xi_i \ge 0, \ \forall i \in [l]. \tag{5.48c}$$

Clearly, according to the new definition of the error:

$$E(R, a, \xi) = R^2 + \sum_{i=1}^{l} \xi_i, \tag{5.49}$$

the empirical and structural errors are combined int a single error functional by introducing slack variables ξ ($\xi_i \ge 0, \forall i \in [l]$). The parameter C represents the tradeoff between the volume of the description and the errors.

In order to transform the original problem into its corresponding dual, it is necessary to compute the Lagrangian function of the primal form which is given by the following equation:

$$L(R, \mathbf{a}, \xi, \boldsymbol{\alpha}, \boldsymbol{\gamma}) = R^2 + C \sum_{i=1}^{l} \xi_i - \sum_{i=1}^{l} \alpha_i \{R^2 + \xi_i - \|\mathbf{x}_i - \mathbf{a}\|^2\} - \sum_{i=1}^{l} \gamma_i \xi_i. \tag{5.50}$$

Here $\boldsymbol{\alpha} = [\alpha_1, \cdots, \alpha_l]^T$ and $\boldsymbol{\gamma} = [\gamma_1, \cdots, \gamma_l]^T$ are the non-negative Lagrange multipliers so that $\alpha_i \ge 0$ and $\gamma_i \ge 0, \forall i \in [l]$. By grouping similar terms, the Lagrangian function in Eq. (5.50) may be rewritten as:

$$L(R, \mathbf{a}, \xi, \boldsymbol{\alpha}, \boldsymbol{\gamma}) = R^2 \left(1 - \sum_{i=1}^{l} \alpha_i\right) + (C - \alpha_i - \gamma_i) \sum_{i=1}^{l} \xi_i + \sum_{i=1}^{l} \alpha_i \|\mathbf{x}_i - \mathbf{a}\|^2, \tag{5.51}$$

which can be equivalently reformulated as:

$$L(R, \mathbf{a}, \xi, \boldsymbol{\alpha}, \boldsymbol{\gamma}) = R^2 \left(1 - \sum_{i=1}^{l} \alpha_i\right) + (C - \alpha_i - \gamma_i) \sum_{i=1}^{l} \xi_i$$

$$+ \sum_{i=1}^{l} \alpha_i \{\langle x_i, x_i \rangle - 2\langle \alpha_i, x_i \rangle + \langle \alpha_i, \alpha_i \rangle\}. \tag{5.52}$$

In this context, the dual optimization problem may be subsequently defined as:

$$\max_{\alpha,\gamma} \min_{R,\mathbf{a},\xi} \ L(R, \mathbf{a}, \xi, \alpha, \gamma) \tag{5.53a}$$

$$\text{s.t} \ \alpha_i \geq 0, \ \forall i \in [l] \tag{5.53b}$$

$$\text{and} \ \gamma_i \geq 0, \ \forall i \in [l] \tag{5.53c}$$

Then, according to the KT theorem, the necessary and sufficient conditions for a normal point $(R^*, \mathbf{a}^*, \xi^*)$ to be an optimum is the existence of a point (α^*, γ^*) such that:

$$\frac{\partial L(\alpha, \gamma, R, \mathbf{a}, \xi)}{\partial R} = 0 \tag{5.54a}$$

$$\frac{\partial L(\alpha, \gamma, R, \mathbf{a}, \xi)}{\partial \mathbf{a}} = \mathbf{0} \tag{5.54b}$$

$$\frac{\partial L(\alpha, \gamma, R, \mathbf{a}, \xi)}{\partial \xi} = \mathbf{0} \tag{5.54c}$$

$$\alpha_i^* \{R^2 + \xi_i - \|\mathbf{x}_i - \mathbf{a}\|^2\} = 0, \forall i \in [l] \tag{5.54d}$$

$$\gamma_i^* \xi_i = 0, \forall i \in [l] \tag{5.54e}$$

$$\alpha_i^* \geq 0, \forall i \in [l] \tag{5.54f}$$

$$\gamma_i^* \geq 0, \forall i \in [l] \tag{5.54g}$$

Equations (5.54a)–(5.54c) yield that:

$$\sum_{i=1}^{l} \alpha_i^* = 1 \tag{5.55a}$$

$$\mathbf{a}^* = \sum_{i=1}^{l} \alpha_i \mathbf{x}_i \tag{5.55b}$$

$$C - \alpha_i^* - \gamma_i^* = 0, \ \forall i \in [l]. \tag{5.55c}$$

Therefore, according to Eqs. (5.55a)–(5.55c), the Lagrangian may be reformulated as:

$$L(R, \mathbf{a}, \xi, \alpha, \gamma) = \sum_{i=1}^{l} \alpha_i \langle \mathbf{x}_i, \mathbf{x}_i \rangle - 2 \sum_{i=1}^{l} \alpha_i \langle \mathbf{a}, \mathbf{x}_i \rangle + \sum_{i=1}^{l} \alpha_i \langle \mathbf{a}, \mathbf{a} \rangle, \tag{5.56}$$

which finally gives that:

$$L(R, \mathbf{a}, \xi, \alpha, \gamma) = \sum_{i=1}^{l} \alpha_i \langle \mathbf{x}_i, \mathbf{x}_i \rangle - \sum_{i=1}^{l} \sum_{j=1}^{l} \alpha_i \alpha_j \langle \mathbf{x}_i, \mathbf{x}_j \rangle. \tag{5.57}$$

According to Eqs. (5.79), (5.55) and (5.57), the dual form of the primal optimization problem may be rewritten as:

$$\max_{\alpha,\gamma} \sum_{i=1}^{l} \alpha_i \langle \mathbf{x}_i, \mathbf{x}_i \rangle - \sum_{i=1}^{l} \sum_{j=1}^{l} \alpha_i \alpha_j \langle \mathbf{x}_i, \mathbf{x}_j \rangle \qquad (5.58a)$$

$$\text{s.t } \sum_{i=1}^{l} \alpha_i \{ R^2 + \xi_i - \|\mathbf{x}_i - \mathbf{a}\|^2 \}, \ \forall i \in [l] \qquad (5.58b)$$

$$\gamma_i \xi_i = 0, \ \forall i \in [l] \qquad (5.58c)$$

$$\gamma_i = C - \alpha_i, \ \forall i \in [l] \qquad (5.58d)$$

$$\alpha_i \geq 0 \qquad (5.58e)$$

$$\gamma_i \geq 0. \qquad (5.58f)$$

The combination of Inequalities (5.58d)–(5.58f) yields that:

$$0 \leq \alpha_i \leq C. \qquad (5.59)$$

The dual optimization problem that was previously formulated presents a well-known quadratic form. Its minimization is an example of a Quadratic Programming problem and there exist standard algorithms in order to obtain its solution. Thus, the optimal values α^* and γ^* for the Lagrangian multipliers are obtained. The Lagrangian formulation of the problem can be utilized in order to provide a further interpretation for the values of α^*. A characteristic of the Lagrange multipliers is that they only play a role when the corresponding constraint should be enforced or violated. In other words, they become active only for cases corresponding to *active constraints* of the form $\{\alpha_i(R^2 + \xi_i - \|\mathbf{x}_i - \mathbf{a}\|^2) = 0; \alpha_i > 0\}$. On the other hand, the contribution of Lagrange multipliers is negligible for *inactive constraints* which may be defined as $\{\alpha_i(R^2 + \xi_i - \|\mathbf{x}_i - \mathbf{a}\|^2) < 0; \alpha_i = 0\}$.

When $\alpha_i^* = 0$, it follows from Eq. (5.58d) that $\gamma_i^* = C \neq 0$ which according to Eq. (5.58c) yields that $\xi_i^* = 0$. This entails that the corresponding training objects $\mathbf{x}_i$ fall within the hypersphere and the constraint $\|\mathbf{x}_i - \mathbf{a}\| \leq R^2$ constitute **inactive constraints**. When $0 < \alpha_i^* < C$, it follows from Eq. (5.58d) that $0 < \gamma_i^* < C$, that is $\gamma_i^* \neq 0$, which according to Eq. (5.58c) yields that $\xi_i^* = 0$. This means that the corresponding training objects $\mathbf{x}_i$ are on the hypersphere boundary defining the set SV^{nbd} and, thus, constitute **active constrains**. Therefore, the set SV^{nbd} of training objects lying on the boundary of the hypersphere can be defined as:

$$SV^{bnd} = \{\mathbf{x}_i \in X^{tr} : \alpha_i^* \in (0, C) \text{ so that } \|\mathbf{x}_i - \mathbf{a}\|^2 = R^2\} \qquad (5.60)$$

Finally, when $\alpha_i^* = C$, it follows from Eq. (5.58d) that $\gamma_i^* = 0$, which according to Eq. (5.58c) yields that $\xi_i^* > 0$. This entails that when the Lagrange multipliers hit the upper bound C, the hyperplane description is not further adjusted in order

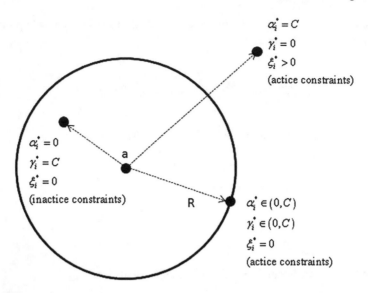

$\alpha_i^* = C$
$\gamma_i^* = 0$
$\xi_i^* > 0$
(actice constraints)

$\alpha_i^* = 0$
$\gamma_i^* = C$
$\xi_i^* = 0$
(inactice constraints)

$\alpha_i^* \in (0, C)$
$\gamma_i^* \in (0, C)$
$\xi_i^* = 0$
(actice constraints)

Fig. 5.3 Support vector data description

to include the corresponding training point $\mathbf{x}_i$. In other words, the corresponding training points will fall outside the hypersphere defining the set SV^{out}. Therefore, this set can be defined as:

$$SV^{out} = \{\mathbf{x}_i \in X^{tr} : \alpha_i^* = C \text{ so that } \|\mathbf{x}_i - \mathbf{a}\|^2 > R^2\} \tag{5.61}$$

These three situations are illustrated in full detail in Fig. 5.3. Having Eq. (5.55b) in mind, the center of the hypersphere $\mathbf{a}$ can be expressed as a linear combination of training objects with weights $\alpha_i^* > 0$. Therefore, for the computation of $\mathbf{a}$, the training objects for which the Lagrangian multipliers are zero may be disregarded. Specifically, during the optimization of the dual problem it is often that a large fraction of the weights becomes 0. Thus, the sum of Eq. (5.55b) is formed by taking into consideration a small fraction of training objects $\mathbf{x}_i$ for which $\alpha_i^* > 0$. These objects are the *support vectors* of the description forming the corresponding set SV which can be expressed as:

$$SV = SV^{bnd} \cup SV^{out}. \tag{5.62}$$

Moreover, the complete set of support vectors can be utilized in order to obtain the optimal values for the slack variables ξ_i^* according to the following equation:

$$\forall \mathbf{x}_i \in SV, \ \xi_i^* = \max(0, \|x_i - a\|^2 - R^2). \tag{5.63}$$

Since it is possible to give an expression for the center of the hypersphere $\mathbf{a}$, it is also possible to test whether a new object $\mathbf{z}$ will be accepted by the description. In particular, this is realized by computing the distance from the object to the center of the hypersphere $\mathbf{a}$. In this context, a test object is accepted when the distance is smaller or equal to the radius:

$$\|\mathbf{z} - \mathbf{a}\|^2 = \langle \mathbf{z}, \mathbf{z} \rangle - 2 \sum_{i=1}^{l} \alpha_i \langle \mathbf{z}, \mathbf{x}_i \rangle + \sum_{i=1}^{l} \sum_{i=1}^{l} \alpha_i \alpha_j \langle \mathbf{x}_i, \mathbf{x}_j \rangle \le R^2. \tag{5.64}$$

By definition, R^2 is the squared distance from the center of the sphere $\mathbf{a}$ to one of the support vectors on the boundary, given by the following equation:

$$R^2 = \langle \mathbf{x}_k, \mathbf{x}_k \rangle - 2 \sum_{i=1}^{l} \alpha_i \langle \mathbf{x}_i, \mathbf{x}_k \rangle + \sum_{i=1}^{l} \sum_{j=1}^{l} \alpha_i \alpha_j \langle \mathbf{x}_i, \mathbf{x}_j \rangle, \ \forall \mathbf{x}_k \in SV^{bnd}. \tag{5.65}$$

This one-class classifier is the so-called Support Vector Data Description, which may be written in the following form:

$$f_{SVDD}(\mathbf{z}; \mathbf{a}, R) = I(\|\mathbf{z} - \mathbf{a}\|^2 \le R^2) \tag{5.66a}$$

$$= I(\langle \mathbf{z}, \mathbf{z} \rangle - 2 \sum_{i=1}^{l} \alpha_i \langle \mathbf{z}, \mathbf{x}_i \rangle + \sum_{i=1}^{l} \sum_{i=1}^{l} \alpha_i \alpha_j \langle \mathbf{x}_i, \mathbf{x}_j \rangle). \tag{5.66b}$$

5.2.2 Flexible Descriptors

The hypersphere is a very rigid model of the data boundary. In general it cannot be expected that this model will fit the data well. If it would be possible to map the data to a new representation, then one might be able to obtain a better fit between the actual data boundary and the hypersphere model. Assuming that there exists a mapping Φ of the data which improves this fit; formally

$$\mathbf{x}^* = \Phi(\mathbf{x}), \tag{5.67}$$

and applying this mapping in Eqs. (5.55) and (5.66), we obtain that

$$L(R, \mathbf{a}, \boldsymbol{\xi}, \boldsymbol{\alpha}, \boldsymbol{\gamma}) = \sum_{i=1}^{l} \alpha_i \langle \Phi(\mathbf{x}_i), \Phi(\mathbf{x}_i) \rangle - \sum_{i=1}^{l} \sum_{j=1}^{l} \alpha_i \alpha_j \langle \Phi(\mathbf{x}_i), \Phi(\mathbf{x}_j) \rangle \tag{5.68}$$

and

$$f_{SVDD}(\mathbf{z}; \mathbf{a}, R) = I\left(\langle \Phi(\mathbf{z}), \Phi(\mathbf{z}) \rangle - 2 \sum_{i=1}^{l} \alpha_i \langle \Phi(\mathbf{z}), \Phi(\mathbf{x}_i) \rangle + \sum_{i=1}^{l} \sum_{i=1}^{l} \alpha_i \alpha_j \langle \Phi(\mathbf{x}_i), \Phi(\mathbf{x}_j) \rangle \right).$$
(5.69)

It is clear from Eq. (5.69) that all mappings $\Phi(\mathbf{x})$ occur only in inner products with other mappings. Therefore, it is once again possible to define a kernel function of the following form:

$$K(\mathbf{x}_i, \mathbf{x}_j) = \langle \Phi(\mathbf{x}_i), \Phi(\mathbf{x}_j) \rangle \rangle$$
(5.70)

in order to replace all occurrences of $\langle \Phi(\mathbf{x}_i), \Phi(\mathbf{x}_j) \rangle$ by this kernel. More details concerning the definition of the kernel function are discusses within the SVM section. The utilization of the kernel function will finally yield that:

$$L(R, \mathbf{a}, \xi, \alpha, \gamma) = \sum_{i=1}^{l} \alpha_i K(\mathbf{x}_i, \mathbf{x}_i) - \sum_{i=1}^{l} \sum_{j=1}^{l} \alpha_i \alpha_j K(\mathbf{x}_i, \mathbf{x}_j)$$
(5.71)

and

$$f_{SVDD}(\mathbf{z}; \mathbf{a}, R) = I\left(K(\mathbf{z}, \mathbf{z}) - 2 \sum_{i=1}^{l} \alpha_i K(\mathbf{z}, \mathbf{x}_i) + \sum_{i=1}^{l} \sum_{i=1}^{l} \alpha_i \alpha_j K(\mathbf{x}_i, \mathbf{x}_j) \right)$$
(5.72)

In this formulation, the mapping $\Phi(\bullet)$ is never used explicitly, but is only defined implicitly by the kernel function $K(\bullet)$, the so-called kernel trick. Another important aspect of this data description method is that it provides a way to estimate the generalization error of a trained SVDD. Specifically, when it is assumed that the training set is a representative sample from the true target distribution, then the number of support vectors may be considered as an indicator of the expected error on the target set. On the other hand, when the training set is a sample which only captures the area in the feature space, but does not follow the true target probability density, it is expected that the error estimate will be far off. Letting f_{SV}^{bnd} be the fraction of training objects that become support vectors on the boundary and f_{SV}^{out} be the fraction of training objects that become support vectors outside the boundary, the error estimate according to the leave one out method becomes:

$$\widetilde{E} \leq f_{SV}^{bnd} + f_{SV}^{out} = f_{SV}.$$
(5.73)

5.2.3 ν - SVC

The basic assumption behind the SVDD method is that concerning the existence of a hypersphere containing all the training objects. In particular, the hypersphere was chosen because it provides a closed boundary around the data. Scholkopf proposed another approach [3], which is called the v-support vector classifier. This approach is focused on placing a hyperplane such that it separates the dataset from the origin with maximal margin. Although, this is not a closed boundary around the data, it is proven in [5] that it provides identical solutions when the data are processed so that they have unit norm. For a hyperplane $\mathbf{w}$ which separates the data $\mathbf{x}_i$ from the origin with maximal margin p, the following equation must hold:

$$\langle \mathbf{w}, \mathbf{x}_i \rangle \geq p - \xi_i, \; \xi_i \geq 0, \; \forall i \in [l]. \tag{5.74}$$

This yields that the function to evaluate a new test object $\mathbf{z}$ becomes:

$$f_{v-SVC} = I(\langle \mathbf{w}, \mathbf{z} \rangle \leq p). \tag{5.75}$$

In order to separate the data from the origin, Scholkopf [3] formulates the following primal optimization problem:

$$\min_{\mathbf{w}, \xi, p} \; \frac{1}{2} \|\mathbf{w}\|^2 + \frac{1}{vl} \sum_{i=1}^{l} \xi_i - p \tag{5.76a}$$

$$\text{s.t} \; \langle \mathbf{w}, \mathbf{x}_i \rangle \geq p - \xi_i, \; \forall i \in [l] \tag{5.76b}$$

$$\xi_i \geq 0, \; \forall i \in [l]. \tag{5.76c}$$

It must be noted that the term $\frac{1}{2}\|\mathbf{w}\|^2$ in the objective function corresponds to the structural error while the empirical error is represented by the sum $\frac{1}{vl} \sum_{i=1}^{l} \xi$. The regularization parameter $v \in (0, 1)$ is a user-defined parameter indicating the fraction of the data that should be accepted by the description. The transformation of the original optimization problem into its corresponding dual requires once again the computation of the Lagrangian function which can be obtained by the following equation:

$$L(\mathbf{w}, \xi, p, \alpha, \beta) = \frac{1}{2} \|\mathbf{w}\|^2 + \frac{1}{vl} \sum_{i=1}^{l} (\xi_i - p) - \sum_{i=1}^{l} \alpha_i \{\langle \mathbf{w}, \mathbf{x}_i \rangle - p + \xi_i\} - \sum_{i=1}^{l} \beta_i \xi_i. \tag{5.77}$$

Therefore, the dual optimization problem may be formulated as:

$$\max_{\alpha, \beta} \min_{\mathbf{w}, \xi, p} \; L(\mathbf{w}, \xi, p, \alpha, \beta) \tag{5.78a}$$

$$\text{s.t} \; \alpha_i \geq 0, \; \forall i \in [l] \tag{5.78b}$$

$$\beta_i \geq 0, \; \forall i \in [l] \tag{5.78c}$$

Then according to KT theorem the necessary and sufficient conditions for a normal point $(\mathbf{w}^*, \xi^*, p^*)$ to be an optimum is the existence of a point (α^*, β^*) such that:

$$\frac{\partial L(\mathbf{w}, \boldsymbol{\xi}, p, \boldsymbol{\alpha}, \boldsymbol{\beta})}{\partial \mathbf{w}} = \mathbf{0} \tag{5.79a}$$

$$\frac{\partial L(\mathbf{w}, \boldsymbol{\xi}, p, \boldsymbol{\alpha}, \boldsymbol{\beta})}{\partial \boldsymbol{\xi}} = \mathbf{0} \tag{5.79b}$$

$$\frac{\partial L(\mathbf{w}, \boldsymbol{\xi}, p, \boldsymbol{\alpha}, \boldsymbol{\beta})}{\partial p} = 0 \tag{5.79c}$$

$$\alpha_i^* \{ \langle \mathbf{w}, \mathbf{x}_i \rangle - p + \xi_i \} = 0, \forall i \in [l] \tag{5.79d}$$

$$\beta_i^* \xi_i = 0, \forall i \in [l] \tag{5.79e}$$

$$\alpha_i^* \geq 0, \forall i \in [l] \tag{5.79f}$$

$$\beta_i^* \geq 0, \forall i \in [l] \tag{5.79g}$$

Equations (5.79a)–(5.79c) yield that:

$$\mathbf{w}^* = \sum_{i=1}^{l} \alpha_i^* \mathbf{x}_i \tag{5.80a}$$

$$\frac{1}{vl} - \alpha_i^* - \beta_i^* = 0, \ \forall i \in [l] \tag{5.80b}$$

$$\sum_{i=1}^{l} \alpha_i^* = 1, \ \forall i \in [l]. \tag{5.80c}$$

Finally, according to Eq. (5.80), the Lagrangian function in Eq. (5.77) may be formulated as:

$$L(\mathbf{w}, \boldsymbol{\xi}, p, \boldsymbol{\alpha}, \boldsymbol{\beta}) = -\frac{1}{2} \sum_{i=1}^{l} \sum_{j=1}^{l} \alpha_i \alpha_j \langle \mathbf{x}_i, \mathbf{x}_j \rangle. \tag{5.81}$$

Having mind Eqs. (5.79d)–(5.79g), (5.80) and (5.81), the dual optimization problem may be re-written in the following form:

$$\max_{\boldsymbol{\alpha}, \boldsymbol{\beta}} \ -\frac{1}{2} \sum_{i=1}^{l} \sum_{j=1}^{j} \alpha_i \alpha_j \langle \mathbf{x}_i, \mathbf{x}_j \rangle \tag{5.82a}$$

$$\text{s.t} \ \sum_{i=1}^{l} \alpha_i \alpha_j \langle \mathbf{x}_i, x_j \rangle = 1 \tag{5.82b}$$

$$\frac{1}{vl} - \alpha_i^* - \beta_i^* = 0, \ \forall i \in [l] \tag{5.82c}$$

$$\alpha_i \{ \langle \mathbf{w}^*, \mathbf{x}_i \rangle - p^* + \xi_i^* \} = 0, \ \forall i \in [l] \tag{5.82d}$$

$$\beta_i^* \xi_i = 0 \tag{5.82e}$$

$$\alpha_i^* \geq 0 \tag{5.82f}$$

$$\beta_i^* \geq 0. \tag{5.82g}$$

Equations (5.82c)–(5.82g) can be combined into a single equation as:

$$0 \leq \alpha_i^* \leq \frac{1}{vl}.$$ (5.83)

The dual optimization problem is, once again, a well-known quadratic optimization problem which can be solved by standard optimization algorithms in order to obtain the optimal values $\boldsymbol{\alpha}^*$, $\boldsymbol{\beta}^*$ for the Lagrangian multipliers. The dual formulation of the original optimization problem can be utilized in order to conduct a further investigation concerning the optimal values of the Lagrangian multipliers. Specifically, when $\alpha_i^* = 0$, then according to Eq. (5.82c) $\beta_i^* = \frac{1}{vl} \neq 0$, which according to Eq. (5.82e) yields that $\xi_i^* \neq 0$. Therefore, the corresponding training objects fulfill **inactive constraints**. These training objects lie further than the hyperplane boundary. On the other hand, when $0 < \alpha_i^* < \frac{1}{vl}$, then Eq. (5.82c) yields that $0 < \beta_i^* < \frac{1}{vl}$, which according to Eq. (5.82e) yields that $\xi_i^* = 0$. This entails that the corresponding training objects fulfill the constraints by lying on the boundary of the hyperplane which maximally separates the data from the origin, that is it will fulfill **active constraints**. Moreover, when $\alpha_i^* = \frac{1}{vl}$, then Eq. (5.82c) yields that $\beta_i^* = 0$, which according to Eq. (5.82e) entails that $\xi_i^* > 0$. Thus, the corresponding training objects fall outside the hyperplane boundary (**active constraints**).

The subset of training objects for which $\alpha_i^* \in (0, \frac{1}{vl})$ are the so-called *boundary support vectors* which may be formally defined as:

$$SV^{bnd} = \left\{ \mathbf{x}_i \in X^{tr} : 0 < \alpha_i^* < \frac{1}{vl} \right\}.$$ (5.84)

On the other hand, the set of training objects for which $\alpha_i^* = \frac{1}{vl}$ define the set of training objects while lie on the hyperplane boundary, that is the set:

$$SV^{out} = \left\{ \mathbf{x}_i \in X^{tr} : \alpha_i^* = \frac{1}{vl} \right\}.$$ (5.85)

The union of sets SV^{bnd} and SV^{out} defines the set of support vectors such that:

$$SV = SV^{bnd} \cup SV^{out}.$$ (5.86)

The sets SV^{bnd} and SV^{out} may be utilized in order to compute the optimal values for the parameter p^* and ξ^*. Specifically, by considering the set of support vectors lying on the boundary of the hyperplane, the following equation holds:

$$\forall \mathbf{x}_i \in SV^{bnd}, \ \langle \mathbf{w}^*, \mathbf{x}_i \rangle - p^* = 0.$$ (5.87)

Therefore, by considering the sum over all boundary support vectors it is possible to write that:

$$\sum_{\mathbf{x}_i \in SV} \langle \mathbf{w}^*, \mathbf{x}_i \rangle - p* = 0 \Leftrightarrow \tag{5.88a}$$

$$\sum_{\mathbf{x}_i \in SV} \langle \mathbf{w}^*, \mathbf{x}_i \rangle = n_{SV}^{bnd} p^* \Leftrightarrow \tag{5.88b}$$

$$p^* = \frac{1}{n_{SV}^{bnd}} \sum_{\mathbf{x}_i \in SV} \langle \mathbf{w}^*, \mathbf{x}_i \rangle. \tag{5.88c}$$

The optimal values for the ξ_i^* parameters may be obtained by considering the complete set of support vectors SV such that:

$$\xi_i^* = \max(0, p^* - \langle \mathbf{w}^*, \mathbf{x}_i \rangle) \tag{5.89}$$

By combining Eqs. (5.80a) and (5.88), it is clear that:

$$p^* = \frac{1}{n_{SV}^{bnd}} \sum_{\mathbf{x}_i \in SV^{bnd}} \sum_{x_j \in SV} \alpha_j^* \langle \mathbf{x}_i, \mathbf{x}_j \rangle. \tag{5.90}$$

Having Eq. (5.90) in mind, it is clear that the maximal margin parameter p^* can be expressed as a function of inner products between training data points. This entails that the kernel trick may also be utilized in order to replace each inner product $\langle \mathbf{x}_i, \mathbf{x}_j \rangle$ occurrence with the equivalent form $K(\mathbf{x}_i, \mathbf{x}_j)$. Common choices for the kernel function according to the literature [6] are the polynomial kernel of power p and the Gaussian kernel given by Eqs. (5.91) and (5.92)

$$K(\mathbf{x}_i, \mathbf{x}_j) = (1 + \langle \mathbf{x}_i, \mathbf{x}_j \rangle)^p \tag{5.91}$$

and

$$K(\mathbf{x}_i, \mathbf{x}_j) = \frac{\|\mathbf{x}_i - \mathbf{x}_j\|^2}{2\sigma^2} \tag{5.92}$$

respectively. The user defined parameters p and σ are to be finely tuned during the training phase through the utilization of a grid searching process [2].

References

1. Cristianini, N., Shawe-Taylor, J.: An Introduction to Support Vector Machines and Other Kernel-Based Learning Methods. Cambridge University Press, Cambridge (2000)
2. Huang, C.-L., Chen, M.-C., Wang, C.-J.: Credit scoring with a data mining approach based on support vector machines. Expert Syst. Appl. 33(4), 847–856 (2007)

3. Schölkopf, B., Williamson, R.C., Smola, A.J., Shawe-Taylor, J., Platt, J.C., et al.: Support vector method for novelty detection. In: NIPS, vol. 12, pp. 582–588. Citeseer (1999)
4. Steinwart, I., Christmann, A.: Support Vector Machines. Springer Science & Business Media, New York (2008)
5. Tax, D.M.J. Concept-learning in the absence of counter-examples:an auto association-based approach to classification. Ph.D. thesis, The State University of NewJersey (1999)
6. Theodoridis, S., Koutroumbas, K.: Pattern Recognition, 3rd edn. Academic Press, Orlando (2006)

Part II
Artificial Immune Systems

Chapter 6
Immune System Fundamentals

Abstract We analyze the biological background of Artificial Immune Systems, namely the physiology of the immune system of vertebrate organisms. The relevant literature review outlines the major components and the fundamental principles governing the operation of the adaptive immune system, with emphasis on those characteristics of the adaptive immune system that are of particular interest from a computation point of view. The fundamental principles of the adaptive immune system are given by the following theories:

- Immune Network Theory,
- Clonal Selection Theory, and
- Negative Selection Theory,

which are briefly analyzed in this chapter.

6.1 Introduction

One of the primary characteristics of human beings is their persistence in observing the natural world in order to devise theories that attempt to describe nature. The Newton's laws of physics and Kepler's model of planetary orbits constitute two major examples of the unsettled human nature which for many years now tries to unravel the basic underpinning behind the observed phenomena. The world, however, need not just be observed and explained, but may also be utilized as inspiration for the design and construction of artifacts, based on the simple principle stating that nature has been doing a remarkable job for millions of years. Moreover, recent developments in computer science, engineering and technology have been determinately influential in obtaining a deeper understanding of the world in general and biological systems in particular. Specifically, biological processes and functions have been explained on the basis of constructing models and performing simulations of such natural systems. The reciprocal statement is also true, that is the introduction of ideas stemming from the study of biology have also been beneficial for a wide range of applications in computing and engineering. This can be exemplified by artificial neural networks, evolutionary algorithms, artificial life and cellular automata. This inspiration from nature is a major motivation for the development of artificial immune systems.

© Springer International Publishing AG 2017

D.N. Sotiropoulos and G.A. Tsihrintzis, *Machine Learning Paradigms*,
Intelligent Systems Reference Library 118, DOI 10.1007/978-3-319-47194-5_6

In this context, a new field of research has emerged under the name of *bioinformatics* referring to the information technology (i.e. computational methods) applied to the management and analysis of biological data. Its implications cover a diverse range of areas from computational intelligence and robotics to genome analysis [2]. The field of biomedical engineering, on the other hand, was introduced in an attempt to encompass the application of engineering principles to biological and medical problems [29]. Therefore, the subject of biomedical engineering is intimately related to bioinformatics.

The bilateral interaction between computing and biology can be mainly identified within the following approaches:

1. *biologically motivated computing* where biology provides sources of models and inspiration in order to develop computational systems (e.g., Artificial Immune Systems and Artificial Neural Networks),
2. *computationally motivated biology* where computing is utilized in order to derive models and inspiration for biology (e.g. Cellular Automata) and
3. *computing with biological mechanisms* which involves the use of the information processing capabilities of biological systems to replace, or at least supplement, current silicon-based computers (e.g. Quantum and DNA computing).

This monograph, however, is explicitly focused on biologically motivated computing within the field of Artificial Immune Systems. In other words, the research presented here does not involve the computational assistance of biology through the construction of models that represent or reproduce biological functionalities of primary interest. On the contrary, the main purpose of this monograph is to utilize biology and, in particular, immunology, as a valid metaphor in order to create abstract and high level representations of biological components or functions. A metaphor uses inspiration in order to render a particular set of ideas and beliefs in such a way that they can be applied to an area different than the one in which they were initially conceived. Paton in [24] identified four properties of biological systems important for the development of metaphors:

1. *architecture* which refers to the form or the structure of the system,
2. *functionality* corresponding to the behavior of the system,
3. *mechanisms* which characterize the cooperation and interactions of the various parts of the system and
4. *organization* referring to the ways the activities of the system are expressed in the dynamics of the whole.

6.2 Brief History and Perspectives on Immunology

Immunology may be defined as the particular scientific discipline that studies the defence mechanisms that confer resistance against diseases. The immune system constitutes the primary bodily system whose main function is to protect our bodies against the constant attack of external *microorganisms*. The immune system specifically recognizes and selectively eliminates foreign invaders by a process known as *immune response*. Individuals who do not succumb to a disease are said to be *immune* to the disease, and the status of a specific resistance to a certain disease is called *immunity*. The immune system has a major contribution in attaining the survival of a living organism by acting efficiently and effectively. The are a large number of distinct components and mechanisms acting on the immune system. Some of theses elements are genetically optimized to defend against a specific invader, while others provide protection against a great variety of *infecting agents*.

The circulation of immune cells, as well as their traffic through the organism, is essential to *immunosurveillance* and to an efficient immune response. Moreover, there exists a great redundancy within the immune system which allows for the activation of many defence mechanisms against a single agent. Studying the various defense mechanisms of the immune system is of primary biological and computational importance since such mechanisms exhibit exceptionally interesting adaptive and memory phenomena. The immune system, in particular, possesses the capability of extracting information from the various agents (environment) and making available for future use in cases of re-infection by the same agent.

Immunology is a relatively new science that was initially conceived in 1796. At that time, Edward Jenner discovered that the introduction of small amounts of *vaccinia* or *cowpox* in an animal would induce future protection against the often lethal disease of *smallpox*. This process is identified by the term *vaccination* and is still used in order to describe the inoculation of healthy individuals with weakened samples of the infectious agents. Therefore, vaccination aims at promoting protection against future infection with a previously encountered disease.

When the process of vaccination was discovered by Jenner, most of the primary functionalities of the immune system remained unknown. Specifically, in the 19th century, Robert Koch proved that infectious diseases are caused by pathogenic microorganisms, each of which is responsible for a *particular infection* or *pathology*. At that time, four major categories of disease were identified which were caused by microorganisms or pathogens, namely: *viruses, bacteria, fungi* and *parasites*.

The initial breakthrough of immunology in the 19th century was followed by the pioneering work of Louis Pasteur who successfully designed a vaccine against *chickenpox* in the second half of the 19th century. Pasteur, although well-versed in the development of vaccines, was unaware of the mechanisms involved in the process of *immunization*, that is the exploitation of these vaccines in order to protect an organism from future infection. Pasteur attributed the phenomenon of immunization to the capability of vaccinia elements in removing nutrients essential to the body so that the growth and proliferation process of disease-causing agents would be interrupted. In 1890, Emil von Behring and Shibashuro Kitasato were the first that demonstrated that protection induced by vaccination was not due to the removal of nutrients, but was rather associated with the appearance of protecting elements in the blood serum of inoculated individuals. Their discovery revealed that people who had been inoculated against diseases contained certain agents that could in some way bind to other infectious agents. These agents were named *antibodies*.

The first major controversy in immunology appeared when Elie Metchnikoff demonstrated in 1882, primarily in invertebrates and later in mammals, that some cells were capable of "eating" microorganisms. These cells were named *phagocytes* and, according to Elie, were the operating constituents of the mechanism that provides resistance against the invading microorganisms. Therefore, based on Elie's perspective, antibodies were of little importance within the immune system. The conflict was resolved in 1904 when Almroth Wright and Joseph Denys demonstrated that antibodies were capable of binding with bacteria and promoting their destruction by phagocytes.

Another important discovery in the history of immunology was the formulation of *side-chain* theory by Paul Ehrlich in the 1890s. The main premise of this theory was that the surfaces of white blood cells, such as B-cells, were covered with several side-chains, or *receptors*. In particular, the receptors on the surfaces of the B-cells form chemical links with the *antigens* encountered. In a broad sense, any molecule that can be recognized by the immune system is called an *antigen*. The variety of the existing receptors on the surface of the various B-cells ensures the existence of at least one receptor with the ability to recognize and bind with a given antigen. Moreover, it was shown that the exposure and binding with a given antigen is followed by an explosive increase in the production of antibodies.

The production of the required antibodies in order to fight the infection from antigens was explained by the formulation of a new theory, the so-called *providential* (or *germinal*) theory. Providential theory suggests that antibodies might be constructed from the collection of *genes*, or *genome*, of the animal. Therefore, it was suggested that the contact of a receptor on a B-cell with a given antigen would be responsible for *selecting* and *stimulating* the B-cell. This results in increasing the production of these receptors, which would then be secreted to the blood stream as antibodies. Therefore, Ehrlich's theory was characterized as *selectivist* based on the selection and stimulation events that dominate the immune response of a living organism. In this context, antigens play the key role of selecting and stimulating B-Cells. The most intriguing aspect of this theory relates to the implied idea that an almost limitless amount of antibodies can be generated on the basis of a finite genome. The theory also suggested that the antigens were themselves responsible for selecting an already existing set of cells through their receptors.

However, in the time period between the years 1914 and 1955, the scientific community remained reluctant in completely adopting a selective theory for the formation of antibodies. The theoretical proposals about such formations were mainly considered as sub-cellular. This entails that the related research was focused on the creation of the antibody molecules produced by cells, with the conclusion that the antigen would bring information concerning the complementary structure of the antibody molecule. This theory is called *template instruction theory* and was originally formulated by Anton Breinl and Felix Haurowitz. This theory was subsequently developed and defended by Nobel Prize winner Linus Pauling. Pauling postulated that all antibodies possess the same *amino acid* sequences, but their particular three-dimensional configuration is determined during synthesis in direct contact with the antigen, which serves as a template.

Therefore, the early immunology was dominated by two controversial theories concerning the structure of the antibody. On the one hand, the germinal or selective theory suggested that the genome was responsible for the definition of the antibody structure, while on the other hand, the instructionist or template theory suggested a fundamental intervention of the antigens in the antibody formation process.

The selective theories of antibody formation were revived in the early 1950s by Niels K. Jerne. Jerne assumed that a diverse population of natural antibodies would appear during development, even in the absence of antigen interaction. The antigen would be matched through the selection of circulating antibodies containing structures complementary to this antigen. Therefore, the quality of an immune response to a given antigen would depend upon the concentration of the circulating antibody of a specific type and could be enhanced by previous exposition to the antigen.

It remained for Burnet to assert that each cell produces and creates on its surface a single type of antibody molecule. The *selective event* is the stimulus given by the antigen, where those cells that produce antibodies complementary to it will proliferate (*clonal expansion*) and secrete antibodies. *Clonal selection* (or *clonal expansion*) theory, formalized by Burnet in 1959, assumed that the generated antibody diversity may be imputed to the activation of random processes in the antibody generation process. According to Burnet, this process takes place during the neo-natal life, such that immediately after birth, the animal would have a fixed repertoire of antibodies. Additionally, this theory postulated the death of any cell possessing antibodies capable of recognizing *self-antigens*, named *self-reactive* cells, during the period of diversity generation.

In 1971, Jerne argued that the elimination of self-reactive cells would constitute a powerful *negative selection* mechanism. This would favor the cellular diversity in cells that could potentially recognize antigens similar to the self. Considerations on how the self-antigens, particularly those of the antibody molecules, named *idiotopes*, could affect the diversity generation and the regulation if the immune responses, lead to the Jerne's proposal of the *immune network theory*. For this work Jerne received the Nobel prize in 1984.

6.3 Fundamentals and Main Components

The immune system constitutes a natural, rapid and effective defense mechanism for a given host against infections. It consists of a two-tier line of defense, namely the *innate immune system* and the *adaptive immune system*, as is illustrated in Fig. 6.1

Both the innate and the adaptive systems depend upon the activity of a large number of immune cells [13, 15, 25] such as *white blood cells*. In particular, innate immunity is mediated mainly by granulocytes and macrophages, while adaptive immunity is mediated by *lymphocytes*. The cells of the innate immune system are immediately available for combatting a wide range of antigens, without requiring previous exposure to them. This reaction will occur in the same way in all normal individuals.

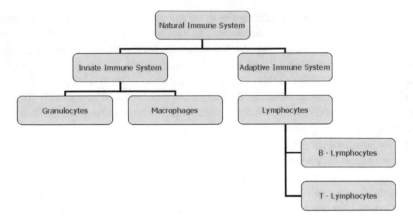

Fig. 6.1 Immune system

The antibody production in response to a determined infectious agent is called a *specific immune response*, also known as *adaptive immune response*. The process of antibody production within the adaptive immune system is explicitly conducted as a response against specific infections. Thus, the presence of certain antibodies in an individual reflects the infections to which this individual has already been exposed. A very interesting aspect of the adaptive immune system is that its cells are capable of developing an immune memory, that is they are capable of recognizing the same antigenic stimulus when it is presented to the organism again. This capability avoids the re-establishment of the disease within the organism. Therefore, the adaptive immune response allows the immune system to improve itself with each encounter of a given antigen.

The immune system consists of a number of components, one of these being lymphocytes. Lymphocytes are the primary mediators of the adaptive immune response and are responsible for the *recognition* and *elimination* of the pathogenic agents. These agents in turn promote immune memory that occurs after the exposition to a disease or vaccination. Lymphocytes usually become active when there is some kind of interaction with an antigenic stimulus leading to the activation and proliferation of lymphocytes. There are two main types of lymphocytes

- *B lymphocytes* or *B-cells* and
- *T lymphocytes* or *T-cells*

which are characterized by the existence of particular antigenic receptors on their surfaces which are highly specific to a given antigenic determinant. In particular, the B-cells have receptors that are able to recognize parts of the antigens that freely circulate in the blood serum. The surface receptors on these B-cells respond to a specific antigen. When a signal is received by the B-cell receptors, the B-cell is activated and will proliferate and differentiate into *plasma cells* that secret antibody molecules in high volumes. These released antibodies, which are a soluble form of the B-cell receptors, are used to neutralize the pathogen, leading to their destruction.

Some of these activated B-cells will remain circulating through the organism for long periods of time, guaranteeing future protection against the same (or similar) antigen that elicited the immune response.

While the adaptive immune response results in immunity against re-infection to the same infectious agent, the innate immune response remains constant along the lifetime of an individual, independent of the antigenic exposure. Altogether, the innate and adaptive immune system contribute to an extremely effective defense mechanism.

6.4 Adaptive Immune System

The ability to present resistance against pathogens is shared by all living organisms. The nature of this resistance, however, defers according to the type of organism. Traditionally, research in immunology has studied almost exclusively the vertebrate (i.e. of animals containing bones) defense reactions and, in particular, the immune system of mammals. Vertebrates have developed a preventive defense mechanism since its main characteristic is to prevent infection against many kinds of antigens that can be encountered, both natural and artificially synthesized.

The most important cells of the adaptive immune system are the *lymphocytes*. They are present only in vertebrates which have evolved a system to proportionate a more efficient and versatile defense mechanism against future infections. On the other hand, the mechanisms of the innate immune system are less evolved. However, the cells of the innate immune system have a crucial role in the initiating and regulating the adaptive immune response.

Each *naive lymphocyte* (i.e. each lymphocyte that has not been involved in an immune response) that enters the blood stream carries antigen receptors of single specificity. The specificity of these receptors is determined by a special mechanism of gene re-arrangement that acts during lymphocyte development in the bone marrow and thymus. It can generate millions of different variants of the encoded genes. Thus, eventhough an individual carries a single specificity receptor, the specificity of each lymphocyte is different. There are millions of lymphocytes circulating throughout our bodies and therefore they present millions of different specificities.

In 1959, Burnet formalized the selective theory which was finally accepted as the most plausible explanation for the behavior of an adaptive immune response. This theory revealed the reasons why antibodies can be induced in response to virtually any antigen and are produced in each individual only against those antigens to which the individual was exposed. Selective theory suggested the existence of many cells that could potentially produce different antibodies. Each of these cells had the capability of synthesizing an antibody of distinct specificity which by binding on the cell surface acted as an antigen receptor. After binding with the antigen, the cell is activated to proliferate and produce a large *clone*. A clone can be understood as a cell or set of cells that are the progeny of a single parent cell. In this context, the clone size refers to the number of offsprings generated by the parent cell. These cells secrete antibodies of the same specificity to its cell receptor. This principle was named *clonal selection theory* or *clonal expansion theory* and constitutes the core of the adaptive immune response. Based upon this clonal selection theory, the lymphocytes can therefore be considered to undergo a process similar to *natural selection* within an individual organism as it was originally formulated by Darwin in [5]. Only those lymphocytes that can interact with the receptor of an antigen are activated to proliferate and differentiate into effector cells.

6.5 Computational Aspects of Adaptive Immune System

The adaptive immune system has been a subject of great research interest because of its powerful information processing capabilities. In particular, the adaptive immune system performs many complex computations in a highly parallel and distributed fashion [8]. Specifically, its overall behavior can be explained as an emergent property of many local interactions. According to Rowe [28], the immune system functionality resembles that of the human brain since it possesses the ability to store memories of the past within the strengths of the interactions of its constituent cells. By doing so, it can generate responses to previously unseen patterns (*antigens*). The primary features of the adaptive immune system which provide several important aspects to the field of information processing may be summarized under the following terms of computation.

6.5.1 Pattern Recognition

From the view point of *pattern recognition* in the immune system, the most important characteristics of B- and T-cells is that they carry *surface receptor molecules* capable of recognizing antigens. B-cell and T-cell receptors recognize antigens with distinct characteristics. The B-cell receptor (BCR) interacts with free antigenic molecules, while the T-cell receptor (TCR) recognizes antigens processed and bound to a surface molecule called *major histocompatibility complex* (MHC). The antigen B-cell receptors are bound to the cell membrane and will be secreted in the form of antibodies when the cell becomes activated. The main role of the B-cell is the production and secretion of antibodies in response to pathogenic agents. Each B-cell produces a single type of antibody, a property called *monospecificity*. These antibodies are capable of recognizing and binding to a determined protein. The secretion and binding of antibodies constitute a form of signaling other cells so that they can ingest, process, and remove the bound substance.

Figure 6.2 illustrates the B-cell detaching the antibody molecule on its surface. The immune recognition occurs at the molecular level and is based on the *complementarity* between the binding region of the receptor and a portion of the antigen called *epitope*. Whilst antibodies present a single type of receptor, antigens might

Fig. 6.2 B- cell detaching the antibody molecule on its surface

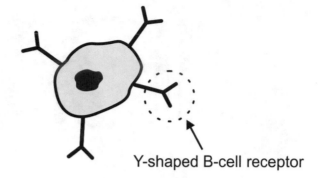

Y-shaped B-cell receptor

Fig. 6.3 The portion of the antigen that is recognized is called epitope

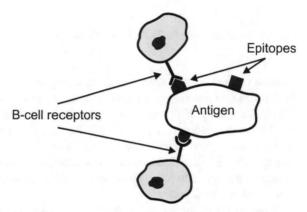

Epitopes

B-cell receptors

Antigen

present several epitopes. This entails that different antibodies can recognize a single antigen, as illustrated in Fig. 6.3.

While B-cells present memory antibody molecules on their surface, and mature within the bone marrow, the maturation process of T-cells occurs in the thymus. T-cell functionality specializes in the regulation of other cells and the direct attack against agents that cause infections to the host organism. T-cells can be divided to two major categories: *helper T-cells*(T_H) and *killer* (or *cytotoxic*) *T-cells*(T_K). T-cell antigenic receptors (TCR) are structurally different when compared to B-cell receptors as illustrated in Fig. 6.4. The TCR recognizes antigens processed and presented in other cells by a surface molecule called major histocompatibility complex or MHC. Figure 6.4 also presents T-cell receptor binding with the MHC complex.

6.5.2 Immune Network Theory

Immune network theory or *idiotypic network theory* was formulated by N.K. Jerne in 1974 [14] in an attempt to provide a conceptually different theory of how the immune system components interact with each other and the environment. This theory is

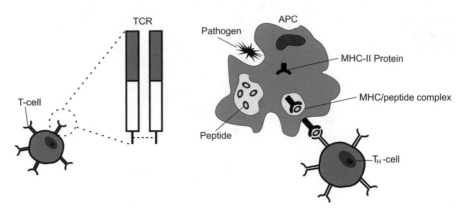

Fig. 6.4 T-cell detaching the TCR on the cell surface

focused on providing a deeper understanding concerning the emergent properties of the adaptive immune system, such as learning and memory, self-tolerance, and size and diversity of the immune repertoires. Immune network theory was based on the demonstration that animals can be stimulated to make antibodies capable of recognizing parts of antibody molecules produced by other animals of the same species or strain. This demonstration led Jerne [14] to realize that, within the immune system of a given individual, any antibody molecule could be recognized by a set of other antibody molecules.

Jerne proposed the following notation in order to describe the immune network. The portion of an antibody molecule which is responsible for recognizing (complementarily) an epitope was named *paratope*. An *idiotype* was defined as the set of epitopes displayed by the variable regions of a set of antibody molecules. An *idiotope* was each single idiotypic epitope. The patterns of idiotopes are determined by the same variable regions of antibody peptide chains that also determine the paratopes. Thus, the idiotopes are located in and around the antigen binding site [11]. Each antibody molecule displays one paratope and a small set of idiotopes [14].

The *immune system* was formally defined as an enormous and complex network of paratopes that recognize sets of idiotopes and of idiotopes that are recognized by sets of paratopes. Therefore, each element could recognize as well as be recognized. This property led to the establishment of a network. As antibody molecules occur both free and as receptor molecules on B-cells, this network intertwines cells and molecules. After a given antibody recognizes an epitope or an idiotope, it can respond either positively or negatively to this recognition signal. A positive response would result in cell activation, cell proliferation, and antibody secretion, while a negative response would lead to tolerance and suppression.

Figure 6.5 illustrates the paratope and idiotope of an antibody molecule as well as the positive and negative responses according to network theory. From a functional point of view, if a paratope interacts with an idiotope, even in the absence of antigens, then the immune system must display a sort of eigen-behavior resulting from this

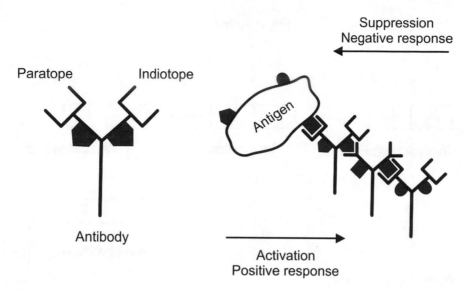

Fig. 6.5 Immune network theory

interaction. This is a fundamental property of the immune system upon which it can achieve a dynamic steady state as its elements interact among themselves in an activating or suppressing manner. In this context, network activation results in increasing the concentration of the antibody molecules while network suppression results in decreasing the concentration of the antibody molecules.

Figure 6.6 provides an illustration of the immune system behavior as it is interpreted through the utilization of the immune network provided by Jerne. When the immune system is primed with an antigen, its epitope is recognized (with various degrees of specificity) by a set of different epitopes, called p_a. These paratopes occur on antibodies and receptor molecules together with certain idiotopes, so that the set p_a of paratopes is associated with the set i_a of idiotopes. The symbol $p_a i_a$ denotes the total set of recognizing antibody molecules and potentially responding lymphocytes with respect to the antigen Ag. Within the immune network, each paratope of the set p_a recognizes a set of idiotopes and the entire set p_a recognizes an even larger set of idiotopes. The set i_b of idiotopes is called the *internal image of the epitope (or antigen)* because it is recognized by the same set p_a that recognized the antigen. The set i_b is associated with a set p_b of paratopes occurring on the molecules and cell receptors of the set $p_b i_b$.

Furthermore, each idiotope of the set $p_a i_a$ is recognized by a set of paratopes, so that the entire set i_a is recognized by an even larger set p_c of paratopes that occur together with a set i_c of idiotopes on antibodies and lymphocytes of the anti-idiotypic set $p_c i_c$. According to this scheme, it is reasonable to assume the existence of ever-larger sets that recognize and or are recognized by previously defined sets within the network. Besides the recognizing set $p_a i_a$ there is a parallel set $p_x i_a$ of antibodies

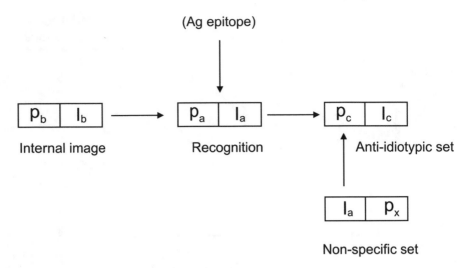

Fig. 6.6 Idiotypic network

that display idiotopes of the set i_a in molecular association with combining sites that do not fit the foreign epitope. The arrows indicate a stimulatory effect when idiotopes are recognized by paratopes on cell receptors and a suppressive effect when paratopes recognize idiotopes on cell receptors.

6.5.3   The Clonal Selection Principle

As each lymphocyte presents a distinct receptor specificity, the number of lymphocytes that might bind to a given antigenic molecule is restricted. Therefore, in order to produce enough specific effector cells to fight against an infection, an activated lymphocyte has to proliferate and then differentiate into these effector cells. This process, called *clonal expansion*, occurs inside the lymph nodes in a micro-environment named *germinal center* (GC) and is characteristic of all immune responses [32, 36].

The *clonal selection principle* (or *clonal expansion principle*) is the theory used to describe the basic properties of an adaptive immune response to an antigenic stimulus. It establishes the idea that only those cells capable of recognizing an antigenic stimulus will proliferate and differentiate into effector cells, thus being selected against those cells that do not. Clonal selection operates on both T-cells and B-cells. The main difference between B-cell and T-cell clonal expansion is that B-cells suffer somatic mutation during reproduction and B-effector cells are active antibody secreting cells. In contrast, T-cells do not suffer somatic mutation during reproduction, and T-effector cells are mainly active secretors of T_K cells. The presence of mutational and selectional events in the B-cell clonal expansion process allow these lympho-

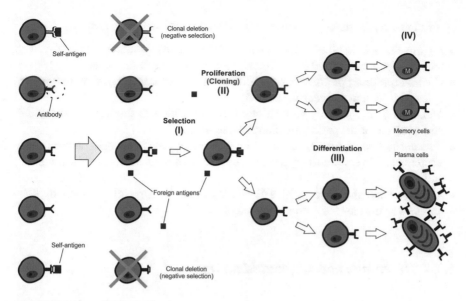

Fig. 6.7 Clonal selection

cytes to increase their repertoire diversity and also to become increasingly better in their capability of recognizing selective antigens. Therefore, the genetic variation, selection, and adaptation capabilities of B-cells provide the reasons for restricting the discussion regarding the clonal selection principle on this specific type of immune cells.

When an animal is exposed to an antigen, the B-cells respond by producing antibodies. Each B-cell secrets a unique kind of antibody, which is relatively specific for the antigen. Figure 6.7 presents a simplified description of the clonal selection principle. Antigenic receptors on a B-cell bind with an antigen (a). When coupled with a *second signal* (or *co-stimulatory signal*) from accessory cells, such as T_H cells, they allow an antigen to stimulate a B-cell. This stimulation of the B-cell causes it to proliferate (divide) (b) and mature into terminal (non-dividing) antibody secretion cells, called plasma cells (c). While plasma cells are the most active antibody secretors, the rapidly dividing B-cells also secrete antibodies, but at a lower rate. The B-cells, in addition to proliferating and differentiating into plasma cells, can differentiate into long-lived B-*memory cells* (d). Memory cells circulate through the blood, lymph, and tissues and when exposed to a second antigenic stimulus they rapidly commence differentiating into plasma cells capable of producing high affinity antibodies. These are pre-selected for the specific antigen that had stimulated the *primary response*. The theory also proposes that potential self-reactive developing lymphocytes are removed from the repertoire previously to its maturation.

In summary, the main properties of the clonal selection theory are:

- *Negative selection*: elimination of newly differentiated lymphocytes reactive with antigenic patterns carried by self components, named self-antigens;
- *Clonal expansion*: proliferation and differentiation on contact of mature lymphocytes with foreign antigens in the body;
- *Monospecificity*: phenotypic restriction of one pattern to one differentiated cell and retention of the pattern by clonal descendants;
- *Somatic hypermutation*: generation of new random genetic changes, subsequently expressed as diverse antibody patterns, by a form of accelerated somatic mutation; and
- *Autoimmunity*: the concept of a *forbidden clone* to early elimination by self-antigens as the basis of *autoimmune diseases*.

6.5.4 Immune Learning and Memory

Antigenic recognition does not suffice to ensure the protective ability of the immune system over extended periods of time. It is important for the immune system to maintain a sufficient amount of resources in order to mount an effective response against pathogens encountered at a later stage. As in typical *predator-prey* situations [33], the size of the lymphocyte sub-population (clone) specific for the pathogen with relation to the size of the pathogen population is crucial to determining the infection outcome. *Learning* in the immune system involves raising the population size and the affinity of those lymphocytes that have proven themselves to be valuable during the antigen recognition phase. Thus, the immune repertoire is biased from a random base to a repertoire that more clearly reflects the actual antigenic environment.

The total number of lymphocytes within the immune system is not kept absolutely constant since there exists a regulation process which increases the sizes of some clones or decreases the sizes of some others. Therefore, if the learning ability of the immune system is uniquely acquired by increasing the population sizes of specific lymphocytes, it must either "forget" previously learnt antigens or increase in size or constantly decrease the portion of its repertoire that is generated at random and that is responsible for responding to novel antigens [27].

In the normal course of the evolution of the immune system, an organism would be expected to encounter a given antigenic pattern repeatedly during its lifetime. The initial exposure to an antigen that stimulates an adaptive immune response is handled by a small number of B-cells, each producing antibodies of different affinity. Storing some high affinity antibody producing cells from the first infection, so as to a large initial specific B-cell sub-population (clone) for subsequent encounters, considerably enhances the effectiveness of the immune response to secondary encounters. These are referred to as memory cells. Rather than "starting from scratch" every time, such a strategy ensures that both the speed and the accuracy of the immune response become successively stronger after each infection. This scheme is reminiscent of a

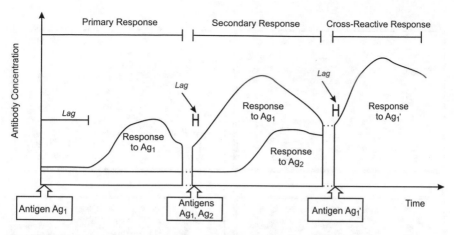

Fig. 6.8 Primary, secondary and cross-reactive immune responses

reinforcement learning strategy, in which the system is continuously learning from direct interaction with the environment.

Figure 6.8 illustrates a particular scenario for the time evolution of an immune response (memory) which is triggered by the introduction of an antigen Ag_1 at time 0. According to this scenario, the initial concentration of the antibodies specific to Ag_1 will not be adequate until after a *lag* has passed so that the specific type of antibodies starts to increase in concentration and affinity up to a certain level. Therefore, in the course of the immune response, the infection is eliminated and the concentration of the antibodies specific to Ag_1 begins to decline. This first phase is known as the *primary response*. When another antigen Ag_2 (different from Ag_1) is introduced, the same pattern of response is presented, but for a kind of antibody with different specificity from the one that recognized Ag_1. This demonstrates the specificity of the adaptive immune response. On the other hand, one important characteristic of the immune memory is that it is associative. This is specifically argued in [31] where the authors elaborate that immunological memory is an associative and robust memory that belongs to the class of *distributed memories*. This class of memories derives its associative and robust nature by sparsely sampling the input space and distributing the data among many independent agents [17]. B-cells adapted to a certain type of antigen Ag_1 can present a faster and more efficient secondary response not only to Ag_1, but also to any structurally related antigen, e.g., Ag_1'. The phenomenon of presenting a more efficient secondary response to an antigen structurally related to a previously seen antigen is called *immunological cross-reaction* or *cross-reactive response*. The immunological cross-reaction is equivalent to the *generalization ability* of machine learning algorithms. The secondary immune response is characterized by a shorter lag phase, a higher rate of antibody production, and longer persistence of antibody synthesis. Moreover, a dose of antigen substantially lower than that required to initiate a primary response can cause a secondary response.

(a)   **(b)**

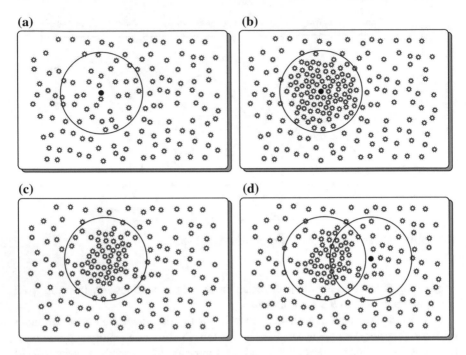

(c)   **(d)**

Fig. 6.9 Primary, secondary and cross-reactive immune responses

The associative nature of the immune memory can be better understood through the series of illustrations within the Fig. 6.9a–d, since they provide a more vivid representation of the previously discussed immune response scenario. This time, however, attention is focused on the immune recognition phenomena that take place at the molecular level during the course of the primary and secondary immune responses. Figure 6.9a presents the initial immune system stimulation by the antigenic pattern Ag_1 where the stimulated B- and T-cells are considered to lie within the *ball of stimulation* of that antigen [26]. Notice that immune molecules and external pathogenic agents are both represented by points in a two dimensional plane where the affinity of interaction between immune cells and antigenic patterns is quantified by their distance.

Figure 6.9b depicts the state of the immune system towards the end of the primary immune response where antibodies of higher affinity have increased in concentration in order to eliminate the majority of the antigenic patterns that circulate in the blood serum. This is the very essence of the clonal selection principle [3] according to which only a specific subset of the available antibody repertoire is selected in order to proliferate in a process similar to natural evolution. At the end of the immune response, the population of antibodies decreases since the antigen that initially triggered the immune response is completely cleared. It is true, however, that a persistent sub-population of memory cells (antibodies of high affinity) remains as it is illustrated in Fig. 6.9c.

So far, the mechanism by which memory cells persist is not fully understood. One theory suggests that a prolonged lifetime is within the nature of memory cells [19] while other theories propose a variety of mechanisms based on the hypothesis that memory cells are re-stimulated at some low level. Such a theory is the idiotypic network theory [14] in which cells co-stimulate each other in a way that mimics the presence of an antigen. Other theories suggest that small amounts of the antigen are retained in lymph nodes [34, 35] or that the existence of related environmental antigens provides cross-stimulation [20]. Although the idiotypic network theory has been proposed as a central aspect of the associative properties of immunological memory [7, 10], it is but one of the possible mechanisms that maintains the memory population and is key to neither the associative nor the robust properties of the immunological memory.

It is true, however, that the persistent population of memory cells is the mechanism by which the immune system remembers. If the same antigen is seen again, the memory population quickly produces large quantities of antibodies and it is an often phenomenon that the antigen is cleared before it causes any disease. This is the so-called secondary immune response according to which a slightly different secondary antigen may overlap partly with the memory population raised by the primary antigen as illustrated in Fig. 6.9d. In other words, this is the cross-reactive response which in the context of computer science could be characterized as associative recall. The strength of the secondary immune response is approximately proportional to the number of memory cells in the ball of stimulation of the antigen. If a subset of the memory population is stimulated by a related antigen, then the response is weaker [6]. Therefore, memory appears to be distributed among the cells in the memory population. Immunological memory is robust because, even when a portion of the memory population is lost, the remaining memory cells persist to produce a response.

6.5.5 *Immunological Memory as a Sparse Distributed Memory*

Sparse Distributed Memories (SDMs) resemble the random access memory in a computer since they share the same reading and writing functionalities. SDMs in particular are written to by providing an address and data and read from by providing an address and getting an output. Unlike random access memory, however, the address space of SDM is enormous, sometimes 1000 bits, yielding 2^{1000} possible addresses. SDMs cannot instantiate such a large number of address-data locations. These instantiated address-data locations are called *hard locations* and are said to sparsely cover the input space [16]. When an address is presented to the memory, hard locations within some threshold Hamming distance of the address are activated. This subset of activated hard locations is called the access circle of the address, as illustrated in Fig. 6.10a.

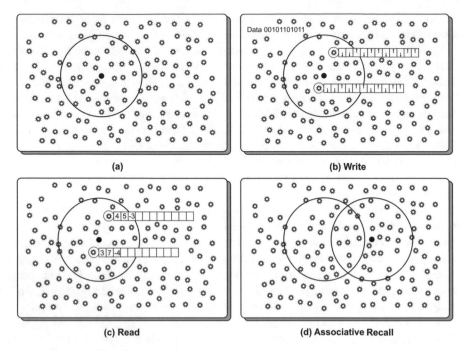

Fig. 6.10 Sparse distributed memories

On a write operation, each bit of the input data is stored independently in a counter in each hard location in the access circle. If the i-th data bit is a 1, the i-th counter in each hard location is incremented by 1. If the i-th data bit is a 0, the counter is decremented by 1, as illustrated in Fig. 6.10b. On a read operation, each bit of the output is composed independently of the other bits. Specifically, the values of the i-th counter of each of the hard locations in the access circle are summed. If the sum is positive, the output for the i-th bit is a 1. If the sum is negative, the output is a 0, as illustrated in Fig. 6.10c.

The distribution of the data among many hard locations makes the memory robust to the loss of some hard locations and permits associative recall of the data if a read address is slightly different from a prior write address. If the access circle of the read address overlaps the write address, then the hard locations that were written to by the write address are activated by the read address and give an associative recall of the write data. This is illustrated in Fig. 6.10d.

The correspondence between the two types of memories is easy to be identified since both SDMs and immunological memory use detectors in order to recognize input. In the case of SDM, hard locations recognize an address, while in the case of immunological memory, B- and T-cells recognize an antigen. In both systems, the number of possible distinct inputs is huge, while, due to resource limitations, the number of detectors is much smaller than the number of possible inputs. Therefore, in both systems, the input recognition is performed through a partial match since the corresponding input spaces are sparsely covered. Partial recognition in the case of SDMs is performed on the basis of the Hamming distance. In the case of immunological memory, partial memory is based on affinity. Thus, in both systems an input pattern activates a subset of detectors, which is the access circle of the input address in the context of SDMs and the ball of stimulation of the antigen in the context of immunological memory.

In both systems, the input-associated information is stored upon the available detectors. In particular, SDM-related information takes the form of exogenously supplied bit string, while immunological memory-related information is based on mechanisms of the immune system determining whether to respond to an antigen and with which particular class of antibodies.

All the available information, in both systems, is distributed across activated detectors. SDMs utilize the input data in order to adjust counters, while immunological memory utilizes a large number of memory cells in order to determine the class of antibodies to be produced. In particular, for the case of B-cells, the information provided by the invading antigenic pattern is used in order to create a highly specific, immune-competent class of antibodies through a genetic reconfiguration process. Because all the information is stored in each detector, each detector can recall the information independently of the other detectors. The strength of the output signal is an accumulation of the information in each activated detector. Thus, as detectors fail, the output signal degrades gracefully and the signal strength is proportional to the number of activated detectors. Therefore, the distributed nature of information storage, in both systems, makes them robust to the failure of individual detectors. In the case of SDM, the data is distributed to hard locations, each of which has different address. In the case of immunological memory, the data is distributed to cells with different receptors. In this context, if the activated subset of detectors of a related input (a noisy address in the case of SDM or mutant strain in the case of immunological memory) overlap the activated detectors of a prior input (Figs. 6.9d and 6.10d), detectors from the prior input will contribute to the output. Such associative recall, as well as the graceful degradation of the signal as inputs differ, are due to the distribution of the data among the activated detectors.

6.5.6 Affinity Maturation

In a T-cell dependent immune response, the repertoire of antigen-activated B-cells is diversified by two major mechanisms, namely

- *hypermutation* and
- *receptor editing*.

Specifically, the pool of memory cells is established by the accumulation of only high-affinity variants of the originally activated B-cells. This maturation process coupled with the clonal expansion takes place within the germinal centers [32].

The set of antibodies that participates in a secondary immune response have, on average, a higher affinity than those of the early primary response. This phenomenon, which is restricted to the T-cell dependent responses, is referred to as the *maturation of the immune response*. This maturation requires that the antigen-binding sites if the antibody molecules in the matured response be structurally different from those present in the primary response. Three different kinds of mutational events have been observed in the antibody V-region [1], namerly

- Point mutations,
- Short deletions and
- Non-reciprocal exchange of sequence following gene conversion (repertoire shift).

The clonal expansion process involves a sequence of random changes, introduced into the variable region (V-region) genes, in order to produce antibodies of increasingly higher affinity. The set of memory cells is subsequently assembled by selecting these higher affinity antibodies. It must be noted that the final antibody repertoire is not only diversified through a hypermutation process. In addition, the immune system utilizes complementary selection mechanisms, in order to preserve a dominating set of B-cells with higher affinity receptors. However, the random nature of the somatic mutation process involves the development of a large portion of nonfunctional antibodies which under the pressure of the hypermutation process can develop harmful anti-self specificities. Therefore, those cells with low affinity receptors, or self-reactive cells, must be efficiently eliminated (or become anergic) so that they do not significantly contribute to the pool of memory cells. This elimination process should be extended to the subset of B-cells that contain damaged and, therefore, disabled antigen-binding sites. The primary functionality of the immune systems that resolves this elimination process is called *apoptosis* and it is very likely to take place within the germinal centers. Apoptosis constitutes an intrinsic property of the immune system dealing with the programmed cell death when a cascade of intercellular events results in the DNA condensation, fragmentation and death [4, 21]. It is clear then, that the two major factors controlling the maturation of the immune response correspond to the processes of somatic hypermutation and clonal selection.

The immune response maturation in the time period between the primary to the secondary exposure to the same or similar antigenic pattern is a continuous process. This is justified by taking into consideration the time evolution of the affinity of the available antibodies, as in Fig. 6.11. Therefore, there are four essential features of the adaptive immune responses:

- *Sufficient diversity* to deal with a universe of antigens,
- *Discrimination of self from non-self*,
- *Adaptability* of the antibodies to the selective antigens and
- *Long lasting immunological memory*.

According to the original formulation of the clonal selection theory by Burnet [3], memory was acquired by expanding the size of an antigen specific clone, with random mutations being allowed in order to enhance affinity. Furthermore, self-reactive

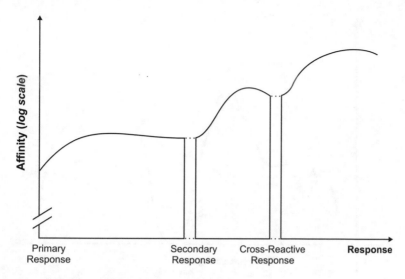

Fig. 6.11 The affinity maturation process

cells would be clonally deleted during development. Recent research developments suggest that the immune system practices molecular selection of receptors in addition to clonal selection of lymphocytes [23]. Instead of the clonal deletion of all the self-reactive cells, occasionally B lymphocytes were found that had undergone *receptor editing*. In other words, these B-cells had deleted their self-reactive receptors and developed new entirely new receptors by genetic re-combination.

George and Gray [9] argued that there should be an additional diversity introducing mechanism during the process of affinity maturation. Their contribution focuses on the idea that receptor editing may be considered as a primary mechanism which offers the ability to escape from *local optima* on an *affinity landscape*. Figure 6.12 illustrates this idea by considering all possible antigen-binding sites depicted in the x-axis, with most similar ones adjacent to each other. The Ag-Ab affinity, on the other hand, is shown on the y-axis.

Figure 6.12, assumes that a particular antibody (Ab_1) is selected during a primary response, so that subsequent point mutation and selection operations allow the immune system to explore local areas around Ab_1 by making small steps towards an antibody with higher affinity and eventually leading to a local optimum (Ab_1^*). As mutations with lower affinity are lost, the antibodies cannot go down the hill. Therefore, receptor editing is a process which allows an antibody to take large steps through the landscape, landing in a local where the affinity might be lower (Ab_2). However, occasionally the leap will lead to an antibody on the side of a hill where the climbing region is more promising (Ab_3), in order to reach the global optimum. For this local optimum, point mutations followed by selection can drive the antibody to the top of the hill (Ab_3^*).

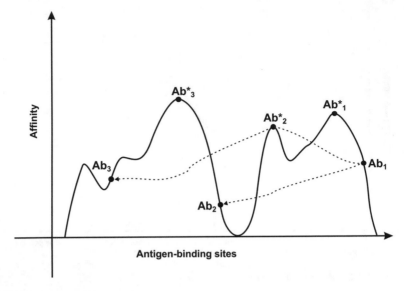

Fig. 6.12 Receptor editing process

6.5.7   Self/Non-self Discrimination

For each of the two main types of cellular components in the lymphoid system (i.e., B-cells and T-cells), there are three possible classes of repertoires:

- *Potential repertoire*: determined by the number, structure and mechanisms of expression of germ-line collections of genes encoding antibodies or T-cell receptors plus the possible somatic variants derived from these. It corresponds to the total repertoire that can be generated,
- *Available (or expressed) repertoire*: defined as the set of diverse molecules that are used as lymphocyte receptors, that is, what can be used and
- *Actual repertoire*: the set of antibodies and receptors produced by effector lymphocytes activated in the internal environment which actually participate in the interactions defining the autonomous activity in any given state.

The recognition ability of the immune system is complete. That is, the antibody molecules and T-cell receptors produced by the lymphocytes of an animal can recognize any molecule, either self or non-self, including those artificially synthesized. Antibody molecules are equipped with *immunogenic idiotopes*; that is, idiotopes that can be recognized by the antigen binding sites on other antibodies. The completeness axiom states that all idiotopes will be recognized by at least one antibody molecule, thus, being the fundamental argument behind the concept of idiotypic networks. The main factors contributing to the repertoire completeness are its *diversity* (obtained by mutation, editing, and gene re-arrangement), its cross-reactivity and its *multispecificity* [12]. Cross-reactivity and multispecificity are considered as the main reasons explaining why a lymphocyte repertoire smaller than the set containing any possible antigenic pattern can in practice recognize and bind to every antigen. The difference between cross-reactivity and multispecificity is that the former indicates the recognition of related antigenic patterns (*epitopes*) while the latter refers to the recognition of very different chemical structures.

The completeness axiom represents a fundamental paradox, since it states that all molecular shapes can be recognized, including our own which are also interpreted as antigens or *self-antigens*. The proper functionality of the immune system relies upon its ability to distinguish between molecules of our own cells (*self*) and foreign molecules (*non-self*), which are a priori indistinguishable. The inability of the immune system to perform this extremely significant pattern recognition task results in the emergence of autoimmune diseases which are triggered by immune responses against self-antigens. Not responding against a self-antigen is a phenomenon called *self-tolerance* or simply *tolerance* [18, 30]. The ability of the immune system to react against valid non-self antigens and to be tolerant against self-antigens is identified by the term self/*non-self discrimination*.

Beyond the large stochastic element in the construction of lymphocyte receptors, an encounter between a lymphocyte receptor and an antigen does not inevitably result in activation of the lymphocyte. Such an encounter may actually lead to its death or inactivation (*anergy*). Therefore, there is a form of *negative selection* that prevents self-specific lymphocytes from becoming auto-aggressive. In contrast, a smaller percentage of cells undergo *positive selection* and mature into *immunocompetent* cells in order to form an individual's repertoire.

Some of the main results of the interaction between a lymphocyte receptor and an antigen are illustrated in Fig. 6.13 and can be listed together with their principal causes:

- *Clonal Expansion*: recognition of non-self antigens in the presence of co-stimulatory signals,
- *Positive Selection*: recognition of a non-self antigen by a mature B-cell,
- *Negative Selection*: recognition of self-antigens in the central lymphoid organs, or peripheral recognition of self-antigens in the absence of co-stimulatory signals, and
- *Clonal Ignorance*: all circumstances in which a cell with potential antigen reactivity fails to react the antigen, e.g., if the antigen concentration is very low or if the affinity of the receptor for the antigen in question is low.

The negative selection process, in particular, concerns the maturation of T-cells within the *thymus* gland. After their generation, T-cells migrate to the protected environment of the thymus gland where a blood barrier prevents them from being exposed to non-self antigens. Therefore, the majority of elements within the thymus are representative of the self class of antigenic patterns rather than the self class. As an outcome, the T-cells containing receptors capable of recognizing these self antigens presented in the thymus are eliminated from the repertoire of T-cells [22]. All T-cells that leave the thymus to circulate throughout the body are said to be *self tolerant* since they do not responde to self antigens. The most important aspect of

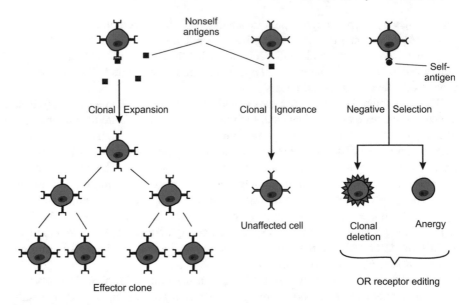

Fig. 6.13 Lymphocyte - Antigen interaction

the negative selection process from an information processing point of view is that it suggests an alternative pattern recognition paradigm by storing information about the complement set of patterns to be recognized.

References

1. Allen, D.: Timing, genetic requirements and functional consequences of somatic hypermutation during b-cell development. Immunol. Rev. **96**, 5–22 (1987)
2. Attwood, T.K., Parry-Smith, D.J.: Introduction to Bioinformatics. Prentice Hall, London (1999)
3. Burnet, F.M.: The Clonal Selection Theory of Immunity. Assoviative Neural Memories: Theory and Implemenations. Cambridge University Press, London (1959)
4. Cohen, J.J.: Apoptosis. Immunol. Today **14**(3), 126–130 (1993)
5. Darwin, C.: On the Origin of Species By Means of Natural Selection, 6th edn. (1859). www.literature.org/authors/darwin
6. East, I.J., Todd, P.E., Leach, S.: Original antigenic sin: experiments with a defined antigen. Molecul. Boiol. **17**, 1539–1544 (1980)
7. Farmer, J.D., Packard, N.H., Perelson, A.S.: The immune system, adaptation, and machine learning. Physica **22D**, 187–204 (1986)
8. Frank, S.A.: The design of natural and artificial adaptive systems (1996)
9. George, A.J.T., Gray, D.: Timing, genetic requirements and functional consequences of somatic hypermutation during b-cell development. Immunol. Today **20**(4), 196–196 (1999)
10. Gibert, C.J., Gibert, T.W.: Associative memory in an immune-based system. In: AAAI 1994: Proceedings of the Twelfth National Conference on Artificial intelligence, vol. 2, pp. 852–857. American Association for Artificial Intelligence, Menlo Park (1994)
11. Hood, L.E., Weissman, I.L., Wood, W.B., Wilson, J.H.: Immunology, 2nd edn. The Benjamin/Cummings Publishing Company, Menlo Park (1984)

12. Inman, J.K.: The antibody combining region: speculations on the hypothesis of general multi-specificity. In: Theoretical Immunology, pp. 243–278 (1978)
13. Jerne, N.K.: The immune system. Sci. Am. **229**(1), 52–60 (1973)
14. Jerne, N.K.: Towards a network theory of the immune system. Annales d'immunologie **125C**(1–2), 373–389 (1974)
15. Jerne, N.K.: The genetic grammar of the immune system. EMBO J. **4**(4), 847–852 (1985)
16. Kanerva, P.: Sparse Distributed Memory. MIT Press, Cambridge (1988)
17. Kanerva, P.: Sparse Distributed Memory and Related Models. Assoviative Neural Memories: Theory and Implemenations. Oxford University Press Inc., New York (1992)
18. Kruisbeek, A.M.: Tolerance. Immunoloigist **3**(5–6), 176–178 (1995)
19. Mackay, C.R.: Immunological memory. Adv. Immunol. **53**, 217–265 (1993)
20. Matziner, P.: Immunological memories are made of this? Nature **369**, 605–606 (1994)
21. McConkey, D.J., Orrenius, S., Jondal, M.: Cellular signalling in programmed cell death (apoptosis). Immunol. Today **11**, 120–121 (1990)
22. Nossal, G.J.V.: Negative selection of lymphocytes. Cell **76**, 229–239 (1994)
23. Nussenweig, M.C.: Immune receptor editing: revise and select. Cell **95**, 875–878 (1998)
24. Paton, R.: Computing with Biological Metaphors. Chapman & Hall, London (1994)
25. Percus, J.K., Percus, O., Perelson, A.S.: Predicting the size of the antibody combining region from consideration of efficient self/non-self discrimination. In: Proceedings of the National Academy of Science, vol. 60, pp. 1691–1695 (1993)
26. Perelson, A.S., Oster, G.F.: Theoretical studies of clonal selection: minimal antibody repertoire size and reliability of self- non-self discrimination. J. Theor. Biol. **81**, 645–670 (1979)
27. Perelson, A.S., Weisbuch, G.: Immumology for physicists. Rev. Modern Phys. **69**(4), 1219–1267 (1997)
28. Rowe, G.W.: Theoretical Models in Biology: The Origin of Life, the Immune System, and the Brain. Oxford University Press Inc., New York (1997)
29. Schwan, H.P.: Biological Engineering. McGrew-Hill, New York (1969)
30. Scjwartz, R.S., Bancherau, J.: Immune tolerance. Immunoloigist **4**(6), 211–218 (1996)
31. Smith, D.J., Forrest, S., Perelson, A.S.: Immunological memory is associative. In: Workshop Notes, Workshop 4: Immunity Based Systems, International Conference on Multiagent Systems, pp. 62–70 (1996)
32. Tarlinton, R.J.: Germinal centers: form and function. Current Opinion Immunol. **10**, 245–251 (1998)
33. Taylor, R.J.: Predation. Chapman and Hall, New York (1984)
34. Tew, J.G., Mandel, T.E.: Prolonged antigen half-life in the lymphoid follicules of antigen-specifically immunized mice. Immunology **37**, 69–76 (1979)
35. Tew, J.G., Philips, P.R., Mandel, T.E.: The maintenance and regulation of the humoral immune response, persisting antigen and the role of follicular antigen-binding dendric cells. Immunol. Rev. **53**, 175–211 (1980)
36. Weissman, I.L., Cooper, M.D.: How the immune system develops. Sci. Am. **269**(3), 33–40 (1993)

Chapter 7
Artificial Immune Systems

Abstract We introduce Artificial Immune Systems by emphasizing on their ability
to provide an alternative machine learning paradigm. The relevant bibliographical
survey is utilized to extract the formal definition of Artificial Immune Systems and
identify their primary application domains, which include:

- Clustering and Classification,
- Anomaly Detection/Intrusion Detection,
- Optimization,
- Automatic Control,
- Bioinformatics,
- Information Retrieval and Data Mining,
- User Modeling/Personalized Recommendation and
- Image Processing.

Special attention is paid on analyzing the Shape-Space Model which provides the
necessary mathematical formalism for the transition from the field of Biology to the
field of Information Technology. This chapter focuses on the development of alternative machine learning algorithms based on Immune Network Theory, the Clonal
Selection Principle and the Theory of Negative Selection. The proposed machine
learning algorithms relate specifically to the problems of:

- Data Clustering,
- Pattern Classification and
- One-Class Classification.

7.1 Definitions

Artificial Immune Systems (AIS) may be defined as *data manipulation, classification, representation and reasoning methodologies which follow a biologically
plausible paradigm, that of the human immune system.* An alternative definition
suggests that *an artificial immune system constitutes a computational system based
upon metaphors of the natural immune system.* In the same spirit, Dasgupta [24,

© Springer International Publishing AG 2017   159
D.N. Sotiropoulos and G.A. Tsihrintzis, *Machine Learning Paradigms*,
Intelligent Systems Reference Library 118, DOI 10.1007/978-3-319-47194-5_7

28] describes AIS as *intelligent methodologies inspired by the immune system toward real-world problem solving*. These definitions, however, do not emphasize the distinction between mathematical theoretical immunology and biologically-inspired computing. Therefore, a more general, complementary definition is necessary in order to incorporate and summarize the definitions already presented. In this context, AIS can be defined as *adaptive systems, inspired by theoretical immunology and observed functions, principles and models, which are applied to problem solving*.

According to the last definition, theoretical immunology and AIS can be differentiated on the basis of their rationales. Artificial Immune Systems are intended to problem solving within a wide range of application areas, whilst theoretical immunology specializes in simulating and/or improving experimental analyzes of the immune system. It must be noted, however, that artificial immune systems are not only related to the creation of abstract or metaphorical models of the biological immune system, but also utilize mathematical models of theoretical immunology in order to address computational problems such as optimization, control and robot navigation. Nevertheless, the majority of AIS exploit only a limited number of ideas and high level abstractions of the immune system. Under the light of these clarifications, an artificial immune system is any system that:

- incorporates, at the minimum, a basic model of an immune component such as *cell*, *molecule* or *organ*,
- is designed upon ideas stemmed from theoretical and/or experimental immunology, and
- its realization is strictly intended to problem solving.

Therefore, simply attributing "immunological terminology" to a given system is not sufficient to characterize it as an AIS. It is essential that a minimal level of immunology is involved, such as a model to perform pattern matching, an incorporated immune principle such as the clonal and/or the negative selection, or an immune network model. In other words, the information processing abilities of the adaptive immune system are the integral part of any methodology suggested by the new computational paradigm of artificial immune systems. The most important information processing properties of the adaptive immune system that should be incorporated, to some degree, by any computational artifact are summarized within the following list:

- *Pattern recognition*: Cells and molecules of the adaptive immune system have several ways of recognizing patterns. For example, there are surface molecules that can bind to an antigen or recognize molecular signals (e.g. lymphokines). Moreover, there exist intra-cellular molecules (e.g. MHC) that have the ability to bind to specific proteins, in order to present them in the cell surface to other immune cells
- *Uniqueness*: Each individual possesses its own immune system, with its particular vulnerabilities and capabilities
- *Self identity*: The uniqueness of the immune system gives rise to the fact that any cell, molecule and tissue that is not native to the body can be recognized and eliminated by the immune system

- *Diversity*: There exist various types of elements (cells, molecules, proteins) that circulate through the body in search of malicious invaders or malfunctioning cells
- *Disposability*: No single cell or molecule is essential for the functioning of the immune system since the majority of the immune cells participate in an endless cycle of death and reproduction during their lifetime. An exception to this mechanism is exhibited by the long-lived memory cells
- *Autonomy*: There is no central "element" controlling the immune system. In other words, it is an autonomous decentralized system which does not require any external intervention or maintenance. Its operation is based upon the straightforward classification and elimination of pathogens. Moreover, the immune system has the ability of partially repairing itself by replacing damaged or malfunctioning cells
- *Multilayeredness*: Multiple layers of different mechanisms that act cooperatively and competitively are combined in order to provide a high overall security
- *Lack of secure layer*: Any cell of the organism can be attacked by the immune system, including those of the immune system itself
- *Anomaly detection*: The immune system can recognize and react to pathogens that had never before been encountered,
- *Dynamically changing coverage*: As the immune system cannot maintain large enough repertoires of cells and molecules to detect all existing pathogens, a trade-off has to be made between space and time. This is accomplished by maintaining a circulating repertoire of lymphocytes that are constantly changing through cell death, production and reproduction
- *Distributivity*: The immune cells, molecules and organs are distributed all over the body with no centralized control,
- *Noise tolerance*: Pathogens are only partially recognized since they are extremely tolerant to molecular noise
- *Resilience*: This property of the immune system corresponds to the ability of the immune system to overcome the disturbances that might reduce its functionality
- *Fault tolerance*: If an immune response has been built up against a given pathogen and the responding cell type is removed, this degeneracy in the immune repertoire is handled by making other cells to respond to that specific antigen. Moreover, the complementary nature of the immune components ensures the reallocation of tasks to other elements in case any of them fails
- *Robustness*: This property of the immune system can be fairly attributed to the sheer diversity and number of the immune components
- *Immune learning and memory*: The molecules of the immune system can adapt themselves to the antigenic challenges both structurally and in number. These adaptation mechanisms are subject to a strong selective pressure, which enforces the elimination of those individuals that demonstrated the minimum ability in fighting invading antigenic patterns. On the other hand, the same procedure qualifies the retainment of those cells that promote faster and more effective responses. These highly adaptive immune cells are the so-called memory cells that undertake the burden of protecting an organism against any given antigenic challenge. Moreover, immune cells and molecules recognize each other even in the absence of exter-

nal pathogens endowing the immune system with a characteristic autonomous eigen-behavior

- *Predator-pray pattern of response*: The vertebrate immune system replicates cell in order to cope with replicating antigens so that the increasing population of these pathogens will not overwhelm the immune defenses. A successful immune response relies upon the appropriate regulation of the immune cells population with respect to the number of the existing antigenic patterns. Therefore, the ideal response to an increasing number of antigenic patterns is the proliferation of the immune competent agents. When the number of pathogens decreases, the repertoires of immune cells returns to a steady state. The collapse of this predator-pray pattern of the immune response is of vital importance since it can lead to the eventual death of the host organism

- *Self organization*: The particular pattern of interaction between antigenic agents and immune molecules during an immune response is not determined a priori. Clonal selection and affinity maturation are the fundamental processes that regulate the adaptation of the available immune cells in order to cope with a particular antigen. Specifically, the selection and expansion procedures involved undertake the responsibility for the immune cells transformation into long living memory cells.

7.2   Scope of AIS

The AIS community has been vibrant and active for almost two decades now, producing a prolific amount of research ranging from modelling the natural immune system, solving artificial or bench-mark problems, to tackling real-world applications, using an equally diverse set of immune-inspired algorithms. Therefore, it is worthwhile to take a step back and reflect on the contributions that this paradigm has brought to the various areas on which it has been applied. Undeniably, there have been a lot of successful stories during its maturation from a naive, alternative, computational paradigm to a standard machine learning methodology. Specifically, the large amount of interesting computational features of the adaptive immune system suggests an almost limitless range of applications, such as:

- Clustering and Classification,
- Anomaly and Intrusion Detection,
- Optimization,
- Control,
- Bioinformatics,
- Information Retrieval and Web Data Mining,
- User Modelling and Personalized Recommendation and
- Image Processing

The wide range of applications of AIS prove the validity of the new computational field as a competitive machine learning paradigm. The literature review presented here provides an overview of the primary application areas to which AIS has currently been applied. This is accomplished by incorporating references from the International Conference on Artificial Immune Systems (ICARIS) in the time period between the years from 2001 to 2009.

A very large number of papers fall under the general heading of *Learning*. Learning can generally be defined as the process of acquiring knowledge from experience and being able to generalize that knowledge to previously unseen problem instances. This generic title applies to the fundamental problems of pattern recognition which involves the unsupervised and supervised versions of machine learning such as clustering and classification. Papers relating to clustering [9, 11, 12, 18, 19, 23, 26, 31, 55, 91, 92, 102] and classification [15, 17, 30, 40, 44, 49, 80, 86, 92, 95, 101, 104, 106, 124, 125, 127, 132] have been separated out from the general learning topic as sub-categories relating particularly to comparing AIS-based *clustering* and *classification* algorithms against conventional techniques. In this context, the relevant papers are primarily interested in formalizing novel clustering or classification algorithms that are subsequently applied on benchmark datasets in order to be compared against state of the art algorithms. This comparison process utilizes the standard accepted quality tests in data-mining such as the classification accuracy. It must be noted that a significant branch of the relevant literature involves the theoretical justification of the related AIS-based machine learning algorithms [16, 42, 111] that may be particularly related to validation of negative selection algorithms and corresponding applications [2, 5, 25, 37, 38, 47, 69, 85, 94, 100, 113, 115] or the mathematical formulation of the underlying shape-spaces [35, 54, 56, 84, 112, 114].

Anomaly detection has been an area of application that has attracted profound interest within the AIS community. Such techniques are required to decide whether an unknown test sample is produced by the underlying probability distribution that corresponds to the training set of normal examples. Typically, only a single class is available on which to train the system. The goal of these immune-inspired systems is to take examples from one class (usually what is considered to be normal operational data) and generate a set of detectors that is capable of identifying a significant deviation from the system's expected behavior. Anomaly detection applications cover a wide range within the AIS literature, such as hardware fault detection [13], fault detection in refrigeration systems [116], misbehavior detection in mobile ad-hoc networks [105], temporal anomaly detection [51] and fault diagnosis for non-linear dynamical systems [50, 83]. Other applications of AIS-based anomaly detection involve the blacklisting of malicious nodes in peer-to-peer networks [120], the detection of anomalies in a foraging swarm robotics system [76] and the identification of variable length unknown motifs, repeated in time series data [130]. Moreover, the authors in [108] articulate the idea of utilizing an AIS for the prediction of bankruptcy of companies while the authors in [10, 93] proposed that anomaly detection in the context of AIS can be a powerful tool for the task of spam detection. Finally, intrusion detection applications and corresponding theoretical papers [1, 3, 52, 107, 117] may be considered as a particular branch of anomaly detection applications specialized for problems related to computer security and virus detection.

A number of publications relate to single [22, 62, 64, 70, 103, 123] and multi-objective [20, 21, 32, 43, 58, 122] constraint function optimization problems, often declaring some success when compared against other state-of-the-art algorithms. The majority of these publications are based on the application of the clonal selection principle, resulting in a number of algorithms such as the CLONALG algorithm [29], AINET [27] and the B-Cell algorithm [118].

A very interesting branch within the general context of AIS-related research concerns the development of general purpose control applications [36] that exploit the ability of AIS to:

- detect changes,
- coordinate agent activities,
- adapt to new information and
- robustly operate on highly complex and indeterminacy environments in a self adaptation manner that handles disturbances.

Therefore, Autonomous Robot Navigation [72, 78, 82, 121, 131] constitutes a leading sub-field of AIS-based control applications. Specifically, the authors in [73, 92] investigate the idea of utilizing an AIS-based robot navigation system in a rescue scenario. Other applications involve the incorporation of the AIS paradigm in order to regulate a multi-vehicle based delivery system in an automated warehouse [74], the development of an AIS-based control framework for the coordination of agents in a dynamic environment [71, 75, 81] and the controlling of the heating of an intelligent home [79]. A very interesting approach is presented by the authors in [57] where they raise the question of whether an AIS-based learning strategy could be utilized in order to develop an architecture that enables robot agents to accomplish complex tasks by building on basic built-in capabilities.

The list of possible AIS-based applications cannot be completed without mentioning the utilization of the AIS paradigm in bioinformatics [8, 45, 109] and image processing [4, 7]. Moreover, a very interesting application domain emerges within the fields of recommendation [14, 87], information retrieval and filtering [77, 89], and user modelling [88, 90].

7.3   A Framework for Engineering AIS

The first issue that needs to be taken into consideration in the development process of an AIS is highly related to the problem domain to which it is going to be applied. It is very important to identify the role assigned to the utilized artificial immune system in order to devise the modelling and simulation details of the relevant biological phenomena. However, the implementation of an artificial immune system may be reduced to the appropriate modelling of its primary mediators independently of the particular application domain. In other words, a special purpose AIS can be realized by particularizing the modelling process of its fundamental components, namely, the B- and T-cells and antibodies. Therefore and according to Timmis [119], a general

Fig. 7.1 Layered framework
for AIS

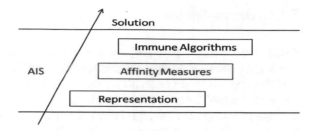

framework for the implementation of an AISW should, at least, contain the following
basic elements:

- *A representation for the components of the system*
- *A set of mechanisms to evaluate the interaction of individual components with
  the environment and each other.* The environment is usually simulated by a set of
  input stimuli and one or more fitness function(s) and
- *Procedures of adaptation that govern the dynamics of the system, that is the particular laws of motion that describe the time evolution of the system behavior.*

This is the basis of the adapted framework in order to design AISs: a representation
to create abstract models of immune organs, cells, and molecules; a set of functions,
termed affinity functions, to quantify the interactions of these "artificial elements",
and a set of general purpose algorithms to govern the dynamics of the AIS.

The adapted framework can be represented by the layered approach appearing in
Fig. 7.1. The application domain forms the basis of every system by determinately
affecting the selection of the representation model to be adapted. This constitutes a
crucial step in the designing process of a valid artificial immune system since the
adaptation of an appropriate representation scheme for its fundamental components
is the first milestone to be placed. The next step in the designing process of an artificial
immune system is the utilization of one or more affinity measures in order to quantify
the interactions of the elements of the system. There are many possible affinity
measures, such as Hamming and Euclidean distances, that are partially dependent
upon the adapted representation model. The final layer involves the use of algorithms
or processes to govern the dynamics (behavior) of the system. In this context, the
description of the framework begins with reviewing a general abstract model of
immune cells and molecules, called *shape-space*. After presenting the shape-space
approach, the framework reviews the wide range of affinity measures that could be
utilized in order to evaluate the interactions of the elements of the AIS. Finally, the
framework presents a general description of the immune inspired algorithms that
were studied for the purposes of the current monograph.

7.3.1   Shape-Spaces

Perelson and Oster in 1979 [99] introduced the concept of *shape-space* ($\mathbb{S}$) by performing a theoretical study on clonal selection which was focused on quantifying
the interactions between molecules of the immune system and antigens. Specifically,
the authors raised the question of how large an immune repertoire has to be for the
immune system to function reliably as a pattern recognition system. Under this perspective, the immune system was seen basically as a pattern (molecular) recognition
system that was especially designed to identify *shapes*. Therefore, the shape-space
approach presents, according to the authors, an ideal formal description in order to
model and evaluate interactions between antigens and antibodies which can be extended to study the binding between any type of cell receptor and the molecules they
bind with.

The affinity between an antibody and an antigen involves several processes, such
as short-range non-covalent interactions based on electrostatic charge, hydrogen
binding, van der Waals interactions, etc. For an antigen to be recognized, the antigen
and antibody molecules must bind complementarily with each other over an ap-
preciable portion of their surfaces. Therefore, extensive *regions of complementarity*
between the molecules are required, as illustrated in Fig. 7.2. The shape and charge
distributions, as well as well as the existence of chemical groups in the appropriate
complementary positions, can be considered as properties of antigens and antibodies
that are important to determine their interactions. This set of features is called the
generalized shape of a molecule.

Assume that it is possible to adequately describe the generalized shape of an
antibody by a set of L parameters (e.g., the length, width, and height of any bump
or groove in the combining site, its charge, etc.). Thus, a point in an L-dimensional
space, called *shape-space* $\mathbb{S}$, specifies the generalized shape of an antigen binding
region of the molecular receptors on the surface of immune cells with regard to its
antigen binding properties. Also, assume that a set of L parameters can be used in
order to describe an antigenic determinant, though antigens and antibodies do not
necessarily have to be of the same length. The mapping from the parameters to their

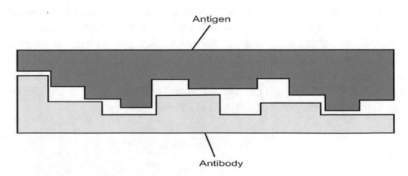

Fig. 7.2 Recognition via regions of complementarity

biological counterparts is not important from a biological standpoint, but will be dictated by the application domain of the AIS.

If an animal has a repertoire of size N, i.e., N antibodies, then the shape-space for that animal contains N points. These points lie within some finite volume V of the shape-space since the range of widths, lengths, charges, etc. is restricted. Similarly, antigens are also characterized by generalized shapes the complements of which lie within the same volume V. If the antigen (Ag) and antibody (Ab) shapes are not quite complementary, then the two molecules may still bind, but with lower affinity.

It is assumed that each antibody specifically interacts with all antigens the complements of which lie within a small surrounding region, characterized by a parameter ϵ named the *cross-reactivity threshold*. The volume V_ϵ resulting from the definition of the cross-reactivity threshold ϵ is called *recognition region*. As each antibody can recognize (cross-react with) all antigens whose complements lie within its reaction region, a finite number of antibodies can recognize a large number of antigens into the volume V_ϵ, depending on the parameter ϵ. If similar patterns occupy neighboring regions of the shape-space, then the same antibody can recognize them, as long as an appropriate ϵ is provided. Figure 7.3 illustrates the idea of a shape-space $\mathbb{S}$, detaching the antibodies, antigens, and the cross-reactivity threshold.

Mathematically, the generalized shape of any molecule m in a shape-space $\mathbb{S}$ can be represented as an attribute string (set of coordinates) of length L. Thus, an attribute string $m = \langle m_1, m_2, \ldots, m_L \rangle$ can be regarded as a point in an L-dimensional shape-space, i.e. $m \in \mathbb{S}^L$. This string might be composed of any type of attributes, such as real values, integers, bits, and symbols. These attributes are usually driven by the problem domain of the AIS and their importance relies upon the quantification of interactions between system components through the adapted affinity measure(s). Most importantly, the type of the attribute will define the particular type of the shape-space to be utilized as follows:

- *Real-valued shape-space*: the attribute strings are real-valued vectors, such that $\mathbb{S} = \mathbb{R}$;

Fig. 7.3 Shape-space model

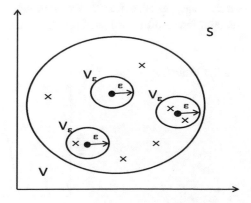

- *Integer shape-space*: the attribute strings are composed of integer values, such that $\mathbb{S} = \mathbb{N}$;
- *Hamming shape-space*: composed of attribute strings built out of a finite alphabet of length M, such that $\mathbb{S} = \Sigma = \{a_1, \ldots, a_M\}$ where Σ denotes a finite alphabet composed of M letters. Usually, hamming shape-spaces are composed of binary attribute strings, such that $\Sigma = \{0, 1\}$;
- *Symbolic shape-space*: usually composed of different types of attribute strings where at least one of them is symbolic, such as a "name", a "category", etc.

7.3.2 Affinity Measures

An affinity measure may be formally defined as a mapping:

$$D : \mathbb{S}^L \times \mathbb{S}^L \to \mathbb{R}^+, \qquad (7.1)$$

which quantifies the interaction between two attribute strings $\mathbf{Ab}, \mathbf{Ag} \in \mathbb{S}^L$ into a single nonnegative real value $D(\mathbf{Ab}, \mathbf{Ag}) \in \mathcal{R}^+$, corresponding to their *affinity* or *degree of match*. In the general case, an antibody molecule is represented by the set of coordinates $\mathbf{Ab} = \langle Ab_1, Ab_2, \ldots, Ab_L \rangle$, while an antigen is given by $\mathbf{Ag} = \langle Ag_1, Ag_2, \ldots, Ag_L \rangle$. Without loss of generality, antibodies and antigens are assumed to be of the same length. Affinity measures are highly important from a computational point of view since they constitute the cornerstone of any AIS-based clustering, classification and recognition (negative selection) algorithm. In other words, the interactions of antibodies or of an antibody and an antigen are evaluated through the utilization of an appropriate *distance measure* which in the context of AIS is termed *affinity measure*.

Given an attribute string (shape) representation of an antigen and a set of antibodies, for each attribute string of an antibody molecule, one can associate a corresponding affinity with the given antigen. Thus, an *affinity landscape* can be defined on the shape-space, as it is illustrated in Fig. 7.4. Usually, the $\mathbf{Ag} - \mathbf{Ab}$ (or $\mathbf{Ab} - \mathbf{Ab}$) recognition status is evaluated through the utilization of a *matching rule*

Fig. 7.4 Affinity landscape

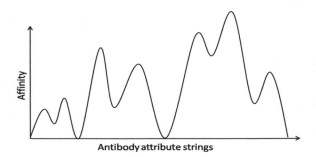

Antibody attribute strings

$$M : \mathbb{S}^L \times \mathbb{S}^L \to \{0, 1\} \tag{7.2}$$

such that:

$$M(\mathbf{Ab}, \mathbf{Ag}) = \begin{cases} I(D(\mathbf{Ab}, \mathbf{Ag}) \geq \theta_{aff}), & \text{Hamming and Symbolic shape-spaces;} \\ I(D(\mathbf{Ab}, \mathbf{Ag}) \leq \theta_{aff}), & \text{Integer- and Real-Valued shape-spaces.} \end{cases} \tag{7.3}$$

In Eq. (7.3), $I(\cdot)$ denotes the indicator function and θ_{aff} is a predefined affinity threshold. In other words, a matching rule with an associated affinity threshold introduces the concept of partial match which is a fundamental control parameter for the generalization ability of any AIS-based machine learning algorithm. Therefore, the components of an artificial immune system are considered to recognize each other whenever their affinity is within (or exceeds) a predefined threshold.

Hamming and Symbolic Shape-Spaces

In Hamming and Symbolic Shape-Spaces, the $\mathbf{Ag} - \mathbf{Ab}$ (or $\mathbf{Ab} - \mathbf{Ab}$) affinity is proportional to the *degree of complementarity* (distance) between the corresponding attribute strings $\mathbf{Ab}$ and $\mathbf{Ag}$ which quantifies the degree of their match. This fact is particularly taken into consideration in Eq. (7.3) stating that two attribute strings match when their distance (affinity) exceeds a predefined threshold. The following list summarizes the most widely used affinity measures that are specifically suitable for finite alphabet and symbolic shape-spaces:

1. **Hamming** or **Edit Distance**: The general form of Hamming or Edit distance for finite alphabet or symbolic shape-spaces is given by the following equation:

$$D(\mathbf{Ab}, \mathbf{Ag}) = \sum_{i=1}^{L} \delta(Ab_i, Ag_i), \tag{7.4}$$

where

$$\delta(Ab_i, Ag_i) = \begin{cases} 1, & Ab_i \neq Ag_i; \\ 0, & \text{otherwise.} \end{cases} \tag{7.5}$$

For binary shape-spaces, the Hamming distance may be formulated as:

$$D(\mathbf{Ab}, \mathbf{Ag}) = \sum_{i=1}^{L} Ab_i \oplus Ag_i \tag{7.6}$$

in which the binary operator $\oplus$ denotes the exclusive OR operation between the bits Ab_i and Ag_i. The Hamming distance of two binary strings $\mathbf{Ab}$ and $\mathbf{Ag}$ may be defined as the minimum number of point mutations required in order to change $\mathbf{Ab}$ into $\mathbf{Ag}$, where point mutations correspond to bit flips at certain positions.

2. **Maximum Number of Contiguous Bits**: This affinity measure can be defined through the following equation:

$$D(\mathbf{Ab}, \mathbf{Ag}) = \max_{\substack{1 \le k \le L \\ k \le m \le L}} \sum_{i=k}^{i=m} Ab_i \oplus Ag_i \tag{7.7}$$

for binary shape-spaces and through the following equation:

$$D(\mathbf{Ab}, \mathbf{Ag}) = \max_{\substack{1 \le k \le L \\ k \le m \le L}} \sum_{i=k}^{i=m} \delta(Ab_i, Ag_i) \tag{7.8}$$

for finite alphabet and symbolic shape-spaces. It was originally proposed by Percus et al. [97] in the form of the *r-contiguous bits rule* which may be formulated as:

$$M(\mathbf{Ab}, \mathbf{Ag}) = I(D(\mathbf{Ab}, \mathbf{Ag}) \ge r) \tag{7.9}$$

where the affinity measure $D(\cdot)$ is provided by Eq. (7.7) or (7.8). The r-contiguous bits (rcb) matching rule requires r contiguous matching symbols in corresponding positions. Considering the fact that perfect matching is rare, the choice of rcb is mainly to simplify mathematical analysis with some flavor of immunology. In fact, the incorporation of a partial matching rule is an integral component of any AIS-based machine learning algorithm since it regulates its ability to generalize from a limited number of training samples. Specifically, the value of r can be used in order to balance between more generalization or more specification.

3. **Maximum Number of Contiguous Bits in a Chunk**: This affinity measure is a variant of the maximum number of contiguous bits, given by the equation:

$$D(\mathbf{Ab}, \mathbf{Ag}, p) = \max_{p \le k \le L} \sum_{i=p}^{i=k} Ab_i \oplus Ag_i \tag{7.10}$$

for binary shape-spaces. It was originally proposed was by Balthrop et al. [6] in the context of the *R-Chunks matching rule* given by the equation:

$$M(\mathbf{Ab}, \mathbf{Ag}) = I(D(\mathbf{Ab}, \mathbf{Ag}, p) \ge r), \tag{7.11}$$

where the affinity measure $D(\cdot)$ is given by Eq. (7.10). It is obvious that the *R-Chunks matching rule* constitutes a simplification of the rcb matching rule since the relevant complementarity requirement involves a portion of the original attribute strings. Two binary strings **Ab** and **Ag** are considered to match, if there exists a sub-range of r complementary bits at corresponding positions within their original bit representations. The required matching degree p may be considered as a window size parameter controlling the generalization ability of the relevant AIS-based machine learning algorithms.

4. **Multiple Contiguous Bits Regions**: Shape-spaces that measure the number of complementary symbols in a pair of attribute strings are biologically more

appealing. Therefore, extensive complementary regions might be interesting for the detection of similar characteristics in symmetric portions of the molecules, and can be useful for specific tasks, such as pattern recognition. An affinity measure that privileges regions of complementarity was proposed by Hunt et al. [63] and may be formulated as follows:

$$D(\mathbf{Ab}, \mathbf{Ag}) = \sum_{i=1}^{L} Ab_i \oplus Ag_i + \sum_{i} 2^{l_i}, \qquad (7.12)$$

where l_i is the length of each complementarity region i with 2 or more consecutive complementary bits. Equation (7.12) is termed the *multiple contiguous bits rule*.

5. **Hamming Distance Extensions**: There is a wide range of affinity measures that can be considered as extensions to the original Hamming distance function. A relevant survey was provided by Harmer et al. [53] reviewing the several approaches that have been employed in order to produce the number of features that match or differ. The corresponding affinity measures are all based on the following basic distance measures:

$$a = a(\mathbf{Ab}, \mathbf{Ag}) = \sum_{i=1}^{l} Ab_i \wedge Ag_i \qquad (7.13)$$

$$b = b(\mathbf{Ab}, \mathbf{Ag}) = \sum_{i=1}^{l} Ab_i \wedge Ag_i^{c} \qquad (7.14)$$

$$c = c(\mathbf{Ab}, \mathbf{Ag}) = \sum_{i=1}^{l} Ab_i^{c} \wedge Ag_i \qquad (7.15)$$

$$d = d(\mathbf{Ab}, \mathbf{Ag}) = \sum_{i=1}^{l} Ab_i^{c} \wedge Ag_i^{c} \qquad (7.16)$$

where the binary operator $\wedge$ corresponds to the logical conjunction between the bits Ab_i and Ag_i. On the other hand, Ab_i^{c} and Ag_i^{c} denote the logical negation of the original bits, assuming that $\mathbf{Ab}, \mathbf{Ag} \in \{0, 1\}^{L}$. The basic distance functions defined in Eqs. (7.13)–(7.16) may be combined into the following affinity measures:

(a) **Russel** and **Rao**:
$$D(\mathbf{Ab}, \mathbf{Ag}) = \frac{a}{a+b+c+d} \qquad (7.17)$$

(b) **Jacard** and **Needham**:
$$D(\mathbf{Ab}, \mathbf{Ag}) = \frac{a}{a+b+c} \qquad (7.18)$$

(c) **Kulzinski**:

$$D(\mathbf{Ab}, \mathbf{Ag}) = \frac{a}{b + c + 1} \qquad (7.19)$$

(d) **Socal** and **Michener**:

$$D(\mathbf{Ab}, \mathbf{Ag}) = \frac{a + d}{a + b + c + d} \qquad (7.20)$$

which is equivalent to:

$$D(\mathbf{Ab}, \mathbf{Ag}) = \sum_{i=1}^{L} Ab_i \overline{\oplus} Ag_i, \qquad (7.21)$$

In Eq. (7.21), $\overline{\oplus}$ corresponds to the logical negation of the *xor* operator between the bits Ab_i and Ag_i.

(e) **Rogers** and **Tanimoto** [46]:

$$D(\mathbf{Ab}, \mathbf{Ag}) = \frac{a + d}{a + b + 2(c + d)}, \qquad (7.22)$$

which is equivalent to:

$$D(\mathbf{Ab}, \mathbf{Ag}) = \frac{\sum_{i=1}^{L} Ab_i \overline{\oplus} Ag_i}{\sum_{i=1}^{L} Ab_i \overline{\oplus} Ag_i + 2 \sum_{i=1}^{L} Ab_i \oplus Ag_i}. \qquad (7.23)$$

This similarity measure was utilized by Harmer et al. [53] in order to evaluate the affinity between two bit-string molecules. The authors suggested that this measure is more selective than the Hamming distance and less selective than the affinity measure corresponding to the *rcb* matching rule.

(f) **Yule**:

$$D(\mathbf{Ab}, \mathbf{Ag}) = \frac{ad - bc}{ad + bc}. \qquad (7.24)$$

6. **Landscape Affinity Measures**: This class of affinity measures was proposed in order to capture ideas of matching biochemical or physical structure, and imperfect matching within a threshold of activation in an immune system. The input attribute strings corresponding to the antibody or antigen molecules for this class of affinity measures are sampled as bytes and converted into positive integer values in order to generate a landscape. The two resulting strings are subsequently compared in a sliding window fashion. This process may be formally described by letting $\mathbf{Ab}'$ and $\mathbf{Ag}'$ be the resulting string representations of the original binary strings $\mathbf{Ab}$ and $\mathbf{Ag}$, after the application of the byte sampling and the integer

conversion processes. Therefore, the modified input strings will be such that $\mathbf{Ab'}, \mathbf{Ag'} \in \{0, \ldots, 255\}^{L'}$ where $L' = \frac{L}{8}$. In this context, the relevant landscape affinity measures will be defined as:

(a) **Difference** affinity measure:

$$D(\mathbf{Ab'}, \mathbf{Ag'}) = \sum_{i=1}^{L'} |Ab'_i - Ag'_i| \tag{7.25}$$

(b) **Slope** affinity measure:

$$D(\mathbf{Ab'}, \mathbf{Ag'}) = \sum_{i=1}^{L'} |(Ab'_{i+1} - Ab'_i) - (Ag'_{i+1} - Ag'_i)|. \tag{7.26}$$

(c) **Physical** affinity measure:

$$D(\mathbf{Ab'}, \mathbf{Ag'}) = \sum_{i=1}^{L'} |Ab'_i - Ag'_i| + 3\mu, \tag{7.27}$$

where $\mu = \min\{\forall i, (Ab'_i - Ag'_i)\}$.

(d) **Statistical Correlation** affinity measure:

$$D(\mathbf{Ab'}, \mathbf{Ag'}) = \frac{\langle \mathbf{Ab'}, \mathbf{Ag'} \rangle}{\|\mathbf{Ab'}\| \cdot \|\mathbf{Ag'}\|}, \tag{7.28}$$

such that $-1 \leq D(\mathbf{Ab'}, \mathbf{Ag'}) \leq 1$. $D(\mathbf{Ab'}, \mathbf{Ag'})$ corresponds to the correlation coefficient value that evaluates the similarity of the two attribute strings. Equation (7.28) is alternatively referred to as the *cosine similarity measure*.

7. **All-Possible Alignments Affinity Measures**: The affinity measures presented so far assume that there is only one possible alignment in which the molecules (attribute strings) react. Nevertheless, from a biological perspective, the molecules are allowed to interact with each other in different alignments. This situation can be modelled by computing the *total affinity* of the two attribute strings by summing the affinity of each possible alignment, as follows:

$$D(\mathbf{Ab}, \mathbf{Ag}) = \sum_{k=0}^{L-1} \sum_{n=0}^{L-K} Ab_{n+k} \oplus Ag_n. \tag{7.29}$$

Another possibility would be to determine all possible alignments for the two attribute strings and subsequently compute the number of alignments that satisfy the *rcb* matching rule, given by the following equation:

$$D(\mathbf{Ab}, \mathbf{Ag}) = \sum_{k=0}^{L-1} I\left(\sum_{n=0}^{L-k} Ab_{n+k} \oplus Ag_n - r + 1\right). \qquad (7.30)$$

An alternative approach would be to compute the maximum affinity over all possible alignments by utilizing the following equation:

$$D(\mathbf{Ab}, \mathbf{Ag}) = \max_{0 \leq k \leq L-1} \sum_{n=0}^{L-k} Ab_{n+k} \oplus Ag_n. \qquad (7.31)$$

8. **Permutation Masks**: Permutation masks were proposed by Balthrop et al. [6] as a significant improvement in the literature relevant to affinity measures and corresponding matching rules for binary shape-spaces. According to the authors, if there is an intrinsic order of the bits in the representation, the specific bit order will hinder the detection of some patterns, since only the consecutive bits are considered to decide a match. Therefore, a different permutation mask, or equivalently a different bit order in the representation, could enable the detection of antigenic patterns that otherwise could not be caught. Whether and how the bit order matters, however, depends on specific applications which is not the case for bit order-free affinity measures, such as the Hamming distance.

A permutation mask can be defined as a bijective mapping π that specifies a particular reordering for all elements $\mathbf{Ag} \in \mathbb{S}^L$, such that $\mathbf{Ag_1} \rightarrow \pi(\mathbf{Ag_1}), \ldots,$ $\mathbf{Ag_{|\mathbb{S}|^L}} \rightarrow \pi(\mathbf{Ag_{|\mathbb{S}|^L}})$. More formally, a permutation $\pi \in \mathbb{S}n$, where $n \in \mathbb{N}$, can be written as $2 \times n$ matrix, as follows:

$$\pi = \begin{pmatrix} \mathbf{x_1}, \ldots, \mathbf{x_n} \\ \pi(\mathbf{x_1}), \ldots, \pi(\mathbf{x_n}) \end{pmatrix}. \qquad (7.32)$$

In the right hand side of Eq. (7.32), the first row presents the original elements while the second row represents the corresponding rearrangements. In this context, an affinity measure that incorporates a particular permutation mask, can be defined by the following equation:

$$D(\mathbf{Ab}, \mathbf{Ag}, \pi) = D(\mathbf{Ab}, \pi(\mathbf{Ag})), \qquad (7.33)$$

so that the corresponding matching rule will take the form:

$$M(\mathbf{Ab}, \mathbf{Ag}, \pi) = I(D(\mathbf{Ab}, \pi(\mathbf{Ag})) \geq r). \qquad (7.34)$$

9. **Fuzzy Affinity Measure**: Kaers et al. [69] suggested an alternative affinity measure by introducing the notions of *fuzzy antibody* and *Major Histocompatibility Complex* (MHC). These new features were proposed in order to incorporate antibody morphology within the definition of an affinity measure. Specifically, antibody morphology was defined as a collection of basic properties of artificial antibodies (detectors), including shape, data-representation and data-ordering,

accounting for the physical and chemical forces that act at the binding site of real biological molecules. A fuzzy antibody may be defined as a string of *fuzzy membership functions* (FMFs)

$$\mathbf{Ab} = \langle F_{1,j_1}, \ldots, F_{L,j_L} \rangle, \ \forall i \in [L], \ \forall j_i \in [n_i] \tag{7.35}$$

where each fuzzy membership function $F_{i,j}$ indicates the degree of recognition of the i-th antigenic attribute by the j-th antibody attribute, such that $F_{i,j} \in [0, 1]$. Moreover, the parameter n_i indicates the number of different fuzzy membership functions for each attribute, that are available for a given antibody population. In addition to this, a matching threshold $\theta_{aff\,i}$ is defined for every attribute such that the affinity measure between an antibody and an antigen molecule may be defined as:

$$D(\mathbf{Ab}, \mathbf{Ag}, p) = \sum_{q=p}^{p+r-1} F_{q,j_q}(Ag_q) - \sum_{q=p}^{p+r-1} \theta_{aff\,i}, \tag{7.36}$$

where $0 \leq D(\mathbf{Ab}, \mathbf{Ag}, p) \leq r$. Here, the parameter p controls the partial matching between the two attribute strings. Therefore, an alternative affinity measure would be to consider the maximum distance over all possible alignments between the given pair of attribute strings, as:

$$D(\mathbf{Ab}, \mathbf{Ag}) = \max_{0 \leq p \leq L-r+1} D(\mathbf{Ab}, \mathbf{Ag}, p). \tag{7.37}$$

This entails that the corresponding matching rule could be formulated as:

$$M(\mathbf{Ab}, \mathbf{Ag}) = I(D(\mathbf{Ab}, \mathbf{Ag}) \geq \theta_{aff}), \tag{7.38}$$

where θ_{aff} is an overall affinity measure. The proposed MHC metaphor is implemented as a permutation mask similar to Balthrop et al. [6] in order to handle the order of attributes.

10. **Value Difference Affinity Measure**: This affinity measure relies upon the *Value Difference Metric* (VDM) which was originally proposed by Stanfill and Waltz [110] in order to provide an appropriate distance function for symbolic shape-spaces and particularly for classification problems. The VDM defines the distance between two values x and y of an attribute a as:

$$vdm_a(x, y) = \sum_{c=1}^{C} |\frac{N_{a,x,c}}{N_{a,x}} - \frac{N_{a,y,c}}{N_{a,y}}| = \sum_{c=1}^{C} |P_{a,x,c} - P_{a,x,y}|. \tag{7.39}$$

In Eq. (7.39):

- $N_{a,x}$: is the number of instances in the training set T that had value x for attribute a;

- $N_{a,x,c}$: is the number of instances in the training set T that had value x for attribute a and an output class c. Note that $N_{a,x}$ is the sum of $N_{a,x,c}$ over all classes given by the following equation:

$$N_{a,x} = \sum_{c=1}^{C} N_{a,x,c} \qquad (7.40)$$

- C: is the number of classes in the problem domain;
- q: is a constant usually set to 1 or 2;
- $P_{a,x,c}$: is the conditional probability that the output class is c given that attribute a has the value x, defined as:

$$P_{a,x,c} = \frac{N_{a,x,c}}{N_{a,x}}. \qquad (7.41)$$

The utilization of the VDM as an affinity measure results in the following formula:

$$D(\mathbf{Ab}, \mathbf{Ag}) = \sum_{i=1}^{L} vdm(Ab_i, Ag_i) weight(Ab_i), \qquad (7.42)$$

where

$$
\begin{aligned}
vdm(Ab_i, Ag_i) &= \sum_{c=1}^{C} (P(c|Ab_i) - P(c|Ag_i))^2 \\
&= \sum_{c=1}^{C} \left(\frac{N(c|Ab_i)}{N(Ab_i)} - \frac{N(c|Ag_i)}{Ag_i} \right)^2 \qquad (7.43)
\end{aligned}
$$

and

$$N(Ab_i) = \sum_{c=1}^{C} N(c|Ab_i) \qquad (7.44)$$

$$N(Ag_i) = \sum_{c=1}^{C} N(c|Ag_i) \qquad (7.45)$$

such that

$$weight(Ab_i) = \sqrt{\sum_{c=1}^{C} P(c|Ab_i)^2} \qquad (7.46)$$

Using the affinity measure provided by the VDM, two values are considered to be closer if they have more similar classifications (i.e. similar correlations with the output classes), regardless of the order in which the values may have be given.

11. **Heterogeneous Value Difference Affinity Measure**: This affinity measure utilizes the *Heterogeneous Value Difference Metric* (HVDM) which is a variation of the original (VDM) particularly suited for mixed data types, given as:

$$hvdm(Ab_i, Ag_i) = \begin{cases} \sqrt{vdm(Ab_i, Ag_i)}, & \text{if } i \text{ is symbolic or nominal;} \\ \frac{|Ab_i - Ag_i|}{A_{i,max} - A_{i,min}}, & \text{if } i \text{ is discrete or real,} \end{cases} \tag{7.47}$$

where $vdm(\cdot, \cdot)$ is given by Eq. (7.43) and $A_{i,max}$, $A_{i,min}$ correspond to the maximum and minimum values for the i-th attribute respectively.

Integer-Valued and Real-Valued Shape-Spaces

Although it is not fully in accordance with the biological concept of shape complementarity, the affinity, in shape-spaces where the variables assume integer or real values, is quantified by utilizing some generalization of the Euclidean distance. This entails that the affinity is inversely proportional to the distance between a given pair of antibody and antigen attribute strings, without reflecting or complementing it. This is particularly taken into consideration in Eq. (7.3), stating that two attribute strings match when their distance (affinity) is below a predefined threshold. Therefore, the utilization of a generalized Euclidean distance function as an affinity metric, implies that there will be a geometric region of points in the shape-space such that all antigens in this region will present the same affinity to a given antibody.

This is illustrated in Fig. 7.5 where all antigens lying within the geometric region a present the same affinity to the antibody **Ab** located at the center of geometric region a. The same situation holds for the antigenic patterns lying within the geometric region b since they all present the same affinity to the given antibody **Ab**. The only difference is that antigenic patterns that lie within the geometric region a present a higher affinity to the antibody **Ab** than those antigens lying within the geometric

Fig. 7.5 Affinity in Euclidean shape-spaces

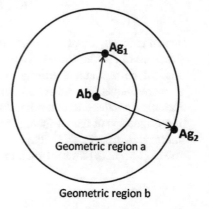

Geometric region a

Geometric region b

region b. This is true since for the given pair of antigens $\mathbf{Ag_1}$ and $\mathbf{Ag_2}$, lying within the geometric regions a and b respectively, it holds that $\|\mathbf{Ab} - \mathbf{Ag_1}\| < \|\mathbf{Ab} - \mathbf{Ag_2}\|$.

Therefore, in most practical applications of AIS, similarity measures are used instead of complementarity ones, eventhough the latter are more plausible from a biological perspective. In such cases, the goal is to search for antibodies with more similar shapes to a given antigenic pattern. The following list summarizes the most widely used affinity measures that are particularly suitable for integer- and real-valued shape-spaces:

1. **Euclidean Distance Affinity Measure**: The Euclidean distance provides the basic metric for the most widely used affinity measure in integer- and real-valued shape-spaces. The Euclidean distance is given by the equation:

$$D(\mathbf{Ab}, \mathbf{Ag}) = \|\mathbf{Ab} - \mathbf{Ag}\| = \sqrt{\sum_{i=1}^{L} |Ab_i - Ag_i|^2}. \qquad (7.48)$$

A modification to this affinity measure was proposed by [47] where the authors express the underlying distance function as a membership function of the antibody, given by the following equation:

$$D(\mathbf{Ab}, \mathbf{Ag}) = \exp\left\{ -\frac{\|\mathbf{Ab} - \mathbf{Ag}\|^2}{2r^2} \right\}. \qquad (7.49)$$

This affinity measure assumes that each antibody recognizes only antigenic patterns that lie within a hypersphere of radius r, the center coordinates of which are given by the antibody attribute string.

2. **Normalized Euclidean Distance Affinity Measure**: Most practical applications, however, utilize a normalized version of the original Euclidean distance, such as the one given by the following equation:

$$D(\mathbf{Ab}, \mathbf{Ag}) = \frac{\|\mathbf{Ab} - \mathbf{Ag}\|}{\sigma_k}, \qquad (7.50)$$

which is mostly employed when the given antigenic patterns are assumed to be elements of disjoint clusters. In this context, σ_k indicates the cluster centroid such that the corresponding affinity measure describes the sparseness of the cluster. A more frequently used version of the normalized Euclidean distance is the one that incorporates a data normalization process. Specifically, given a real-valued representation of the attribute strings corresponding to the antibody and antigen molecules, $\mathbf{Ab}, \mathbf{Ag} \in \mathbb{R}^L$, the i-th attribute Ab_i or Ag_i is considered to lie within the $[A_{i,min}, A_{i,max}]$ interval such that:

$$A_{i,min} \leq Ab_i, Ag_i \leq A_{i,max} \Leftrightarrow$$
$$A_{i,min} - A_{i,min} \leq Ab_i - A_{i,min}, Ag_i - A_{i,min} \leq A_{i,max} - A_{i,min} \Leftrightarrow$$
$$\frac{A_{i,min} - A_{i,min}}{A_{i,max} - A_{i,min}} \leq \frac{Ab_i - A_{i,min}}{A_{i,max} - A_{i,min}}, \frac{Ag_i - A_{i,min}}{A_{i,max} - A_{i,min}} \leq \frac{A_{i,max} - A_{i,min}}{A_{i,max} - A_{i,min}} \Leftrightarrow$$
$$0 \leq \frac{Ab_i - A_{i,min}}{A_{i,max} - A_{i,min}}, \frac{Ag_i - A_{i,min}}{A_{i,max} - A_{i,min}} \leq 1.$$

$$(7.51)$$

Therefore, the incorporation of the data normalization process within the distance computation yields that:

$$D(\mathbf{Ab}, \mathbf{Ag}) = \sqrt{\sum_{i=1}^{l} \left(\frac{Ab_i - A_{i,min}}{A_{i,max} - A_{i,min}} - \frac{Ag_i - A_{i,min}}{A_{i,max} - A_{i,min}} \right)^2}$$

$$= \sqrt{\sum_{i=1}^{L} \left(\frac{Ab_i - Ag_i}{A_{i,max} - A_{i,min}} \right)^2}. \qquad (7.52)$$

Let $\mathbf{Ab}' = \langle Ab'_1, \ldots, Ab'_L \rangle$ and $\mathbf{Ag}' = \langle Ag'_1, \ldots, Ag'_L \rangle$ such that:

$$Ab'_i = \frac{Ab_i - A_{i,min}}{A_{i,max} - A_{i,min}} \qquad (7.53)$$

$$Ag'_i = \frac{Ag_i - A_{i,min}}{A_{i,max} - A_{i,min}}. \qquad (7.54)$$

Then, Eq. (7.52) can be rewritten as:

$$D(\mathbf{Ab}', \mathbf{Ag}') = \|\mathbf{Ab}' - \mathbf{Ag}'\| = \sqrt{\sum_{i=1}^{L} (Ab'_i - Ag'_i)^2}, \qquad (7.55)$$

which, according to Eq. (7.51), implies that the distance between any given pair of antibody and antigen attribute strings satisfies the following inequality:

$$0 \leq D(\mathbf{Ab}', \mathbf{Ag}') \leq \sqrt{L}. \qquad (7.56)$$

A wide range of applications require that the utilized affinity measure be also within a certain range (i.e. the [0, 1] interval), which modifies Eq. (7.52) into

$$D(\mathbf{Ab}', \mathbf{Ag}') = \frac{1}{\sqrt{L}} \|\mathbf{Ab}' - \mathbf{Ag}'\|. \qquad (7.57)$$

3. **Manhattan Distance Affinity Measure**: Another widely used affinity measure is provided by the Manhattan distance, given by

$$D(\textbf{Ab}, \textbf{Ag}) = \sum_{i=1}^{L} |Ab_i - Ag_i|. \tag{7.58}$$

This can be normalized as:

$$D(\textbf{Ab}', \textbf{Ag}') = \frac{1}{L} \sum_{i=1}^{L} |Ab_i' - Ag_i'| \tag{7.59}$$

where the Ab_i's and Ab_i's are given by Eqs. (7.53) and (7.54).

4. **Generalized Euclidean Distance Affinity Measure**: Euclidean and Manhattan distances are not a panacea, especially when all the dimensions of the input pattern space do not equally contribute to classification. Such cases can benefit from the utilization of the generalized version of Euclidean and Manhattan distances given by the λ-norm Minkowski distance:

$$D_\lambda(\textbf{Ab}, \textbf{Ag}) = \left\{ \sum_{i=1}^{L} |Ab_i - Ag_i|^\lambda \right\}^{\frac{1}{\lambda}} \tag{7.60}$$

for any arbitrary λ or the corresponding normalized version formulated as:

$$D_\lambda(\textbf{Ab}', \textbf{Ag}') = L^{-\frac{1}{\lambda}} \cdot \left\{ \sum_{i=1}^{L} |Ab_i' - Ag_i'|^\lambda \right\}^{\frac{1}{\lambda}}. \tag{7.61}$$

Equation (7.60) may be used in order to define the infinity norm Minkowski distance-based affinity measure

$$D_\infty(\textbf{Ab}, \textbf{Ag}) = \lim_{\lambda \to \infty} D_\lambda(\textbf{Ab}, \textbf{Ag}) = \max_{1 \le i \le L} \{|Ab_i - Ag_i|\} \tag{7.62}$$

The utilization of the generalized Euclidean distance-based affinity measures, $D_\lambda(\cdot)$, results in different shaped detection regions around a given antibody, located at the center of the corresponding area. The term detection region identifies the subrange of the shape-space that is recognized by a given antibody, such that $\mathcal{R}_\lambda(\textbf{Ab}) = \{\textbf{Ag} \in \mathbb{S} : D_\lambda(\textbf{Ab}, \textbf{Ag}) \le R\}$. In other words, $\mathcal{R}_\lambda(\textbf{Ab})$ contains the subset of the antigenic patterns that are within a predefined distance from **Ab**, parameterized by λ and R, where R is the radius of the detection region. Figure 7.6a–e illustrate five different detection regions in the shape-space $\mathbb{S} = [0, 1] \times [0, 1]$ for five different values of the parameter λ for an antibody located at the center of the corresponding shape-space, such that $\textbf{Ab} = [0.5\ 0.5]$.

5. **Heterogeneous Euclidean Overlap Metric**: This affinity measure can handle mixed data with both continuous and nominal values, given by the equation:

$$D(\mathbf{Ab}, \mathbf{Ag}) = \sqrt{\sum_{i=0}^{L} heom(Ab_i - Ag_i)^2}, \qquad (7.63)$$

where

$$heom(Ab_i, Ag_i) = \begin{cases} \delta(Ab_i, Ag_i), & \text{if } i \text{ is symbolic or nominal;} \\ \frac{|Ab_i, Ag_i|}{A - A}, & \text{if } i \text{ is discrete or real.} \end{cases} \qquad (7.64)$$

6. **Discretized Value Difference Metric Affinity Measure (DVDM)**: This affinity measure allows for the use of VDM on real-valued data. The fundamental pre-requisite for the application of DVDM is the discretization of each attribute range into s equal-width intervals. Therefore, the width w_i of the discretized interval for the i-th attribute will be given by:

$$w_i = \frac{A_{i,min} - A_{i,max}}{s}. \qquad (7.65)$$

In this context, the DVDM affinity measure may be defined as:

$$D(\mathbf{Ab}, \mathbf{Ag}) = \sum_{i=1}^{L} |vdm(disc(Ab_i), disc(Ag_i))|^2, \qquad (7.66)$$

where

$$disc(Ab_i) = \begin{cases} \lfloor \frac{Ab_i - A_{i,min}}{w_i} \rfloor + 1, & \text{if } i \text{ is continuous;} \\ Ab_i, & \text{if } i \text{ is discrete.} \end{cases} \qquad (7.67)$$

and

$$disc(Ag_i) = \begin{cases} \lfloor \frac{Ag_i - A_{i,min}}{w_i} \rfloor + 1, & \text{if } i \text{ is continuous;} \\ Ag_i, & \text{if } i \text{ is discrete.} \end{cases} \qquad (7.68)$$

7.3.3   Immune Algorithms

The artificial immune system-based machine learning algorithms covered in this monograph pertain to three major categories:

1. Clustering,
2. Classification and
3. One Class Classification.

Therefore, the following sections are dedicated to thoroughly describing each algorithm and providing the theoretical justification for its validity.

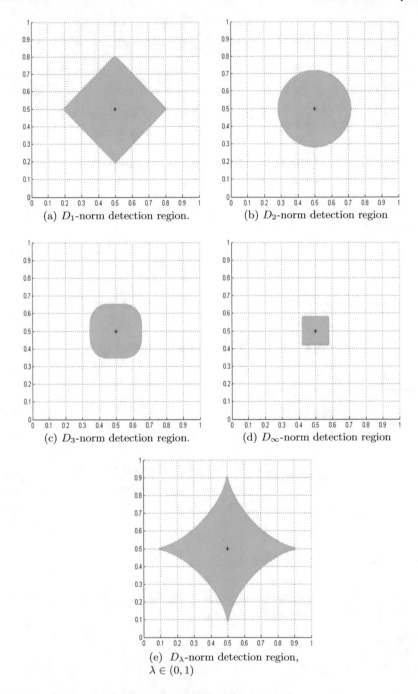

(a) D_1-norm detection region.   (b) D_2-norm detection region

(c) D_3-norm detection region.   (d) D_∞-norm detection region

(e) D_λ-norm detection region,
$\lambda \in (0, 1)$

Fig. 7.6 Generalized Euclidean distances detection regions

7.4 Theoretical Justification of the Machine Learning Ability of the Adaptive Immune System

Pattern recognition constitutes the fundamental machine learning capability of the adaptive immune system which is of primary importance for the survival of any living organism. The immune system of the vertebrates, in particular, is a highly evolved biological system whose functionality is focused on the identification and subsequent elimination of foreign material. In order to accomplish this task, the immune system must be able to distinguish between foreign and self-molecules. Therefore, memorizing and learning from the encountered antigenic patterns are essential prerequisites for the pattern recognition functionalities of the immune system. Specifically, novel antigenic shapes can be recognized by employing genetic mechanisms for change, similar to those utilized in biological evolution. In the immune system, however, these operations run on a time scale that can be as short as a few days, making the immune system an ideal candidate for the study and modelling of adaptive processes.

In this context, Farmer et al. [39] proposed a mathematical model, based on the immune network theory of Jerne [65], which utilizes genetic operators on a time scale fast enough to observe experimentally in a computer. Specifically, the relevant model attributes the ability of the immune system to recognize novel antigenic patterns without pre-programming, on the basis of its self-organized, decentralized and dynamic nature. According to [39], the learning and memory phenomena exhibited by the adaptive immune system can be reduced to the particular patterns (i.e. the predator-pray pattern of immune response) of its dynamic behavior. That is, the time evolution of its fundamental components governs the emergence of the high level pattern recognition characteristics of the immune system.

The ideal mathematical framework for describing the time evolution of such a dynamical system is provided by the associated state vector and the corresponding set of differential equations, determining the interactions between the elements of the state vector. Specifically, the state vector $\mathbf{S}$ may be formally defined by the following equation:

$$\begin{aligned}
\mathbf{S} &= [\mathbf{X}(t), \mathbf{Y}(t)], \text{ where} \\
\mathbf{X}(t) &= \langle X_1(t), \ldots, X_N(t) \rangle, \text{ and} \\
\mathbf{Y}(t) &= \langle Y_1(t), \ldots, Y_M(t) \rangle.
\end{aligned} \tag{7.69}$$

In Eq. (7.69), the variables $\{X_i(t), \ \forall i \in [N]\}$ and $\{Y_i(t), \ \forall i \in [M]\}$ correspond to the instantaneous concentrations of the existent antibody and antigen molecules, respectively. In particular, variable $X_i(t)$ represents the concentration of the i-th type of antibodies $\mathbf{Ab}_i$ at time t, whereas, variable $Y_i(t)$ represents the concentration of the i-th type of antigens $\mathbf{Ag}_i$ at time t. Moreover, the different types of antibody and antigen molecules present within the immune system at a particular moment are given by the parameters N and M. Therefore, the state vector $\mathbf{S}$ provides qualitative and quantitative information concerning the constituent elements of the immune system at a given time instance.

The motion equations for each constituent of the state vector $\mathbf{S}$ are given by the following set of differential equations:

$$\frac{dX_i(t)}{dt} = F_i(\mathbf{X}(t), \mathbf{Y}(t)) = F_i(t), \; \forall i \in [N] \tag{7.70}$$

$$\frac{dY_i(t)}{dt} = G_i(\mathbf{X}(t), \mathbf{Y}(t)) = G_i(t), \; \forall i \in [M], \tag{7.71}$$

where the particular form of the functions $\{F_i : \mathbb{R}^{N+M} \to \mathbb{R}\}$ and $\{G_i : \mathbb{R}^{N+M} \to \mathbb{R}\}$ will be thoroughly described in the context of the various modes of interaction between the elements of the immune system. Specifically, by letting

$$\mathbf{F}(t) = [F_1(t), \ldots, F_N(t)] \tag{7.72}$$
$$\mathbf{G}(t) = [G_1(t), \ldots, G_M(t)] \tag{7.73}$$

the time evolution of the state vector $\mathbf{S}(t)$ according to Eqs. (7.70) and (7.71) may be written in the following form:

$$\frac{d\mathbf{S}^T(t)}{dt} = \begin{bmatrix} \frac{d\mathbf{X}^T(t)}{dt} \\ \frac{d\mathbf{Y}^T(t)}{dt} \end{bmatrix} = \begin{bmatrix} \mathbf{F}^T(t) \\ \mathbf{G}^T(t). \end{bmatrix} \tag{7.74}$$

An essential element of the model is that the list of antibody and antigen types is dynamic, changing as new types of molecules are added or removed. Thus, it is more appropriate to consider the quantities N and M in Eq. (7.69) as functions of time, such that $N = N(t)$ and $M = M(t)$. In this context, the dimensionality of the state vector may also be expressed as a function of time according to the following equation:

$$dim(\mathbf{S})(t) = N(t) + M(t) \tag{7.75}$$

However, the quantities $N(t)$ and $M(t)$ are being modified in a slower time scale than the quantities $X_i(t)$ and $Y_i(t)$, so that it is possible to integrate Eq. (7.74) for a series of simulations, ensuring that the exact composition of the system is examined and updated as needed. This updating is performed on the basis of a minimum concentration threshold, so that a particular variable and all of its reactions are eliminated when the concentration drops below that threshold. This is an important property of the model since new space is needed in a finite animal or computer memory.

Antibody and antigen molecules are treated as being composed of elementary binary strings corresponding to epitopes or paratopes. Therefore, the utilized shape-space for the paratope or epitope representation is the binary Hamming space, such that $\mathbb{S} = \Sigma^L$, where $\Sigma = \{0, 1\}$. Specifically, an antibody molecule may be represented as an ordered pair of paratope and epitope binary strings, such that $\mathbf{Ab} = \langle \mathbf{p}, \mathbf{e} \rangle$, where $\mathbf{p}, \mathbf{e} \in \mathbb{S}$. Accordingly, an antigenic pattern may be represented by the corresponding epitope binary string as $\mathbf{Ag} = \langle \mathbf{e} \rangle$, where $\mathbf{e} \in \mathbb{S}$.

The model proposed by Farmer et al. [39] considers two modes of interaction amongst the elements of the immune system. The first mode concerns the antibody-antibody interaction while the second one focuses on the antibody-antigen interaction. It must be noted, that these modes of interaction are essential for the derivation of the system motion equations governed by the functions $\{F_i : i \in [N]\}$ and $\{G_i : i \in [M]\}$ that can be analytically defined in the following way:

- The set of functions defining vector $\mathbf{F}$ captures both modes of interaction since it describes the time evolution of the antibody molecule concentrations. Specifically, each vector element corresponds to a function F_i, given by the following equation:

$$F_i(t) = c \left[\sum_{j=1}^{N} d_{ji} X_i(t) X_j(t) - k_1 \sum_{j=1}^{N} d_{ij} X_i(t) X_j(t) + \sum_{j=1}^{M} d_{ji} X_i(t) Y_j(t) \right] - k_2 X_i(t). \quad (7.76)$$

The first term represents the stimulation of the paratope of a type i antibody, $\mathbf{Ab}_i = \langle \mathbf{p}_i, \mathbf{e}_i \rangle$, by the epitope of a type j antibody, $\mathbf{Ab}_j = \langle \mathbf{p}_j, \mathbf{e}_j \rangle$. The second term represents the suppression of a type i antibody when its epitope is recognized by the paratope of a type j antibody. The form of these terms is dictated by the fact that the probability of collision between a type i antibody and a type j antibody is proportional to the product $X_i \cdot X_j$. The parameter c corresponds to a rate constant depending on the number of collisions per unit time and the rate of antibody production stimulated by a collision. The term d_{ij} represents the affinity measure between a type i epitope and a type j paratope, given by the following equation:

$$d_{ij} = D(\mathbf{e}_i, \mathbf{p}_j), \quad (7.77)$$

where $D(\cdot)$ is given by Eq. (7.30). The constant k_1 represents a possible inequality between stimulation and suppression which is eliminated when $d_{ij} = d_{ji}$. This immune network model, however, is driven by the presence of antigenic patterns, represented by the third term. Therefore, this term corresponds to the stimulation of a type i antibody when its paratope recognizes the epitope of a type j antibody, which is proportional to the quantity $X_i Y_j$, regulated by the factor $d_{ji} = D(\mathbf{e}_j, \mathbf{p}_i)$. This term changes in time as antigens grow or are eliminated. The final term models the tendency of the cells to die in the absence of any interaction, at a rate determined by the parameter k_2.

- The set of functions defining vector $\mathbf{G}$, on the other hand, is uniquely based on the antibody-antigen mode of interaction describing the time evolution of the antigen molecules concentration. Specifically, each vector element corresponds to a function G_i, given by the following equation:

$$G_i = -k_3 \sum_{j=1}^{M} d_{ij} X_j(t) Y_i(t) \quad (7.78)$$

representing the suppression of the type i antigen when it is recognized by the type j antibody.

The model proposed by Farmer et al. [39] is a dynamical model in which antibody and antigenic molecules are constantly being added or removed from the immune system. As it was previously mentioned, the elimination process consists of removing all molecules and their corresponding interactions when their concentration drops below a predefined threshold. However, reviewing the fundamental generation mechanisms for the new antibody and antigen molecules is of greater importance since they are essential for the emergence of the learning capabilities of the immune system. According to Farmer et al. [39] new antibody types are generated through the application of genetic operators to the paratope and epitope strings, such as crossover, inversion and point mutation. *Crossover* is implemented by randomly selecting two antibody types, randomly choosing a position within the two strings, and interchanging the pieces on the side of the chosen position in order to produce two new types. Paratopes and epitopes are crossed over separately. *Point mutation* is implemented by randomly changing one of the bits in a given string and *inversion* is implemented by inverting a randomly chosen segment of the string. Random choices are weighted according to concentration whenever appropriate. Antigens, on the other hand, can be generated either randomly or by design. Therefore, the learning ability of the adaptive immune system can be primarily attributed in raising the population size and affinity of those antibody types that present the maximum complementarity to a given antigenic pattern. Increasing the population size of selected antibodies is particularly captured by Eq. (7.76) while the affinity maturation process is captured through the application of genetic operators in order to generate highly evolved immune competent antibodies.

The authors' findings indicate that the ability of the immune system to learn and effectively eliminate antigens depends upon the detailed regulatory rules governing the operation of the system, as it is demonstrated by Eqs. (7.76) and (7.78). Specifically, the fundamental hypothesis behind the immune network theory of Jerne [65], stating that internal recognition events between antibodies play crucial regulatory role even in the absence of external antigenic stimulation, is experimentally justified. In particular, it is argued that antibodies whose paratopes match epitopes are amplified at the expense of other antibodies. According to the authors' experimentation, when $k_1 = 1$ (equal stimulation and suppression) and $k_2 > 0$, every antibody type eventually dies due to the damping term. On the other hand, letting $k_1 < 1$ favors the formation of reaction loops which form the fundamental underpinning behind the ability of the immune system to remember certain states, even when the system is disturbed by the introduction of new types.

The formation of an immunological cycle initiated by the antigen $\mathbf{Ag}_0 = \langle \mathbf{e}_0 \rangle$ is illustrated in Fig. 7.7, where the presence of the epitope $\mathbf{e}_0$ increases the concentration of all antibodies that recognize that particular epitope. Letting this set of antibodies be $\mathbf{Ab}_1 = \langle \mathbf{p}_1, \mathbf{e}_1 \rangle$, then the concentrations of all antibodies that recognize epitopes on $\mathbf{Ab}_1$ will also increase. If this set of antibodies is identified by $\mathbf{Ab}_2 = \langle \mathbf{p}_2, \mathbf{e}_2 \rangle$, then the application of the same domino-recognition logic results in a sequence of

Fig. 7.7 Immunological
cycle

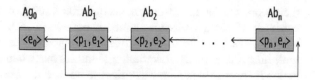

antibody molecules $\mathbf{Ab}_1$, $\mathbf{Ab}_2$, ..., $\mathbf{Ab}_n$, where the epitope $\mathbf{e}_{i-1}$ of the $(i-1)$-th type antibody is recognized by the paratope $\mathbf{p}_i$ of the i-th type antibody, for $i \in \{2, \ldots, n\}$. In this context, if $\mathbf{Ab}_n$ resembles the original antigen, its epitope $\mathbf{e}_n$ will be recognized by the paratope $\mathbf{p}_1$ of $\mathbf{Ab}_1$, forming a cycle or autocatalytic loop of order n, even in the absence of the original antigenic shape. Therefore, memorizing a given antigenic pattern can be achieved on the basis of an immunological cycle containing at least one antibody type that resembles the original antigen.

The network approach adopted by Farmer et al. [39] is particularly interesting for the development of computational tools for clustering and classification. This is true since this model naturally provides an account for emergent properties such as learning, memory, self-tolerance, size and diversity of cell populations, and network interactions with the environment and the self components. In general terms, the structure of most network models can be described similarly to the way suggested by Perelson [98]:

$$\begin{array}{c} \text{Rate of} \\ \text{population} \\ \text{variation} \end{array} = \begin{array}{c} \text{Network} \\ \text{stimulation} \end{array} - \begin{array}{c} \text{Network} \\ \text{supression} \end{array} + \begin{array}{c} \text{Influx of} \\ \text{new} \\ \text{elements} \end{array} - \begin{array}{c} \text{Death of} \\ \text{unstimulated} \\ \text{elements} \end{array} \quad (7.79)$$

Perelson and Oster [99], in particular, studied the immune system from the view point of a molecular recognition device which is designed in order to identify "foreign shapes". According to the authors, learning within the immune system involves raising the population size and affinity of those antibodies that have proven to be immune-competent in previous encounters of antigenic patterns. In this context, the authors investigate the relationship between the antibody repertoire size N_{Ab} and the probability P of recognizing a random foreign molecule. Specifically, it is argued that the probability of recognizing a foreign molecule increases with the antibody repertoire size. This is the very essence of the clonal selection principle which was originally formulated in order to explain how a great variety of different antigenic patterns could elicit specific cellular or humoral immune responses. The basic premise of the clonal selection principle is that the lymphocytes of a vertebrate organism have the necessary information to synthesize the receptors to recognize an antigen without using the antigen itself as a template. This is true, since during the course of an immune response a selection mechanism is triggered by the antigens, enforcing the proliferation of those antibodies with receptors complementary to the given antigenic pattern.

The mathematical analysis provided by Perelson and Oster [99] interprets the reliability of the immune system in terms of its antibody repertoire size. In particular, their work attributes the pattern recognition ability of the immune system to the existence of a large variety of antibody molecules with different three-dimensional binding sites, specific for the different "antigenic determinants" found on antigen molecules. Specifically, their findings indicate that when the size N_{Ab} of the available antibody repertoire is below a critical value then the immune system could become very ineffectual. On the other hand, increasing the antibody repertoire size beyond a critical value yields diminishing small increases in the probability of recognizing a foreign molecule.

In order to estimate the number of distinct antibody shapes required to reliably recognize any given antigenic shape, Perelson and Oster [99] assumed that antibody shapes are randomly distributed throughout some volume V within the Euclidean shape-space $\mathbb{S}$ with uniform density p_{Ab}. The restriction of the antibody binding sites and antigenic determinant shapes within a volume V is justified by the fact that both are characterized by finite size and finite values of other physical parameters such as charge. The choice of a random distribution for the shapes within V, on the other hand, is based solely on the complete lack of knowledge about the actual distribution. In any particular organism, however, the distribution is certainly not random since subregions of V corresponding to shapes of self antigens would have been deleted from the available antibody repertoire through the mechanism of negative selection. On the contrary, shapes that correspond to particular antigenic determinants of infectious agents, may have increased in population due to the process of clonal selection.

In order to calculate the number of different antibody shapes required to cover the sub-portion V of the shape-space $\mathbb{S}$, it is assumed that each antibody shape can bind all antigenic patterns that are within a distance ϵ from its center, represented by the corresponding attribute string. In other words, each attribute string may be viewed as a point in the L-dimensional vector space $\mathbb{S} = \Sigma^L$, where $\Sigma = \mathbb{R}$ and the distance between a given pair of points is measured via the Euclidean metric. This is particularly illustrated in Fig. 7.8 where each antibody string is assumed to be surrounded by a ball or sphere of radius ϵ recognizing any antigenic pattern whose shape happens to fall within that ball. Therefore, a limited number of antibodies

Fig. 7.8 Euclidean space shape-space model where • represents an antibody combining site and **x** represents an antigenic determinant shape

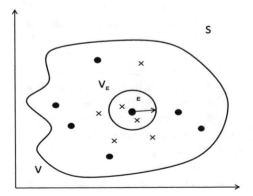

results in the occupation of a small fraction of the volume V by the associated recognition balls so that a random antigen will have a high probability of escaping detection by any antibody. Conversely, the existence of a large enough number of antibodies in the shape-space $\mathbb{S}$ ensures that the corresponding recognition balls will asymptotically fill the volume V, entailing that all antigens can be detected.

Each L-dimensional ball of radius ϵ takes up a volume $c_L \epsilon^L$, where c_L is a constant which depends upon the dimensionality of $\mathbb{S}$. For example, with a Euclidean metric, for $L = 2$, we have $c_2 = \pi$. Similarly, $L = 3$ gives $c_3 = \frac{4}{3}\pi$ and an arbitrary L gives $c_L = \frac{2\pi^{\frac{L}{2}}}{L\Gamma(\frac{L}{2})}$, where $\Gamma(\cdot)$ is the Gamma function. If there are a total of N_{Ab} antibodies within the repertoire, one would expect them to occupy a volume somewhat less than $N_{Ab}c_L\epsilon^L$ since balls would overlap. In order to compare this volume to the total volume of the shape-space occupied by both antibodies and antigens, it is assumed that V is a sphere of radius R centered at a point $\mathbf{p}$ representing a typical shape in $\mathbb{S}$. The total volume of V is then $c_L R^L$. Consequently, if

$$N_{Ab}c_L\epsilon^L \gg c_L R^L \tag{7.80}$$

or, equivalently, if

$$N_{Ab} \gg \left(\frac{\epsilon}{R}\right)^{-L}, \tag{7.81}$$

the balls surrounding each antibody would overlap considerably and cover most of V. Therefore, Eq. (7.81) determines the minimum size of the available antibody repertoire required in order to ensure the reliability of the immune system.

Assuming that the number of antibodies in each small subregion of V is distributed according to the Poisson distribution, the probability of no antibodies within distance ϵ of a randomly chosen antigenic shape is given by the exponential distribution. Having in mind that the mean number of antibodies in a volume $c_L\epsilon^L$ of the shape-space is $p_{Ab}c_L\epsilon^L$, the probability, $P_0(\epsilon, p_{Ab}, L)$, that no antibody is within distance ϵ of the antigen in the L-dimensional shape-space is given by

$$P_0(\epsilon, p_{Ab}, L) = e^{-p_{Ab}c_L\epsilon^L}. \tag{7.82}$$

Hence the probability, $P(\epsilon, p_{Ab}, L)$, that there are one or more antibodies within distance ϵ of the antigen is given by:

$$P(\epsilon, p_{Ab}, L) = 1 - e^{-p_{Ab}c_L\epsilon^L}. \tag{7.83}$$

In order to study the variation of P with ϵ, L and p_{Ab}, it is convenient to non-dimensionalize all the parameters. Thus, expressing p_{Ab} in the following form:

$$p_{Ab} = \frac{N_{Ab}}{c_L R^L} \tag{7.84}$$

yields that P may be written as:

$$P = 1 - e^{-N_{Ab}\hat{\epsilon}^L} \tag{7.85}$$

where $\hat{\epsilon} \triangleq \frac{\epsilon}{R}$ is non-dimensional and $0 \le \hat{\epsilon} \le 1$. The number of distinct antibodies in a vertebrate's repertoire can conceivably vary from 1 to many millions. From Eq. (7.85), it is obvious that P increases monotonically with N_{Ab}, when ϵ and L are held constant and it is essentially 1 when $N_{Ab} = 10\hat{\epsilon}^{-L}$. The parameter $\hat{\epsilon}$ may be considered as a measure of the antibody specificity. Specifically, when $\hat{\epsilon}$ is large, each antibody can combine with a multitude of different shapes. On the other hand, when $\hat{\epsilon}$ is small, each antibody is assumed to combine with a small fraction of antigens which have shapes very nearly complementary to that of the antibody. The antigens within a distance $\hat{\epsilon}$ of an antibody all cross-react. Thus, $\hat{\epsilon}$ can also be viewed as a measure of the cross-reactivity of random antigens with a given antibody. For a fixed repertoire size, when $\hat{\epsilon}$ is small, only a small fraction of shape space is covered by the balls of radius $\hat{\epsilon}$ surrounding each antibody shape, and, thus, P is small. Conversely, when $\hat{\epsilon}$ is large, the balls surrounding each antibody overlap and cover all of V, so that $P \approx 1$.

The number of parameters required to specify a point in shape-space, L, can be interpreted as a measure of antibody and/or antigenic determinant complexity. A simple shape can be described by very few parameters, while a complicated antigenic determinant might require many parameters to specify its shape. With this interpretation of L, Eq. (7.85) predicts that in order to maintain a given level of performance, increases in antigen complexity, as measured by L, must be accompanied by large increases in repertoire size. Moreover, there is a critical complexity, L^*, above which the immune system cannot function effectively. Letting P^* be the minimum value of P required for the efficient functioning of the immune system, then Eq. (7.85) yields that the critical complexity will be given by the following equation:

$$L^* = \frac{\ln\{\ln[1 - P^*]^{-\frac{1}{N_{Ab}}}\}}{\ln \hat{\epsilon}}. \tag{7.86}$$

For $L \ll L^*$, $P \approx 1$, whereas for $L \gg L^*$, $P \approx 0$. Having in mind that the immune system of a vertebrate organism has evolved to the point where $P \ge P^*$, then it is assumed that it can tolerate a fraction $f, f \le f^* \triangleq 1 - P^*$, of randomly encountered antigens not being complementary to any antibody in its repertoire. Once again, Eq. (7.85) may be utilized in order to derive that:

$$N_{Ab} \ge N_{Ab}^* \triangleq -\hat{\epsilon}^{-L} \ln(1 - P^*) \tag{7.87}$$

for $P \ge P^*$. Thus, the minimum repertoire size, N_{Ab}^*, varies as the logarithm of $(1 - P^*)^{-\hat{\epsilon}^{-L}}$ and grows without bound as P^* approaches 1.

7.5 AIS-Based Clustering

The generic immune network model proposed by Perelson [98] formed the basis for the conception of many AIS-based machine learning algorithms, such as the Artificial Immune NETwork (aiNet) clustering algorithm formulated by de Castro and Von Zuben [29]. aiNet constitutes an immune network learning algorithm exploiting the computational aspects of the Clonal Selection Principle. In particular, the network is initialized with a small number of randomly generated elements. Each network element corresponds to an antibody molecule which is represented by an attribute string in an Euclidean shape-space. The next stage in the course of the algorithm execution involves the presentation of the antigenic patterns. Each antigenic pattern is presented to each network cell and their affinity is measured according to Eq. (7.48). Subsequently, a number of high affinity antibodies are selected and reproduced (*clonal expansion*) according to their affinity. In particular, the number of clones produced for each antibody is proportional to the measure of its affinity with a given antigenic pattern. The clones generated undergo somatic mutation inversely proportional to their antigenic affinity. That is, the higher the affinity, the lower the mutation rate. A number of high affinity clones is selected to be maintained in the network, constituting what is defined as a clonal memory. The affinity between all remaining antibodies is determined. Those antibodies whose affinity is less than a given threshold are eliminated from the network (*clonal suppression*). All antibodies whose affinity with the currently presented antigen are also eliminated from the network (*metadynamics*). The remaining antibodies are incorporated into the network, and their affinity with the existing antibodies is determined. The final step of the algorithm is the elimination of all antibody pairs whose affinity is less than a predefined threshold.

7.5.1 Background Immunological Concepts

The aiNet learning algorithm, in terms of the utilized biological concepts, can be summarized as follows:

1. *Initialization*: Create an initial random population of network antibodies
2. *Antigenic presentation*: For each antigenic pattern do:

    (a) *Clonal selection and expansion*: For each network element, determine its affinity with the currently presented antigen. Select a number of high affinity elements and reproduce (clone) them proportionally to their affinity;

    (b) *Affinity maturation*: Mutate each clone inversely proportionally to each affinity with the currently presented antigenic pattern. Re-select a number of highest affinity clones and place them into a clonal memory set;

    (c) *Metadynamics*: Eliminate all memory clones whose affinity with the currently presented antigenic pattern is less than a pre-defined threshold;

(d) *Clonal interactions*: Determine the network interactions (affinity) among all the elements of the clonal memory set;

(e) *Clonal suppression*: Eliminate those memory clones whose affinity with each other is less than a pre-specified threshold;

(f) *Network construction*: Incorporate the remaining clones of the clonal memory with all network antibodies;

3. *Network interactions*: Determine the similarity between each pair of network antibodies;
4. *Network suppression*: Eliminate all network antibodies whose affinity is less than a pre-specified threshold;
5. *Diversity*: Introduce a number of new randomly generated antibodies into the network;
6. *Cycle*: Repeat Steps 2–5 until a pre-specified number of iterations is reached.

7.5.2   The Artificial Immune Network (AIN) Learning Algorithm

The AIN is developed is as an edge-weighted graph composed of a set of nodes, called memory antibodies and sets of node pairs, called edges, with an assigned weight or connection strength that reflects the affinity of their match. In order to quantify immune recognition, we consider all immune events as taking place in shape-space $\mathbb{S}$ which constitutes a multi-dimensional metric space where each axis stands for a physico-chemical measure characterizing a molecular shape. For the purposes of the current monograph, we utilized a real valued shape-space where each element of the AIN is represented by a real valued vector of L elements. This implies that $\mathbb{S} = \mathbb{R}^L$. The affinity/complementarity level of the interaction between two elements of the AIN was computed on the basis of the Euclidean distance between the corresponding vectors in $\mathbb{R}^L$. The antigenic patterns set to be recognized by the AIN will be composed of the set of M L-dimensional feature vectors while the produced memory antibodies can be considered as an alternative compact representation of the original feature vector set. Each feature vector encapsulates the relevant information content depending on the particular problem space.

The following notation will be adapted throughout the formulation of the learning algorithm:

- **Ab** $\in M_{N \times L}$ is the matrix storing the available antibody repertoire, such that **Ab** = **Ab**$_{\{m\}}$ $\cup$ **Ab**$_{\{d\}}$.
- **Ab**$_{\{m\}}$ $\in M_{m \times L}$ is the matrix storing the total memory antibodies repertoire, where $m \leq N$.
- **Ab**$_{\{d\}}$ $\in M_{d \times L}$ is the matrix storing the d new antibody feature vectors to be inserted within the **Ab** pool such that $d = N - m$.
- **Ag** $\in M_{M \times L}$ is the matrix storing the antigenic population that is initially fed within the algorithm.

- **F** $\in M_{N \times M}$ is the matrix storing the affinities between any given antibody and any given antigen. In particular, the $\mathbf{F}_{ij}$ element of the matrix corresponds to the affinity between the i-th antibody and the j-th antigen feature vectors. The affinity measures that can be utilized are thoroughly described within Sect. 7.3.2.
- **S** $\in M_{N \times N}$ is the matrix storing the similarities between any given pair of antibody feature vectors. Specifically, the $\mathbf{S}_{ij}$ element of the matrix corresponds to the similarity between the i-th and j-th antibody feature vectors.
- **C** $\in M_{N_c \times L}$ is the matrix storing the population of N_c clones generated from **Ab**.
- **C*** $\in M_{N_c \times L}$ is the matrix storing the population **C** after the application of the affinity maturation process.
- **D**$_j$ $\in M_{1 \times N_c}$ is the vector storing the affinity measure among every element from **C*** with the currently presented antigen **Ag**$_j$, such that $\mathbf{D}_j = \{\|\mathbf{C}_k^* - \mathbf{Ag}_j\|, k \in [N_c]\}$
- ζ is the percentage of mature antibodies to be selected.
- **M**$_j$ $\in M_{[\zeta \cdot N_c] \times L}$ is the matrix storing the memory clones for antigen Ag_j.
- **S**$_m$ $\in M_{[\zeta \cdot N_c] \times [\zeta \cdot N_c]}$ is the matrix storing the similarities between any given pair of memory antibody feature vectors.
- **M**$_j^*$ $\in M_{M_c' \times L}$ is the matrix storing the M_c' resulting memory clones after the application of the clonal suppression process such that $M_c' < [\zeta \cdot N_c]$.
- σ_d represents the AIN apoptosis criterion. It is the *natural death* threshold value that corresponds to the maximum acceptable distance between any antibody and the currently presented antigen **Ag**$_j$.
- σ_s represents the AIN suppression criterion. It is the threshold that corresponds to the minimum acceptable distance between any pair of memory antibodies.

The aiNet learning algorithm aims at building a memory set that recognizes and represents the data structural organization. The more specific the antibodies, the less parsimonious the network (low compression rate). On the other hand, the more generalist the antibodies, the more parsimonious the network with respect to the number of produced memory antibodies (improved compression rate). The suppression threshold σ_s controls the specificity level of the produced memory antibodies, the clustering accuracy and network plasticity. In this context, the general form of the AIN training algorithm appears in Fig. 7.9 while its analytical description involves the following steps:

1. At each iteration, do:

   (a) For each antigenic pattern **Ag**$_j$ $\in M_{1 \times L}$, where $j \in [M]$, do:

   i. Determine its affinity $\mathbf{F}_j(i) \in M_{N \times 1}$ to all the available antibodies **Ab**$_i$ $\in M_{1 \times L}$, $i \in [N]$ such that $\mathbf{F}_j(i) = \{\|\mathbf{Ab}_i - \mathbf{Ag}_j\|, \forall i \in [N]\}$.

   ii. Select a subset **Ab**$_{\{n\}}$ from **Ab**, composed of the n highest affinity antibodies.

   iii. The n selected antibodies are going to proliferate inversely proportional to their antigenic affinity $\mathbf{F}_j(i)$ generating a set **C** of clones. The higher the affinity, the larger the clone size for each for each of the n selected antibodies.

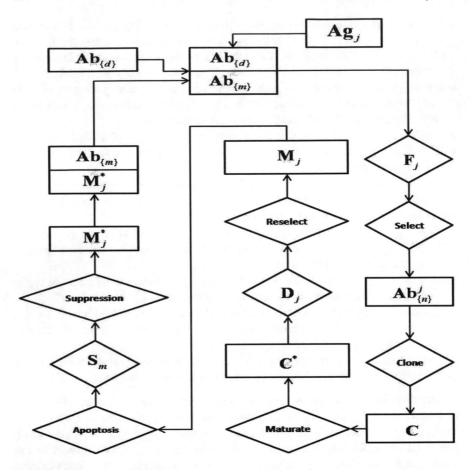

Fig. 7.9 AIN learning algorithm

iv. The set **C** is submitted to a directed affinity maturation process generating a maturated set **C***. During the affinity maturation process each
antibody C_k from **C** will suffer a number of guided mutation steps in
order to approach the currently presented antigenic pattern $\mathbf{Ag}_j$ within a
neighborhood of radius r where $r < \sigma_d$. In each mutation step, antibody
C_k from **C** will be modified at a rate a_{kj} according to the equation:

$$\mathbf{C}_k^* = \mathbf{C}_k + a_{kj} \cdot (\mathbf{Ag}_j - \mathbf{C}_k), \qquad (7.88)$$

where a_{kj} is proportional to the antigenic affinity $\mathbf{F}_j(p_k)$ of its parent
antibody and p_k corresponds to the parent antibody index of $\mathbf{C}_k$ within
the original antibody matrix **Ab**, such that $a_{kj} \propto \mathbf{F}_j(p_k)$ and $k \in [N_c]$,
$p_k \in [n], j \in [M]$.

     v. Determine the affinity matrix $\mathbf{D}_j$ among the currently presented antigen $\mathbf{Ag}_j$ and all the elements of $\mathbf{C}^*$. Specifically, the element $\mathbf{D}_j(k)$ of $\mathbf{D}$ corresponds to affinity measure between the k-th element of $\mathbf{C}^*$ and the j-th element of $\mathbf{Ag}$ such that $\mathbf{D}_j(k) = \|C_k^* - Ag_j\|$, where $k \in [N_c]$ and $j \in [M]$.

    vi. From $\mathbf{C}^*$ re-select a percentage z of the antibodies that present the lowest distance to the currently presented antigen $\mathbf{Ag}_j$ and put them into a matrix $\mathbf{M}_j$ of clonal memory such that $\mathbf{M}_j \in M_{[\zeta \cdot N_c] \times L}$.

    vii. *Apoptosis*: eliminate all the memory clones from $\mathbf{M}_j$ whose affinity to the currently presented antigenic pattern is greater than a pre-specified threshold sd such that $\mathbf{D}_j(k) > \sigma_d$.

    viii. Determine the affinity matrix $\mathbf{S}_m$ storing the distances between any given pair of memory clones such that $\mathbf{S}_m = \{\|\mathbf{M}_j(i) - \mathbf{M}_j(k)\|, i \in [\zeta \cdot N_c], k \in [\zeta \cdot N_c]\}$.

    ix. *Clonal Suppression*: Eliminate all memory clones for which $\mathbf{S}_{ik} < \sigma_s$ so that the resulting memory matrix $\mathbf{M}_j^*$ is obtained.

    x. Concatenate the total antibody memory matrix $\mathbf{Ab}_{\{m\}}$ with the resultant clonal memory matrix $\mathbf{M}_j^*$ for the currently presented antigenic pattern such that:

$$\mathbf{Ab}_{\{m\}} \leftarrow [\mathbf{Ab}_{\{m\}}; \mathbf{M}_j^*] \tag{7.89}$$

(b) Determine the affinity matrix $\mathbf{S}$ storing the affinities between all possible pairs of memory antibodies such that $\mathbf{S} = \{\|\mathbf{Ab}_{\{m\}}^i - \mathbf{Ab}_{\{m\}}^k\|, i \in [m], k \in [m]\}$.

(c) *Network suppression*: Eliminate all memory antibodies such that $\mathbf{S}_{ik} < \sigma_s$.

(d) Build the total antibody matrix by introducing d randomly generated antibody feature vectors according to the following equation:

$$\mathbf{Ab} \leftarrow [\mathbf{Ab}_{\{m\}}; \mathbf{Ab}_{\{d\}}] \tag{7.90}$$

2. Test the stopping criterion.

The selection of the n highest affinity antibodies is performed by sorting the distances of the available antibodies with the currently presented antigenic pattern in ascending order. Therefore, the amount N_c of clones generated for all these n selected antibodies will be given by the following equation:

$$N_c = \sum_{i=1}^{n} \left[\frac{\beta N}{i} \right] \tag{7.91}$$

where β is a multiplying factor and N is the total amount of antibodies. Each term of this sum corresponds to the clone size of each selected antibody. For example, if $N = 100$ and $\beta = 1$ then the highest affinity antibody ($i = 1$) will produce 100 clones, while the second highest affinity antibody produces 50 clones, and so on.

    The primary characteristic of the AIN learning algorithm is the elicitation of a clonal immune response to each antigenic pattern presented to the network. Moreover, this algorithm incorporates two suppression steps, namely the *clonal suppression* and *network suppression*, respectively. The clonal suppression ensures the elimination of intra-clonal self recognizing antibodies that are the by product of the generation of a different clone for each antigenic pattern. Network suppression, on the other hand, is required in order to search for similarities between different sets of clones. The resulting memory antibodies may be considered as the internal images of the antigens (or group of antigens) that have been presented to the network during the learning phase. Specifically, the network outputs can be taken to be the matrix of antibody feature vectors $\mathbf{Ab}_{\{m\}}$ and their affinity matrix $\mathbf{S}$. While matrix $\mathbf{Ab}_{\{m\}}$ represents the network internal images of the antigens presented to the aiNet, matrix $\mathbf{S}$ is responsible for determining which network antibodies are connected to each other, describing the general network structure.

The Artificial Immune Network convergence can be evaluated on the basis of several alternative criteria such as the following:

1. Stop the iterative process after a predefined number of iteration steps;
2. Stop the iterative process when the network reaches a pre-defined number of antibodies;
3. Evaluate the average error between all the antigens and the network memory antibodies, according to the following equation:

$$E = \frac{1}{m \cdot M} \sum_{i=1}^{m} \sum_{j=1}^{M} \|\mathbf{Ab}_{\{m\}}^{i} - \mathbf{Ag}_{j}\|, \qquad (7.92)$$

and stop the iterative process if this average error is larger than a pre-specified threshold. This strategy is particularly useful for obtaining less parsimonious solutions;
4. The network is supposed to have converged if its average error rises after k consecutive iterations.

7.5.3   AiNet Characterization and Complexity Analysis

The aiNet learning model can be classified as a *connectionist, competitive* and *constructive network*, where the resulting memory antibodies correspond to the network nodes while the antibody concentration and affinity may be viewed as their internal states. Moreover, the utilized learning mechanisms govern the time evolution of the antibody concentration and affinity. The connections among the memory antibodies, represented by the elements of matrix $\mathbf{S}$, correspond to the physical mechanisms that measure their affinity in order to quantify the immune network recognition. The aiNet graph representation, in particular, provides a description of its architecture, with the final number and spatial distribution of the identified clusters. The network dynamics employed regulates the plasticity of the aiNet, while the meta-dynamics is responsible for a broader exploration of the search-space. The competitive nature of the AIN stems from the fact that its antibodies compete with each other for antigenic recognition and, consequently, survival. Finally, the aiNet is plastic in nature, in the sense that its architecture since the number and shape of the immune cells are completely adaptable to the problem space.

The aiNet general structure is different from the neural network models [59] if one focuses on the function of the nodes and their connections. In the aiNet case, the nodes work as internal images of ensembles of patterns (thus, representing the acquired knowledge) and the connection strengths describe the similarities among these ensembles. On the other hand, in the neural network case, the nodes correspond to processing elements while the connection strengths may represent the acquired knowledge. More importantly, the immune clonal selection pattern of antigenic response may be considered as a microcosm of Darwinian evolution. The processes of simulated evolution proposed by Holland [61] try to mimic some aspects of the original theory of evolution. Regarding the aiNet learning algorithm, it is possible to identify several features in common with the process of simulated evolution which are mostly encountered within the context of evolutionary algorithms. These features relate to the fact that the aiNet is a population-based algorithm which operates on a set of randomly chosen initial candidate solutions that are subsequently refined through a series of genetic modifications and selection processes.

Analyzing an algorithm refers to the process of deriving estimates for the time and space requirements for its execution. *Complexity of an algorithm*, on the other hand, corresponds to the amount of time and space needed until its termination. Notice that determining the performance parameters of a computer program is not an easy task since it is highly dependent on a number of parameters such as the computer model being used, the way the data are represented and the programming language in which the code is implemented. In this section, the learning algorithm for the construction of an AIN will be generally evaluated for its complexity in terms of the computational cost per generation and its memory (space) requirements. Specifically, the necessary time for an algorithm to converge can be measured by counting the number of instructions executed, which is proportional to the maximum number of iterations conducted. Regardless of the particular affinity measure employed, the fundamental parameters characterizing the computational cost of the algorithm refer to the dimension $N \times L$ of the available antibody repertoire, the total number of clones N_c, the amount M of antigens to be recognized, the amount m of the produced memory antibodies and the amount n and percentage ζ of the selected antibodies for reproduction or storage as memory antibodies.

The AIN learning algorithm implemented has the following three main processing steps:

1. Defining the affinity of antibodies (Steps: 1(a) i, 1(a) v, 1(a) viii and 1(b));
2. Eliminating antibodies the affinities of which are below or above a pre-specified threshold through the mechanisms of:

   - Selection (Step 1(a) ii);
   - Re-selection (Step 1(a) vi);
   - Apoptosis (Step 1(a) vii).

- Clonal Suppression (Step 1(a) ix);
- Network Suppression (Step 1(c));

3. Hyper-mutating the population (Step 1(a) iv).

The usual way of selecting n individuals from a population is by sorting their affinities in the corresponding storing vector and subsequently extracting the first n elements of the sorted vector. This operation can be performed in $O(n)$ time, where n is the number of elements in the affinities vector. Moreover, the elimination processes that take place within the learning phase of the AIN algorithm involve the comparison of each element of a square affinity matrix with a pre-defined threshold. This operation may be conducted in $O(n^2)$ time where n is the number of elements in each matrix dimension. Finally, mutating the N_c clones demands a computational time of order $O(N_c L)$.

Table 7.1 summarizes the individual complexities for each step of the AIN training algorithm. Therefore, by summing up the computational time required for each of these steps, it is possible to determine the computational time for each individual iteration of the algorithm according to the following equation:

$$O(N + N_c + [\zeta \cdot N_c] + [\zeta \cdot N_c]^2 + m^2 + N_c L) = O(m^2) \qquad (7.93)$$

since $m \gg [\zeta \cdot N_c]$. The total time complexity of the algorithm can be estimated by taking into consideration the fact that these steps have to be performed for each of the M antigenic patterns. Thus, the overall time complexity may be given by the following equation:

$$\textbf{Required Time} = O(Mm^2). \qquad (7.94)$$

It is important to note that m may vary along the learning iterations, such that at each generation the algorithm has a different cost. Therefore, variable m in Eq. (7.94) may be considered as the average number of memory antibodies generated at each generation. On the other hand, the required memory to run the algorithm, can be estimated by summing the individual space requirement for each one of the utilized variables $\textbf{Ag}$, $\textbf{Ab}$, $\textbf{F}_j$, $\textbf{C}$ and $\textbf{C}^*$, $\textbf{M}_j$ and $\textbf{M}_j^*$, and $\textbf{S}_m$. In particular, the overall space requirement of the algorithm can be given by the following equation:

$$\textbf{Required Memory} = (M + N + N_c)L + N + N_c + [\zeta \cdot N_c] + m^2. \qquad (7.95)$$

Table 7.1 AIN learning algorithm complexity

Description	Step	Time Complexity
Selection	1(a) ii	$O(N)$
Reselection	1(a) vi	$O(N_c)$
Apoptosis	1(a) vii	$O([\zeta \cdot N_c])$
Clonal suppression	1(a) xi	$O([\zeta \cdot N_c]^2)$
Network suppression	1(c)	$O(m^2)$
Hypermutation	1(a) iv	$O(N_c L)$

7.6   AIS-Based Classification

Experimentation with the AIN learning algorithm revealed that the evolved artificial immune networks, when combined with traditional statistical analysis tools, were very efficient in extracting interesting and useful clusters from data sets. Moreover, the AIN learning algorithm may be utilized as a data compression technique since the number of resulting memory antibodies in aiNet is significantly less than the initial set of antigenic patterns in a minimal average distance sense. Therefore, the evolved memory antibodies may be considered as providing an alternative data representation that revelates their original space distribution. In addition, the AIN learning algorithm may be exploited as a dimension reduction technique when the utilized distance function can serve as a correlation measure between a given a pair of feature vectors. These feature vectors, however, correspond to a column-wise interpretation of the original data set, such that the resulting memory antibodies may be considered as an alternative feature set for the original data points.

Therefore, the exceptional success of the AIN training algorithm in dealing with unsupervised learning tasks, led Watkins et al. [126] to the conception of an Artificial Immune Recognition System (AIRS) which is particularly suited for supervised learning problems. The goal of the AIRS algorithm is to develop a set of memory cells, through the utilization of an artificial evolutionary mechanism, that may be used in order to classify data. The primary mechanism for providing evolutionary pressure to the population of the initial antibody molecules is the competition for system wide resources. This concept is the means by which cell survival is determined and reinforcement for high quality classification is provided.

The primary objective of the AIRS algorithm is to develop a set of immune competent memory cells, evolved from an initial population of randomly generated antibody molecules. Therefore, the incorporation of a fitness concept is essential in order to evaluate the pattern recognition ability of any given individual. In the AIRS algorithm, fitness is initially determined by the stimulation response of a particular antibody molecule to the currently presented antigenic pattern. According to this fitness value, a finite number of resources is allocated to each antibody molecule. In this context, by purging the antibody molecules that exhibited the least ability to acquire resources, the algorithm enforces a great selective pressure toward a place in the search space that will provide the highest reward. Otherwise stated, this process ensures the survival of those individuals in the population that will eventually provide the best classification results.

The antibody molecules surviving the phase of competition for resources are further rewarded by being given the opportunity to produce mutated offsprings. Once again, this competition for survival can be seen in a truly evolutionary sense. That is, while the fittest individuals in a given round of antigenic exposure might not survive to become a memory cell, their offspring might. Therefore, it it the survival of species that the algorithm promotes.

7.6.1   Background Immunological Concepts

The artificial memory cells that are generated by the AIS-based classification algorithm, embody several characteristics that are encountered in the natural immune systems. Primarily, the produced memory cells are based on memory B-Cells that have undergone a maturation process in the body. In mammalian immune systems, these antibody memory cells are initially stimulated by invading pathogens and subsequently undergo a process of somatic hyper-mutation as a response to the original signals indicating the antigenic recognition. The embodiment of this concept is seen in the function of memory cells in the AIRS algorithm. The artificial memory cells may also be thought of as taking on the role of T-Cells and Antigen Presenting Cells to some degree. In natural immune systems T-Cells tend to be associated with a specific population of B-Cells. When a T-Cell recognizes an antigen, it then presents this antigen to the B-Cells associated with it for further recognition. In the context of the AIRS algorithm this process may be encountered in the initial stages involving the identification of the matching memory cell which in turn develops a population of closely related antibody clones.

Table 7.2, in particular, presents the fundamental immune principles employed by the AIRS algorithm which, in terms of the utilized immune principles, may be summarized as follows:

1. *Initialization*:

   (a) *Data normalization*: All antigenic attribute strings are normalized such that the distance between any given pair of antigens lies strictly in the [0, 1] interval.
   (b) *Affinity threshold computation*: After normalization the affinity threshold is calculated corresponding to the average affinity value over all training data.

Table 7.2 Mapping between the immune system and AIRS

Immune system	AIRS
Antibody	Combination of feature vector and vector class
Shape-space	Type and possible values of the data vector
Clonal expansion	Reproduction of highly stimulated antibody molecules
Antigens	Combination of training feature vector and training vector class
Affinity maturation	Random antibody mutation and removal of the least stimulated
Immune memory	Memory set of artificial antibody molecules
Metadynamics	Continual removal and creation of antibody and memory cells

(c) *Memory cells initialization*: The set of available memory cells for each class of patterns pertaining to the training data is set to be null.

(d) *Antibody cells initialization*: The set of available antibody molecules for each class of patterns pertaining to the training data is set to be null.

2. *Antigenic presentation*: for each antigenic pattern do:

(a) *Memory cell identification*: Find the memory cell that presents the highest stimulation level to the currently presented antigenic pattern. If the set of memory cells of the same classification as the currently presented antigen is empty, then this antigen should be incorporated within the set of memory cells. Additionally, the current antigen will be denoted as the matching memory cell.

(b) *Antibody molecules generation*: Once the matching memory antibody has been identified, it is subsequently utilized in order to generate offsprings. These offsprings are mutated clones of the matching memory cell that will eventually be incorporated within the available antibodies repertoire. Specifically, the number of mutated offsprings to be generated is proportional to the stimulation level of the matching memory cell to the currently presented antigen. The mutation process, in particular, is performed element-wise for each constituent of the antibody attribute string to be mutated. Moreover, the decision concerning the mutation or not of a specific element is randomly taken such that an average number of *MutationRate* elements will be finally modified for each attribute string. The mutation range for each element to be altered is also proportional to the stimulation level of the matching memory cell, while the mutation magnitude is a value randomly chosen according to the uniform distribution in the [0, 1] interval.

(c) *Stimulations computation*: Each antibody molecule in the available repertoire is presented with the current antigen in order to determine the corresponding stimulation level.

(d) *Actual learning process*: The goal of this portion of the algorithm is to develop a candidate memory cell which is the most successful in correctly classifying the currently presented antigen. Therefore, this algorithmic step strictly concerns the antibodies pertaining to the same class with the current antigen. The following sub-steps will be executed until the average stimulation level of the available antibodies becomes greater than or equal to a predefined threshold:

i. *Stimulations normalization*: The stimulation level for each antibody is normalized across the whole antibody population based on the raw stimulation value.

ii. *Resource allocation*: Based on this normalized stimulation value, each antibody is allocated a finite number of resources.

iii. *Competition for resources*: In case the number of resources allocated across the population exceed a predefined maximum value, then resources are removed from the weakest (least stimulated) antibody until the total number of resources in the system returns to the number of resources allowed. Moreover, those antibodies that have zero resources are completely removed from the available antibody population.

    iv. *Stimulations re-computation*: Re-estimate the stimulation levels for each available antibody molecule to the currently presented antigen.

    v. *Candidate memory cell identification*: The antibody molecule with the highest stimulation level to the current antigen is identified as the candidate memory cell.

    vi. *Mutation of surviving antibody molecules*: Regardless of whether the stopping criterion is met or not, all antibodies in the repertoire are given the opportunity to produce mutated offsprings.

(e) *Memory cell introduction*: The final stage in the learning routine is the potential introduction of the just-developed candidate memory cell into the set of the existing memory cells. It is during this stage that the affinity threshold calculated during initialization becomes critical since it dictates whether the candidate memory cell replaces the matching memory cell that was previously identified. The candidate memory cell is added to the set of memory cells only if it is more stimulated by the currently presented training antigen than the matching memory cell. If this test is passed, then if the affinity between the matching memory cell and the candidate memory cell is less than the product of the affinity threshold and the affinity threshold scalar, then the candidate memory cell replaces the matching memory cell in the set of memory cells.

3. *Classification*: The classification is performed in a k-nearest neighbor approach. Each memory cell is iteratively presented with each data item for stimulation. The system classification of a data item is determined by using a majority vote of the outputs of the k most stimulated memory cells.

7.6.2   The Artificial Immune Recognition System (AIRS) Learning Algorithm

The evolutionary procedure of developing memory antibodies lies within the core of the training process of the proposed AIS-based classifier applied on each class of antigenic patterns. The evolved memory cells provide an alternative problem domain representation since they constitute points in the original feature space that do not coincide with the original training instances. However, the validity of this alternative representation follows from the fact that the memory antibodies produced recognize the corresponding set of training patterns in each class in the sense that their average affinity to them is above a predefined threshold.

To quantify immune recognition, all immune events are considered as taking place in a shape-space $\mathbb{S}$, constituting a multi-dimensional metric space in which each axis stands for a physico-chemical measure characterizing molecular shape. Specifically, a real-valued shape-space was utilized in which each element of the

AIS-based classifier is represented by a real-valued vector of L elements, such that $\Sigma = \mathbb{R}$ and $\mathbb{S} = \Sigma^L$. The affinity/complementarity level of the interaction between two elements of the constructed immune-inspired classifier was computed on the basis of the Euclidean distance between the corresponding vectors in $\mathbb{S}$. The antigenic pattern set to be recognized by the memory antibodies produced during the training phase of the AIS-based classifier is composed of the set of representative antibodies, which maintain the spatial structure of the set of all data in the music database, yet form a minimum representation of them.

The following notation will be adapted throughout the formulation of the AIRS learning algorithm:

- AT denotes the Affinity Threshold.
- ATS denotes the Affinity Threshold Scalar.
- CR denotes the Clonal Rate.
- MR denotes the Mutation Rate.
- HCR denotes the Hyper Clonal Rate.
- TR denotes the number of Total Resources, which is the maximum number of resources allowed within the system.
- RA denotes the number of Resources Allocated within the system during the training process on a particular antigenic pattern.
- NRR is the Number of Resources to be Removed in case $RA > TR$.
- C denotes the number of classes pertaining to a given classification problem.
- $\mathbf{Ag} \in M_{M \times L}$ is the matrix storing the complete set of training antigenic patterns. This matrix may be considered as being constructed by the following concatenation of matrices:

$$\mathbf{Ag} = [\mathbf{Ag}^{(1)}; \ldots; \mathbf{Ag}^{(C)}] \tag{7.96}$$

where $\mathbf{Ag}^{(k)} \in M_{M_k \times L}$ denotes the sub-matrix storing the training instances for the k-th class of patterns, such that:

$$M = \sum_{k=1}^{C} M_k \tag{7.97}$$

- $\mathbf{Ab} \in M_{N \times L}$ is the matrix storing the available antibody repertoire for the complete set of classes pertaining to the classification problem under investigation. This matrix may be considered as being constructed by the following concatenation of matrices:

$$\mathbf{Ab} = [\mathbf{Ab}^{(1)}; \ldots; \mathbf{Ab}^{(C)}] \tag{7.98}$$

where $\mathbf{Ab}^{(k)} \in M_{N_k \times L}$ denotes the sub-matrix storing the available antibodies for the k-th class of patterns, such that:

$$N = \sum_{k=1}^{C} N_k \tag{7.99}$$

- $\mathbf{S} \in M_{1 \times N}$ is the vector storing the stimulation levels to the currently presented antigenic pattern for the complete set of available antibodies in the repertoire. This vector may be considered as being constructed by the following concatenation of matrices:

$$\mathbf{S} = [\mathbf{S}^{(1)}, \ldots, \mathbf{S}^{(C)}] \tag{7.100}$$

where $\mathbf{S}^{(k)} \in M_{1 \times N_k}$ denotes the sub-vector storing the stimulation levels of the available antibodies for the k-th class of patterns to the current antigen which is also of the same classification.

- $\mathbf{R} \in M_{1 \times N}$ is the vector storing the resources allocated for the complete set of available antibodies in the repertoire after the presentation of the current antigenic instance. This vector may be considered as being constructed by the following concatenation of matrices:

$$\mathbf{R} = [\mathbf{R}^{(1)}, \ldots, \mathbf{R}^{(C)}] \tag{7.101}$$

where $\mathbf{R}^{(k)} \in M_{1 \times N_k}$ denotes the sub-vector storing the resources allocated for the available antibodies of the k-th class of patters after the presentation of the current antigenic instance which is also of the same classification.

- $\mathbf{M} \in M_{m \times L}$ is the matrix storing the memory antibodies for each class of patterns pertaining to a given classification problem. This matrix may be thought of as being constructed by the following concatenation of matrices:

$$\mathbf{M} = [\mathbf{M}^{(1)}; \ldots; \mathbf{M}^{(C)}] \tag{7.102}$$

where $\mathbf{M}^{(k)} \in M_{m_k \times L}$ denotes the sub-matrix storing the memory antibodies for the k-th class of patterns such that:

$$m = \sum_{k=1}^{C} m_k \tag{7.103}$$

- $\mathbf{s} \in M_{1 \times C}$ denotes the vector storing the average stimulation level of the available antibodies for each class of patterns, such that:

$$\mathbf{s} = [s(1), \ldots, s(C)] \tag{7.104}$$

In this context, the $s_j(k)$ element denotes the average stimulation level for the antibodies of the k-th class of patterns after the presentation of the j-th training antigen pertaining to the same class.

The analytical description of the AIRS algorithm, illustrated in Figs. 7.10, 7.11 and 7.12, involves the following steps:

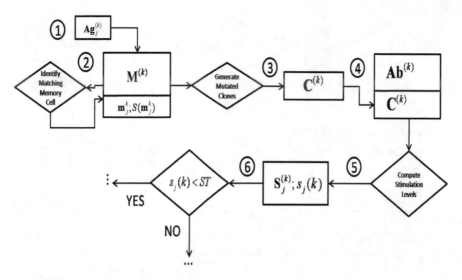

Fig. 7.10 AIRS learning algorithm 1

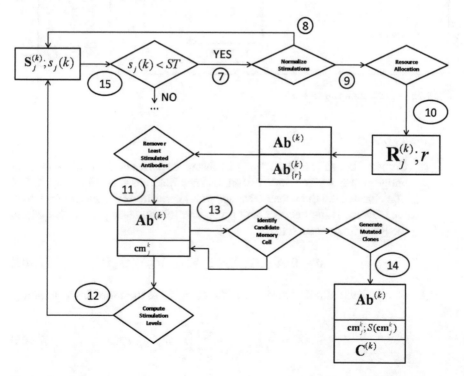

Fig. 7.11 AIRS learning algorithm 2

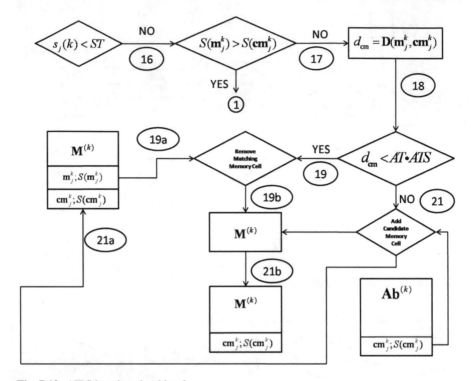

Fig. 7.12 AIRS learning algorithm 3

1. *Initialization*:

   (a) Compute the matrix $\mathbf{D} \in M_{M \times M}$ of distances between all possible pairs of antigens in $\mathbf{Ag}$, where the utilized distance function is given by Eq. (7.57). The Normalized Euclidean Distance provides an affinity measure for which the distance value between any given pair of antigenic pattern lies within the [0, 1] interval such that:

$$0 \leq \mathbf{D}(\mathbf{Ag}_i, \mathbf{Ag}_j) \leq 1, \ \forall i \in [M], \forall j \in [M] \qquad (7.105)$$

   (b) Compute the affinity threshold (AT) according to the following equation:

$$AT = \frac{2}{M(M-1)} \sum_{i=1}^{M} \sum_{j=1}^{M} \mathbf{D}(\mathbf{Ag}_i, \mathbf{Ag}_j) \qquad (7.106)$$

   (c) Initialize matrices $\mathbf{Ab}$, $\mathbf{M}$ and vectors $\mathbf{S}$, $\mathbf{R}$ according to the following equations:

$$\mathbf{Ab}^{(k)} \leftarrow [], \ \forall k \in [C] \tag{7.107}$$

$$\mathbf{S}^{(k)} \leftarrow [], \ \forall k \in [C] \tag{7.108}$$

$$\mathbf{R}^{(k)} \leftarrow [], \ \forall k \in [C] \tag{7.109}$$

$$\mathbf{M}^{(k)} \leftarrow [], \ \forall k \in [C] \tag{7.110}$$

such that:

$$N_k \leftarrow 0, \ \forall k \in [C] \tag{7.111}$$

$$m_k \leftarrow 0, \ \forall k \in [C] \tag{7.112}$$

where [] denotes the empty matrix or the empty vector.

2. For each class $k \in [C]$, of patterns do:

- For each antigenic pattern $\mathbf{Ag}_j^{(k)} \in M_{1 \times L}$, where $j \in [M_k]$, do:
  (a) Determine the matching memory cell $(\hat{\mathbf{m}}_j^k)$

$$\hat{\mathbf{m}}_j^k = \begin{cases} \mathbf{Ag}_j^{(k)}, & j = 1; \\ \arg\max_{\mathbf{m} \in \mathbf{M}_j^{(k)}} stimulation(\mathbf{Ag}_j^{(k)}, \mathbf{m}), & \text{otherwise}. \end{cases} \tag{7.113}$$

and corresponding stimulation level $(\hat{s}_j^k)$

$$\hat{s}_j^k = stimulation(\mathbf{Ag}_j^{(k)}, \hat{\mathbf{m}}_j^k) \tag{7.114}$$

The stimulation level for a given pair of vectors $\mathbf{x}, \mathbf{y}$ is given by the following equation:

$$stimulation(\mathbf{x}, \mathbf{y}) = 1 - \mathbf{D}(\mathbf{x}, \mathbf{y}) \tag{7.115}$$

such that the distance $\mathbf{D}(\mathbf{x}, \mathbf{y})$ is once given by Eq. (7.57).

(b) Compute the matrix $\mathbf{C}^{(k)} \in M_{N_c \times L}$ of mutated clones, originating from the matching memory cell $\hat{\mathbf{m}}_j^k$, where the number of produced antibodies N_c is given by the following equation:

$$N_c = HCR \cdot CR \cdot \hat{s}_j^k. \tag{7.116}$$

The computation of matrix $\mathbf{C}^{(k)}$ requires the computation of two auxiliary random matrices $\mathbf{P}_j, \mathbf{Q}_j \in M_{N_c \times L}$ such that their elements $\mathbf{P}_j(l, p)$ and $\mathbf{Q}_j(l, p)$ are uniformly distributed in the [0, 1] interval, where $l \in [N_c]$ and $p \in [L]$. Matrix $\mathbf{C}^{(k)}$, then, may be computed according to the following equation:

$$\mathbf{C}^{(k)}(l, p) = \begin{cases} low(p) + \mathbf{P}_j(l, p) \cdot \delta_j^k, & \mathbf{Q}_j(l, p) < MR; \\ \hat{m}_j^k, & \text{otherwise}, \end{cases} \tag{7.117}$$

where $\delta_j^k = 1 - \hat{s}_j^k$ and $low(p) = \max(0, \hat{\mathbf{m}}_j^k(p) - \frac{1}{2}\delta_j^k)$. Finally, matrix $\mathbf{Ab}^{(k)}$ and variable N_k are updated according to the following equations:

$$\mathbf{Ab}^{(k)} \leftarrow [\mathbf{Ab}^{(k)}; \mathbf{C}^{(k)}] \tag{7.118}$$

$$N_k \leftarrow N_k + Nc. \tag{7.119}$$

(c) Compute the sub-matrix $\mathbf{S}_j^{(k)} \in M_{1 \times N_k}$ containing the stimulations of the available antibodies for the k-th class of patterns after the presentation of the j-th antigen of the k-th class $\mathbf{Ag}_j^{(k)}$ according to the following equation:

$$\mathbf{S}_j^{(k)}(i) = stimulation(\mathbf{Ag}_j^{(k)}, \mathbf{Ab}_i^{(k)}), \; \forall i \in [N_k]. \tag{7.120}$$

Compute the average stimulation level for the k-th class of patterns after the presentation of the j-th antigenic instance originating from the same class according to the following equation:

$$s_j(k) = \frac{1}{N_k} \sum_{i=1}^{N_k} \mathbf{S}_j^{(k)}(i). \tag{7.121}$$

(d) *Actual learning process*: While $s_j(k) < ST$, do:
   i. Normalize stimulations of the antibodies according to the following equations:

$$\mathbf{S}_{j,min}^{(k)} = \min_{i \in [N_k]} \mathbf{S}_j^{(k)}(i) \tag{7.122}$$

$$\mathbf{S}_{j,max}^{(k)} = \max_{i \in [N_k]} \mathbf{S}_j^{(k)}(i) \tag{7.123}$$

$$\mathbf{S}_j^{(k)}(i) \leftarrow \frac{\mathbf{S}_j^{(k)}(i) - \mathbf{S}_{j,min}^{(k)}}{\mathbf{S}_{j,max}^{(k)} - \mathbf{S}_{j,min}^{(k)}}. \tag{7.124}$$

   ii. Compute the sub-vector $\mathbf{R}_j^{(k)} \in M_{1 \times N_k}$ of the available antibody resources for the k-th class of patterns after the presentation of the j-th antigen from the same class, according to the following equation:

$$\mathbf{R}_j^{(k)}(i) = \mathbf{S}_j^{(k)}(i) \cdot CR, \; \forall i \in [N_k]. \tag{7.125}$$

   iii. Compute the number of resources allocated by the complete set of antibodies for the current class and antigen according to the following equation:

$$RA = \sum_{i=1}^{N_k} \mathbf{R}_j^{(k)}(i). \tag{7.126}$$

iv. Compute the number of resources to be removed according to the following equation:

$$NRR = RA - TR. \tag{7.127}$$

v. Re-order the elements in matrix $\mathbf{Ab}^{(k)}$ and vectors $\mathbf{S}^{(k)}$ and $\mathbf{R}^{(k)}$ according to the permutation $\pi : [N_k] \rightarrow [N_k]$, such that:

$$\mathbf{Ab}^{(k)} \leftarrow \pi(\mathbf{Ab}^{(k)}) \tag{7.128}$$

$$\mathbf{S}_j^{(k)} \leftarrow \pi(\mathbf{S}_j^{(k)}) \tag{7.129}$$

$$\mathbf{R}_j^{(k)} \leftarrow \pi(\mathbf{R}_j^{(k)}). \tag{7.130}$$

This permutation rearranges the elements in $\mathbf{R}_j^{(k)}$, so that:

$$\mathbf{R}_j^{(k)}(i) \leq \mathbf{R}_j^{(k)}(i+1), \ \forall i \in [N_k - 1] \tag{7.131}$$

vi. Compute the vector $\mathbf{I}_j^{(k)} \in M_{1 \times \mu}$, where $\mu \leq N_k$, such that:

$$\mathbf{I}_j^{(k)} = \left\{ r \in [N_k] : \sum_{i=1}^{r} \mathbf{R}_j^{(k)}(i) < NRR \right\} \tag{7.132}$$

vii. Compute the optimal value $\hat{r}$, given by the following equation:

$$\hat{r} = \arg \max_{i \in \mathbf{I}_j^{(k)}} \sigma_j^k(r) \tag{7.133}$$

where $\sigma_j^k(r)$ denotes the partial sum of the r first elements in $\mathbf{R}_j^{(k)}$ given by the following equation:

$$\sigma_j^k(r) = \sum_{i=1}^{r} \mathbf{R}_j^{(k)} \tag{7.134}$$

viii. Compute the remaining number of allocated resources according to the following equation:

$$RA = \sum_{i=\hat{r}+1}^{N_k} \mathbf{R}_j^{(k)} \tag{7.135}$$

This number corresponds to the amount of system wide resources after the removal of the $\hat{r}$ least stimulated antibodies. This value, however, yields a remaining number of resources to be removed which is, once again, given by Eq. (7.127) such that:

$$0 < NRR < \mathbf{R}_j^{(k)}(\hat{r} + 1) \tag{7.136}$$

ix. Eliminate a number NRR of resources from the $\hat{r} + 1$-th antibody, such
that:

$$\mathbf{R}_j^{(k)}(\hat{r} + 1) \leftarrow \mathbf{R}_j^{(k)}(\hat{r} + 1) - NRR \tag{7.137}$$

which finally yields $NRR = 0$ and $\mathbf{R}_j^{(k)}(\hat{r} + 1) > 0$.

x. Remove the $\hat{r}$ least stimulated elements from the available antibodies
repertoire corresponding to the current class and antigen according to
the following equation:

$$\mathbf{Ab}^{(k)} \leftarrow \mathbf{Ab}^{(k)} \setminus \{\mathbf{Ab}_1^{(k)}, \dots, \mathbf{Ab}_{\hat{r}}^{(k)}\} \tag{7.138}$$

and re-estimate the number of antibodies for the current class of patterns
as:

$$N_k \leftarrow N_k - \hat{r} \tag{7.139}$$

xi. Re-estimate vector $\mathbf{S}_j^{(k)}$ containing the stimulations for the available
antibodies against the current antigen and associated class according
to Eq. (7.120). Subsequently, re-estimate the corresponding average
stimulation level $s_j(k)$ according to Eq. (7.121).

xii. Determine the candidate memory cell $\tilde{\mathbf{m}}_j^k$ and corresponding stimulation level $\tilde{s}_j^k$ according to the following equations:

$$\tilde{\mathbf{m}}_j^k = \arg \max_{\mathbf{m} \in \mathbf{Ab}^{(k)}} stimulation(\mathbf{Ag}_j^{(k)}, \mathbf{m}) \tag{7.140}$$

$$\tilde{s}_j^k = stimulation(\mathbf{Ag}_j^{(k)}, \tilde{\mathbf{m}}_j^k) \tag{7.141}$$

xiii. Compute the matrix $\tilde{\mathbf{C}}^{(k)} \in M_{N_c \times L}$ of mutated offsprings corresponding
to the surviving antibodies, such that:

$$N_c = \sum_{i=1}^{N_k} N_c(i) \tag{7.142}$$

Estimating the number of clones for each surviving antibody requires
the computation of an auxiliary random vector $\mathbf{z}_j \in M_{1 \times N_k}$ such that
each vector element $z_j(i)$, $\forall i \in [N_k]$ is uniformly distributed in the
$[0, 1]$ interval. Therefore, the number of mutated clones to be produced
from each surviving antibody will be given by:

$$N_c(i) = \begin{cases} \mathbf{S}_j^{(k)}(i) \cdot CR, & \mathbf{S}_j^{(k)}(i) < z_j(i); \\ 0, & \text{otherwise} . \end{cases} \tag{7.143}$$

In this context, matrix $\tilde{\mathbf{C}}^{(k)}$ of mutated clones may be considered as constructed through the following concatenation process:

$$\tilde{\mathbf{C}}^{(k)} = [\tilde{\mathbf{C}}_1^{(k)}; \ldots; \tilde{\mathbf{C}}_{N_k}^{(k)}]. \tag{7.144}$$

This involves the sub-matrices of mutated clones $\tilde{\mathbf{C}}_i^{(k)}$, $\forall i \in [N_k]$ produced for the i-th antibody of the k-th class $\mathbf{Ab}_i^{(k)}$, such that $\tilde{\mathbf{C}}_i^{(k)} \in M_{N_c(i) \times L}$, where

$$\tilde{\mathbf{C}}_i^{(k)} = \begin{cases} \hat{\mathbf{C}}_i^{(k)}, & \mathbf{S}_j^{(k)}(i) < z_j(i); \\ [], & \text{otherwise.} \end{cases} \tag{7.145}$$

In other words, the elements $\hat{\mathbf{C}}_i^{(k)} \in M_{N_c(i) \times L}$ denote the non-empty sub-matrices of $\tilde{\mathbf{C}}^{(k)}$, such that $N_c(i) \neq 0$. Determining sub-matrices $\hat{\mathbf{C}}_i^{(k)}$, $\forall i : N_c(i) \neq 0$ requires the computation of two random matrices $\mathbf{P}_i, \mathbf{Q}_i \in M_{N_c(i) \times L}$ such that their elements $\mathbf{P}_i(l, p)$ and $\mathbf{Q}_i(l, p)$ are uniformly distributed in the $[0, 1]$ interval where $l \in [N_c(i)]$ and $p \in [L]$. The elements of matrix $\hat{\mathbf{C}}_i^{(k)}$ may be then computed by the following equation:

$$\hat{\mathbf{C}}_i^{(k)}(l, p) = \begin{cases} low(p) + \mathbf{P}_i(l, p) \cdot \delta_j^k(i), & \mathbf{Q}_i(l, p) < MR); \\ \mathbf{Ab}_i^{(k)}(p), & \text{otherwise .} \end{cases} \tag{7.146}$$

where $\delta_j^k(i) = 1 - \mathbf{S}_j^{(k)}(i)$ and $low(p) = \max(0, \mathbf{Ab}_i^{(k)}(p) - \frac{1}{2}\delta_j^k(i))$ Finally, update matrix $\mathbf{Ab}^{(k)}$ and variable N_k according to the following equations:

$$\mathbf{Ab}^{(k)} \leftarrow [\mathbf{Ab}^{(k)}; \tilde{\mathbf{C}}^{(k)}] \tag{7.147}$$

$$N_k \leftarrow N_k + N_c \tag{7.148}$$

(e) *Memory cell introduction*: If $\tilde{s}_j^k > \hat{s}_j^k$, then:
   i. $d_{cm} = \mathbf{D}(\tilde{\mathbf{m}}_j^k, \hat{\mathbf{m}}_j^k)$
   ii. If $d_{cm} < AT \cdot ATS$, then:
      – Update the sub-matric $\mathbf{M}^{(k)}$ of memory cells for the class of the currently presented matrix and the corresponding number of memory cells m_k according to the following equations:

$$\mathbf{M}^{(k)} \leftarrow \mathbf{M}^{(k)} \setminus \{\hat{\mathbf{m}}_j^k\} \tag{7.149}$$

$$m_k \leftarrow m_k - 1 \tag{7.150}$$

These operations correspond to the substitution of the matching memory cell with the candidate memory cell.

iii. Incorporate the candidate memory cell within the sub-matrix of memory cells pertaining to the same class as the currently presented antigenic pattern according to the following equations:

$$\mathbf{M}^{(k)} \leftarrow [\mathbf{M}^{(k)}; \tilde{\mathbf{m}}_j^k] \qquad (7.151)$$

$$m_k \leftarrow m_k + 1. \qquad (7.152)$$

7.6.3   Source Power of AIRS Learning Algorithm and Complexity Analysis

The AIRS learning algorithm was originally conceived in an attempt to demonstrate that Artificial Immune Systems are amenable to the task of classification. AIRS formulation, in particular, is based on the principles of resource-limited AISs in which the competition for system-wide resources forms the fundamental framework in order to address supervised learning problems. As it will be thoroughly described within the next section:

1. AIRS has been extremely effective in a broad array of classification problems, including paradigms with high dimensional feature spaces, multi-class classification problems and real-valued feature domains.
2. General purpose classifiers such as Neural Networks, although highly competent, perform poorly until an appropriate architecture is determined, and the search for the most suitable architecture may require substantial effort. AIRS performance, on the contrary, has been experimentally tested to be within a couple of percentage points of the best results for the majority of problems to which it has been applied. Moreover, the best results obtained by AIRS are highly competent.
3. The primary feature of the AIRS learning algorithm is its self-adjusting nature, justifying its ability to consistently perform over a wide range of classification problems.

Therefore, it is of crucial importance to reveal the source of the classification power exhibited by this alternative machine learning paradigm. The authors in [48] investigate the hypothesis that the evolutionary process of deriving candidate memory cells from an initial population of antibody molecules is the fundamental underpinning behind the AIRS algorithm. Specifically, it is argued that the memory cell pool produced as a final result may serve as an internal image of the important aspects of the problem domain the capturing of which is essential for a high classification performance. This conjecture, in particular, is experimentally tested by replacing that part of the algorithm with a function which directly generates candidate memory cells from some suitably constrained probability density function that resembles the original space distribution of the training data. The results obtained indicate that this replacement does not significantly affect the classification performance of the algorithm, thus, providing evidence supporting the initial hypothesis.

Finally, it is very important to estimate the temporal and spatial complexity of the AIRS classification algorithm since this complexity provides essential information concerning its overall performance when compared against state of the art supervised learning paradigms. Regardless of the particular affinity measure employed, the fundamental parameters characterizing the computational cost of the algorithm refer to the dimension $N_c \times L$ of the mutated antibody clones to be produced, the total number N of antibodies in the repertoire, the amount M of training antigens to be recognized and the amount m of the produced memory antibodies.

The most time consuming computational steps of the AIRS learning algorithm appear in the following list:

- Memory Cell Identification: (Step 2(a));
- Mutated Clone Generation - Phase A: (Step 2(b));
- Stimulation Level Normalization: (Step 2(d) i);
- Least Stimulated Antibody Identification: (Step 2(d) v);
- Candidate Memory Cell Identification: (Step 2(d) xii);
- Mutated Clone Generation - Phase B: (Step 2(d) xiii).

The previous are time-consuming steps as they involve sorting operations on a given vector of elements or the generation of a matrix of mutated clones. It is important to note that Steps 2(a) and 2(b) are executed once per antigenic pattern while Steps 2(d) i, v, xii and xiii are executed until a pre-defined condition is met for each antigenic pattern.

Table 7.3 summarizes the overall time complexity for each individual processing step of the AIRS learning algorithm. Therefore, by summing up the computational time required for each individual step, it is possible to estimate the overall time complexity of the algorithm according to the following equation:

$$O(m) + O(mL) + O(N) + O(N) + O(N) + O(NL) = O(NL) \qquad (7.153)$$

since $N \gg M$. In fact, the total number of produced antibodies may be expressed as $N = \hat{k}M$, where $\hat{k}$ is the average number of antibodies produced per antigenic pattern. In this context, the overall time complexity may be given by the following equation:

Table 7.3 AIRS learning algorithm complexity

Description	Time complexity
Memory cell identification	$O(m)$
Mutated clone generation - Phase A	$O(mL)$
Stimulation level normalization	$O(N)$
Least stimulated antibodys identification	$O(N)$
Candidate memory cell identification	$O(N)$
Mutated clones generation - Phase B	$O(NL)$

$$\textbf{Required Time} = O(\hat{k}ML), \tag{7.154}$$

where $\hat{k}$ is strictly dependent on the stimulation threshold ST, so that $\hat{k} = f(ST)$ and f is a monotonically increasing function. Specifically, when the stimulation threshold approaches unity (1), the time complexity of the algorithm is extremely increased. However, the generalization ability of the classifier relies upon tuning the stimulation threshold at values that are significantly lesser than its maximum value. The required memory to run the algorithm, on the other hand, can be estimated by summing up the individual space requirement for each one of the utilized variables **Ag**, **Ag**, **D**, **C**, $\tilde{\textbf{C}}$, **S**, **R**, **M** and **s**. In particular, the overall space requirement of the algorithm can be given by the following equation:

$$\textbf{Required Space} = (M + N + m + 2N_c)L + 2N + C + M^2. \tag{7.155}$$

7.7   AIS-Based Negative Selection

Negative Selection (NS) was the first among the various mechanisms in the immune system what were explored as a computational paradigm in order to develop an AIS. Specifically, it was introduced in 1994 by Forrest et al. in [41] as a loose abstract model of biological negative selection that concentrates on the generation of change detectors. It is often referred to as *negative detection* since the set of generated detectors are intended to recognize events indicating that a given set of self attribute strings deviate from an established norm. Although the original version of the NS algorithm has significantly evolved and diversified over the years since its original conception, its main characteristics remain unchanged forming the backbone for any negative selection based algorithm.

The initial formulation of the NS algorithm by Forrest et al. involves the definition of a collection S of self-attribute strings that have equal lengths over a finite alphabet Σ such that $S \subset \mathbb{S} = \Sigma^L$. This collection of self strings is usually considered to be protected or monitored against changes. For example, S may be a program, data file or normal pattern of activity, which is segmented into equal-sized substrings. The actual process of the algorithm is completed in two phases, namely the training and testing phase as follows:

- *Training Phase*: this phase is usually referred to as the *censoring* or the *generation* phase of the NS algorithm involving the generation of a set R of detectors that fail to match any self-attribute string in S. Specifically, during the generation phase of the algorithm, which is illustrated in Fig. 7.13, an initial set of detectors R_0 is generated by some random process and is subsequently censored against the set S of self samples resulting in a set R of valid non-self detectors. A crucial factor of the NS algorithm is that it does require an exact or perfect matching between a given pair of attribute strings. On the contrary, the degree of match is evaluated in a partial manner so that two attribute strings are considered to match when their

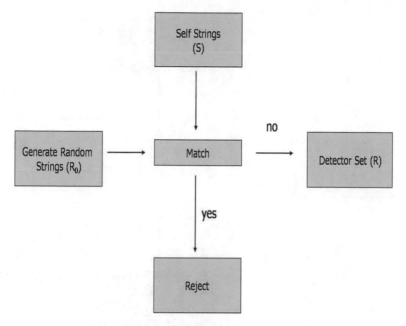

Fig. 7.13 Negative selection algorithm: censoring phase

affinity exceeds a predefined threshold. In case a binary alphabet Σ is chosen for the data representation such that $\Sigma = \{0, 1\}$, the notion of partial match is realized through the utilization of the r-contiguous bits matching rule where r is a suitably chosen parameter.

- *Testing Phase*: this phase is usually referred to as the *monitoring* or the *detection* phase during which a subsequent version S^* of the self-set S is continuously checked against the detector set R for changes, indicating the presence of noise or corruption within the initial data set. Specifically, during the detection phase of the algorithm, which is illustrated in Fig. 7.14, each element of the set S^* is checked against each element of the detector set R for matching. If any non-self detector matches a self-string, then a change is known to have occurred since the detectors are designed so as not to match any of the original strings in S.

The most important aspects of the NS algorithm justifying its validity as an alternative anomaly detection approach appear in the following list:

1. Each copy of the detection algorithm is unique. Most detection algorithms need to monitor multiple sites (e.g., multiple copies of software or multiple computers on a network). In these distributed environments, any single detection scheme is unlikely to be effective, since detection failure on a particular site results in an overall system failure. The NS algorithm approach, on the other hand, involves

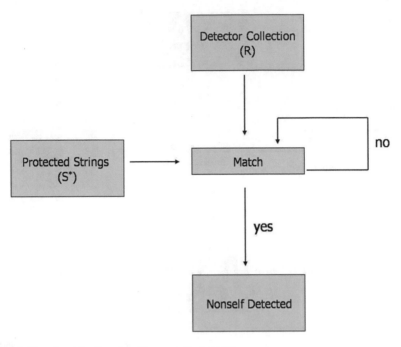

Fig. 7.14 Negative selection algorithm: monitoring phase

the protection of each particular site with a unique set of detectors. Therefore, even if one site is compromised, other sites will remain protected.

2. Detection is probabilistic. One consequence of using different sets of detectors to protect each entity is that probabilistic detection methods are feasible. This is because an intrusion at one site is unlikely to be successful at multiple sites. The utilization of probabilistic detection methods ensures high system-wide reliability at relatively low processing and memory cost. The price, of course, is a somewhat higher chance of intrusion at each particular suite.

3. The detection functionality of the NS algorithm relies upon its ability to probabilistically recognize any foreign activity rather than looking for known patterns of anomaly. Traditional protection systems, such as virus or intrusion detection programs operate by scanning for unique patterns (e.g., digital signatures) that are known at the time the detection software is distributed, thus, leaving systems vulnerable to attack by novel means. The NS algorithm approach, on the contrary, learns the self-defining patterns of behavior so that it will gain the ability to probabilistically identify any deviation from self.

Finally, the most significant characteristics of this alternative classification paradigm include:

1. The negative representation of information, since from a machine learning point of view the NS algorithm outputs the complementary concept of the real target

concept. Although it is obviously distinct from any learning mechanism in the positive space, its strength and applicability has been proven over a wide range of applications.

2. The utilization of some form of detector set as the detection/classification mechanism. Generating an appropriate set of detectors constitutes an essential element of any NS algorithm regulating the non-self coverage and the resulting classification/generalization performance of the algorithm.

3. The most important feature of the NS algorithm in the context of this monograph is its validity as an alternative one-class classification paradigm that can efficiently address the class imbalance problem. This is true, since the ultimate purpose of the NS algorithm is to discriminate between two classes of patterns based on samples that originate uniquely from the minority class of self/normal patterns. In other words, NS algorithms in general, have proven to be extremely competitive against classification problems involving the identification of a class of patterns that occupies only an insignificant amount of volume in the complete space of patterns.

7.7.1 Background Immunological Concepts

The NS algorithm proposed by Forrest et al. constitutes an alternative approach to the problem of change detection based on the principles of self/non-self discrimination in the natural immune system. This discrimination is achieved in part by T-cells, which have receptors on their surface that can detect foreign proteins (antigens). During the generation of T-cells, receptors are generated via a pseudo-random genetic rearrangement process. Then they undergo a censoring process, called negative selection, in the *thymus* gland where T-cells reacting against self-proteins are destroyed, so that only those that do bind to self-proteins are allowed to leave the thymus. These matured T-cells then circulate throughout the body to perform immunological functions protecting the host organism against foreign antigens. The negative selection algorithm works on similar principles, generating detectors randomly, and eliminating the ones that detect self, so that the remaining T-cells can detect any non-self instance. Table 7.4, in particular summarizes the immunological principles employed by the artificial NS algorithm.

7.7.2 Theoretical Justification of the Negative Selection Algorithm

The main feature of the NS algorithm proposed by Forrest et al. is concerned with the utilization of a binary shape-space $\mathbb{S}$ such that all immunological events may be considered as taking place within $\mathbb{S} = \Sigma^L$, where $\Sigma = \{0, 1\}$. Therefore, self

Table 7.4 Mapping between biological and artificial negative selection

Biological NS	Artificial NS
T-Cell	Detector
Shape-space	Type and possible values of attribute strings
Self space	Attribute stings indicating normal behavior (to be monitored)
Non-self space	Attribute stings indicating abnormal behavior (to be recognized)
Affinity measure	Distance/similarity measure
Partial recognition	R-contiguous bits matching rule

($\mathcal{S}$) and non-self ($\mathcal{N}$) spaces may be defined as subsets of the original binary string space $\mathbb{S}$ such that $\mathbb{S} = \mathcal{S} \cup \mathcal{N}$ and $\mathcal{S} \cap \mathcal{N} = \emptyset$ where $|\mathcal{N}| \gg |\mathcal{S}|$. In this context, the artificial counterparts, i.e. detectors and training self-samples, of the original biological molecules, i.e. T-cells and antigens, are all represented as binary strings of equal length. The primary objective of the NS algorithm consists of randomly generating a set of detectors

$$\mathcal{D} \subset \mathcal{N} = \{\mathbf{d}_1, \ldots, \mathbf{d}_m\} \tag{7.156}$$

that can recognize almost any binary string in the non-self space $\mathcal{N}$ on the basis of a limited-size self-space sample

$$\hat{\mathcal{S}} \subset \mathcal{S} = \{\mathbf{s}_1, \ldots, \mathbf{s}_n\} \tag{7.157}$$

without recognizing any string in it. More formally, the NS algorithm aims at developing a set of detectors $\mathcal{D}$ given a self-sample $\hat{\mathcal{S}}$ such that:

$$\forall \mathbf{s} \in \hat{\mathcal{S}}, \forall \mathbf{d} \in \mathcal{D} \, \neg match(\mathbf{s}, \mathbf{d}) \tag{7.158}$$

and the probability of not recognizing any sting in $\mathcal{N}$ is given by a predefined threshold P_f according to the following equation:

$$P(\exists \mathbf{n} \in \mathcal{N} : \forall \mathbf{d} \in \mathcal{D}, \, \neg match(\mathbf{n}, \mathbf{d})) = P_f \tag{7.159}$$

The matching rule utilized by the NS is the *rcb* matching rule defined in Eq. (7.9), taking into special consideration the fact that a perfect match between two random binary strings of equal length is an extremely rare event. Specifically, the probability P_M that two random strings of equal length L over a finite alphabet of m elements match at least r contiguous locations is given by the following equation [96, 97]:

$$P_M \approx m^{-r} \frac{(L - r)(m - 1)}{m + 1} \tag{7.160}$$

under the assumption $m^{-r} \ll 1$.

The most important issue concerning the validity of the NS algorithm as an alternative machine learning paradigm, is that it is mathematically feasible despite its counter-intuitive nature. That is, the vast majority of the non-self space can be efficiently covered by a limited number of detectors that one can accurately estimate having in mind the particular requirements of a given classification problem. In other words, the number of necessary detectors that need to be randomly generated in order to ensure a maximum probability of failure (P_f), may be formally defined as a function of the cardinality of the self-space (N_S), the probability of matching between two random strings in $\mathcal{S}$ (P_M) and the probability threshold P_f. Specifically, we make the following definitions and calculations:

- N_{R_0}: Number of initial detector strings before the censoring phase.
- N_R: Number of detectors after the censoring phase corresponding to the size of the repertoire.
- N_S: Number of self strings.
- P_M: Matching probability for a random pair of strings.
- f: the probability of a random string not matching any of the N_S self stings, such that $f = (1 - P_M)^{N_s}$.
- P_f: Probability that N_R detectors fail to detect an abnormal pattern.

Under these assumptions, the number of required detectors may be estimated according to the following procedure: Assuming that P_M is small and N_S is large, the following equations hold:

$$f \approx e^{-P_M N_S}, \tag{7.161}$$

$$N_R = N_{R_0} \cdot f, \tag{7.162}$$

$$P_f = (1 - P_M)^{N_R}. \tag{7.163}$$

Moreover, if P_M is small and N_R is large, then:

$$P_f \approx e^{-P_M N_R} \tag{7.164}$$

Thus,

$$N_{R_0} = N_R \cdot f = \frac{-\ln P_f}{P_M} \tag{7.165}$$

Therefore, by solving Eqs. (7.161) and (7.162) with respect to N_{R_0}, the following formula is obtained:

$$N_{R_0} = \frac{-\ln P_f}{P_M \cdot (1 - P_M)^{N_S}}. \tag{7.166}$$

It is important to note that the number of initial detectors to be generated is minimized by choosing a matching rule such that

$$P_M \approx \frac{1}{N_S}.$$

(7.167)

Having in mind the above analysis, the following observations can be made on the NS algorithm:

1. It is tunable, since by choosing a desired probability threshold (P_f), the number of required detector strings can be estimated through the utilization of Eq. (7.166).
2. N_R is independent of N_S for fixed P_M and P_f per Eq. (7.165). That is, the size of the detector set does not necessarily grow with the number of strings being protected. Therefore, it is possible to efficiently protect very large data sets!
3. The probability of detection increases exponentially with the number of independent detection algorithms. If N_t is the number of copies of the algorithm, then the probability of the overall system to fail is given by the following equation:

$$P = P_f^{N_t}.$$

(7.168)

4. Detection is symmetric, since changes to the detector set are detected by the same matching process that notices changes to self. This implies that the set of training self strings may be considered as change detectors for the set of generated detectors.

Finally, the above analysis may be utilized in order to estimate the computational cost of the NS algorithm, having in mind, that its execution relies upon the following basic operations:

- Phase I: generating a random string of fixed length and
- Phase II: comparing two strings in order to check whether they meet the matching criterion.

Assuming that these operations take constant time, then the time complexity of the algorithm will be proportional to the number of strings in R_0 (i.e., N_{R_0}) and the number of strings in S (i.e., N_S) in Phase I and proportional to the number of strings in R (i.e., N_R) in Phase II. Therefore, the time complexity of the NS algorithm may be given by the following equation:

$$\textbf{Required Time} = O\left(N_S \cdot \frac{-\ln P_f}{P_M \cdot (1 - P_M)^{N_S}}\right).$$

(7.169)

7.7.3 Real-Valued Negative Selection with Variable-Sized Detectors

Most research in negative selection algorithms utilizes the general framework provided by binary shape-spaces. Binary representation is usually a natural choice since it provides a finite problem space that is easier to analyze by enumerative combinatorics and is more straightforward to use for categorized data. However, the vast majority of real-world applications require the utilization of real-valued, continuous shape-spaces extending the representation ability of the discrete binary space. Specifically, there is a wide range of classification problems that are naturally real-valued, imposing the adaptation of real-valued representation, which is more suitable for the interpretation of the results and the establishment of a more stable algorithm by maintaining affinity within the representation space.

Real-valued negative selection algorithms operate on the basis of the same principles as the original NS algorithm in the binary shape-space. In particular, detectors are randomly generated and subsequently eliminated according to the philosophy of the initial algorithm, with the only difference being the utilization of an infinite size continuous alphabet, such that $\Sigma = \mathbb{R}$ and $\mathbb{S} = \Sigma^L$. Therefore, the generated detectors correspond to real-valued attribute stings of equal size that may be represented as points in a multi-dimensional vector space.

Independently of the particular affinity measure utilized and corresponding matching rule, detectors usually share some common characteristics such as the number of contiguous bits required for a matching in a binary string space or the distance threshold deciding recognition in a real-valued representation. Generally, there is a common affinity threshold θ_{aff}, as formulated in Eq. (7.3), among the detectors determining the recognition status between a given pair of attribute strings. However, detector features can reasonably be extended to overcome this limitation. This is the primary insight that led Ji and Dasgupta [66–68] to the conception of a Real-Valued Negative Selection algorithm with Variable-Sized Detectors, which is usually referred to as *V-Detector*.

The key idea behind this algorithm consists of assigning a different detection region at each detector, parameterized by the corresponding affinity threshold. In other words, each detector string in $\mathbb{S} = \mathbb{R}^L$ may be considered as the center of a hypersphere in the relevant L-dimensional Euclidean vector space, with an associated radius, defining the volume of the corresponding detection region. It is a reasonable choice to make the radius variable considering that the non-self regions to be covered by detectors are more likely to be in different scales. However, a variable radius is not the only possibility provided by the V-Detector algorithm. Detector variability can also be achieved by utilizing different affinity measures such as the Generalized Euclidean distance. Employing alternative affinity measures results in different detection region shapes such as hyper-rectangles or hyper-hyperbolas, which constitute generalizations of the two-dimensional detection regions illustrated in Fig. 7.6a–e.

Figure 7.15 illustrates the core idea of variable-sized detectors in 2-dimensional space. The white-colored circles represent the actual self-region, which is usually given through the training data. The grey-colored circles, on the other hand, are the possible detectors covering the non-self region. Figure 7.15a, b illustrate the case for constant-sized and variable-sized detectors, respectively. The issues related to the existence of holes [5, 33, 34, 60, 128, 129] are represented by the black areas within the two graphs corresponding to the portions of the non-self space that are not covered by the set of generated detectors. Figure 7.15a, in particular, depicts the

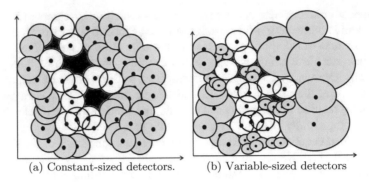

(a) Constant-sized detectors.   (b) Variable-sized detectors

Fig. 7.15   Main concepts of real-valued negative selection and V-detector

existence of three major areas of the non-self space that the real-valued NS algorithm with constant-sized detectors fails to cover, since it is impossible to place a detector within those regions without matching the neighboring self elements. This, however, is not the case when utilizing variable sized-detectors, as illustrated in Fig. 7.15b, where the portion of not-covered non-self space is minimized by the utilization of variable-sized detectors. Specifically, the ability to use detectors of different radii provides the flexibility to cover the space that was originally occupied by holes with detectors of smaller size. Moreover, using variable-sized detectors, as illustrated in Fig. 7.15b, gives us the opportunity to significantly reduce the total number of detectors required in order to adequately cover the larger area of the non-self space.

The issue of holes was primarily identified by D'haeseleer [33, 34] as a major detector coverage problem concerning the impossibility of constructing a detector set capable of recognizing any string in the non-self space $\mathcal{N}$. The existence of not-covered holes within the non-self space was originally conceived by D'haeseleer within the context of binary shape-spaces. Consider, for example, the case of a binary shape-space $\mathbb{S} = \{0, 1\}^L$ where the training set of self samples $\hat{S}$ contains the self-stings s_1, s_2 such that $s_1 = \langle 00111100 \rangle$ and $s_2 = \langle 11111111 \rangle$. Then, according to D'haeseleer, any non-self string $\mathbf{n} \in \mathcal{N}$ such that $\mathbf{n} = \langle b_1 b_2 1111 b_3 b_4 \rangle$, $b_1, b_2, b_3, b_4 \in \{0, 1\}$, cannot be detected by any detector since it will match the self-strings s_1, s_2 according to the *rcb* matching rule. In this context, D'haeseleer proved that the utilization of a matching rule with a constant matching probability P_M and a self set of size $N_S = P_M \cdot |\mathbb{S}|$ always suffices to induce holes.

Finally, it must be noted that the estimated coverage of the non-self space can be used as a control parameter of V-Detector since it has the ability to automatically evaluate the estimated coverage during its generation phase.

7.7.4 AIS-Based One-Class Classification

A very important property of computational NS is that it provides an alternative theoretical framework for the development of one-class classification algorithms. This is true, since negative selection algorithms deal with highly imbalanced problem domains where the only training samples originate from the minority class of positive/target patterns. This entails that the complementary space of negative/outlier patterns remains completely unknown. Therefore, NS algorithms are expected to be useful with extremely skewed binary classification problems where the target class of interest occupies only a small fragment of the complete pattern space. The fundamental difference, however, from traditional one-class classification paradigms is concerned with the fact that NS algorithms do not operate by trying to estimate a sufficient boundary around the subspace of positive patterns. On the contrary, the learning process of a NS algorithm consists of generating an appropriate set of detectors that almost completely covers the subspace of negative patterns. In other words, the one-class learning paradigm inspired by the NS algorithm, instead of trying to obtain a compact description of the target space, is focused on maximally covering the complementary space of outliers with an appropriate set of detectors.

A formal description of the AIS-based one-class classification approach, in the context of the utilized NS algorithm, may be conducted by initially setting the underlying binary classification problem. Specifically, the V-Detector algorithm employs a real-valued shape-space $\mathbb{S}$ which is assumed to be partitioned into two disjoint subsets $\mathcal{S}$ and $\mathcal{N}$, corresponding to the subspaces of self and non-self patterns respectively, such that $|\mathcal{N}| \gg |\mathcal{S}|$ and:

$$\mathbb{S} = [0, 1]^{L}, \tag{7.170}$$

$$\mathbb{S} = \mathcal{S} \cup \mathcal{N}, \tag{7.171}$$

$$\mathbb{S} \cap \mathcal{N} = \emptyset, \tag{7.172}$$

Equation (7.170) yields that all artificial immune molecules, training self-cells and generated detectors are modelled as finite-sized attribute strings over the infinite alphabet $\Sigma = [0, 1]$. Specifically, training self-samples and the centers of the generated detectors may be considered as normalized points in the L-dimensional vector space $\mathbb{S}$. The application of the same normalization philosophy on the affinity measure results in the utilization of the normalized Euclidean distance, given by Eq. (7.57), so that:

$$\forall \, \mathbf{x}, \mathbf{y} \in \mathbb{S} \quad \mathbf{D}(\mathbf{x}, \mathbf{y}) \in [0, 1]. \tag{7.173}$$

All immune recognition events in $\mathbb{S}$ are modelled through the notion of the detection region. Specifically, each artificial immune molecule $\mathbf{s} \in \mathbb{S}$ is associated with a detection radius $r(\mathbf{s})$, such that:

$$r : \mathbb{S} \rightarrow [0, 1] \tag{7.174}$$

and

$$\forall \mathbf{s} \in \mathcal{S}, \ r(\mathbf{s}) = R_{self}, \tag{7.175}$$

where R_{self} is the constant radius value associated with each training self-sample.

Therefore, the detection region assigned to an immune molecule $\mathbf{s} \in \mathbb{S}$ with an associated detection radius $r(\mathbf{s})$ will be given by the following equation:

$$\mathcal{R}(\mathbf{s}, r(\mathbf{s})) = \{\mathbf{x} \in \mathbb{S} : \mathbf{D}(\mathbf{x}, \mathbf{s}) \leq r(\mathbf{s})\}, \tag{7.176}$$

which corresponds to a hyper-sphere of radius $r(\mathbf{s})$ centered at the point $\mathbf{s} \in \mathbb{S}$. In this context, the matching rule for a given pair of immune molecules may be expressed in the following form:

$$M(\mathbf{x}, \mathbf{y}) = I(\mathbf{y} \in \mathcal{R}(\mathbf{x}, r(\mathbf{x})) \lor \mathbf{x} \in \mathcal{R}(\mathbf{y}, r(\mathbf{y})). \tag{7.177}$$

Having in mind the previous definitions, the ultimate purpose of AIS-based one-class classification may be defined as follows:

Definition 1 Given a set $\hat{\mathcal{S}} \subset \mathcal{S}$ of training self-samples where $\hat{\mathcal{S}} = \{\mathbf{s}_1, \ldots, \mathbf{s}_n\}$ with an associated self detection radius R_{self}, develop a set of detectors $\mathcal{D} \subset \mathcal{N}$, where $\mathcal{D} = \{\mathbf{d}_1, \ldots, \mathbf{d}_m\}$ so that the associated detection radii be given by the set $R = \{r(\mathbf{d}) : \mathbf{d} \in \mathcal{D}\}$ and the following constraints be satisfied:

1.

$$\forall \mathbf{d} \in \mathcal{D}, \mathbf{d} \notin \mathcal{R}_{\hat{\mathcal{S}}}, \tag{7.178}$$

where $\mathcal{R}_{\hat{\mathcal{S}}}$ denotes the region of positive patterns given by the following equation:

$$\mathcal{R}_{\hat{\mathcal{S}}} = \bigcup_{\mathbf{s} \in \hat{\mathcal{S}}} \mathcal{R}(\mathbf{s}, r(\mathbf{s})). \tag{7.179}$$

2.

$$P(\exists \mathbf{n} \in \mathcal{N} : \mathbf{n} \notin \mathcal{R}_{\mathcal{D}}) = 1 - C_0, \tag{7.180}$$

where $\mathcal{R}_{\hat{\mathcal{D}}}$ denotes the region of negative patterns given by the following equation:

$$\mathcal{R}_{\mathcal{D}} = \bigcup_{\mathbf{d} \in \mathcal{D}} \mathcal{R}(\mathbf{d}, r(\mathbf{d})) \tag{7.181}$$

and C_0 the estimated percentage of coverage of the non-self space $\mathcal{N}$.

Definition 2 The detector coverage C_0 of a given detector set may be defined as the ratio of the volume of the non-self region that can be recognized by any detector in the detector set to the volume of the entire non-self region. In general, this quantity may be expressed according to the following equation:

Fig. 7.16 Self/non-self coverage

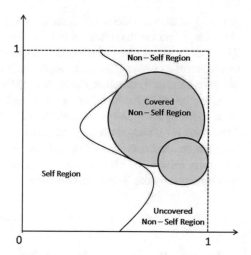

$$C_0 = \frac{\int_{\mathbf{x} \in \mathcal{D}} d\mathbf{x}}{\int_{\mathbf{x} \in \mathcal{N}} d\mathbf{x}}. \tag{7.182}$$

In case the shape-space in question is discrete and finite, Eq. (7.182) may be re-expressed in the following form:

$$C_0 = \frac{|\mathcal{D}|}{|\mathcal{N}|}. \tag{7.183}$$

Figure 7.16 illustrates the three regions in question, namely, self-region, covered non-self region and uncovered non-self region for the particular case involving a two-dimensional shape-space, such that $\mathbb{S} = [0, 1]^2$. In statistical terms, the points of the non-self region form the population of interest which is generally of infinite size. In this context, the probability of each point to be covered by detectors has a binomial distribution. Therefore, the detector coverage is the same as the proportion of the covered points, which equals to the probability that a random point in the non-self region is a covered point. Assuming that all points from the entire non-self space are equally likely to be chosen in a random sampling process, the probability of a sample point being recognized by detectors is thus equal to C_0.

7.7.5 V-Detector Algorithm

According to the original conception of the V-Detector algorithm, its operation consists of randomly generating a set of detectors until the estimated coverage of the non-self space is approximately equal to C_0 and at the same time the estimated coverage of the self-space is below a predefined threshold C_{self}. The input arguments to the algorithm are the set $\hat{S}$ of training self-samples, the corresponding detection radius R_{self} and the maximum number of detectors T_{max} that should not be exceeded. The estimated coverage of the self and non-self spaces constitute the primary control parameters of the V-Detector algorithm since they providing additional stopping criterions for its generation phase. Estimated coverage is a by-product of variable detectors based on the idea that, when sampling k points in the space under consideration and only one point is not covered, then the estimated coverage would be $1 - \frac{1}{k}$. Therefore, if m random tries are executed without finding an uncovered point, then it is reasonable to conclude that the estimated coverage is at least $a = 1 - \frac{1}{k}$. Thus, the necessary number of tries in order to ensure that the estimated coverage is at least a will be given by the following equation:

$$k = \frac{1}{1 - a}. \tag{7.184}$$

The generation phase of the algorithm, as it is illustrated in Fig. 7.17 consists of continuously sampling points in the shape-space under consideration according to the uniform distribution, until the set of generated detectors reaches a predefined size. However, the generation phase of the algorithm may be terminated before reaching

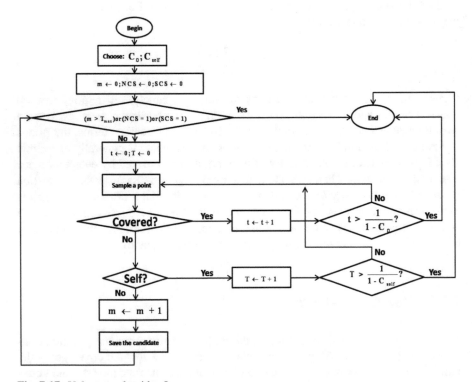

Fig. 7.17   V-detector algorithm I

the pre-specified number of detectors in case the coverage of self or non-self spaces exceeds a given threshold. Specifically, during the generation phase of the algorithm, each candidate detector is checked for matching against the sets of training self-samples and the non-self detectors that have already been accepted. Each time a candidate detector is found to lie within the already covered spaces of self or non-self patterns, this fact is recorded in a corresponding counter in order to obtain the current estimate for the coverage of the self or non-self spaces respectively. Therefore, in case the current coverage of the self or non-self spaces exceeds the given threshold values, the generation phase of the algorithm terminates. On the other hand, if the candidate detector, is neither an already covered non-self instance nor a matched self-sample, it is saved as a valid non-self detector and the corresponding detection radius is based on the minimum distance to each detector that is going to be retained. It is important to note that exceeding the pre-specified level of self-space coverage indicates an abnormal algorithm termination. This may be the case when the given training self data cover almost completely the considered shape-space or the self-radius is set to be too large so that the whole space is assigned to the positive class of patterns.

The following notation will be adapted throughout the formulation of the V-Detector I one-class learning algorithm:

- $S \in M_{n \times L}$ is the matrix storing the training self patterns such that S_i denotes the i-th element of **S**, where $i \in [n]$.
- $D \in M_{m \times L}$ is the matrix storing the generated detectors such that D_i denotes the i-th element of **D**, where $i \in [m]$.
- $\mathbf{R} \in M_{1 \times m}$ is the vector storing the radiuses of the generated detectors such that $\mathbf{R}(i)$ is the radii associated with detector $\mathbf{D}_i$.
- $\mathbf{D}_d \in M_{1 \times m}$ is the vector storing the distances from the currently generated detector to all the previously accepted detectors.
- $\mathbf{I}_d \in M_{1 \times \hat{m}}$, $\hat{m} \leq m$ is the vector storing the indices of the already accepted detectors whose distances from the currently sampled detector are less than the corresponding radii.
- $\mathbf{D}_s \in M_{1 \times n}$ is the vector storing the distances from the currently generated detector to all the training self samples.
- *SCS* (Self Coverage Status) is a boolean variable indicating whether the estimated coverage of the self space is above a predefined threshold.
- *NCS* (Non-Self Coverage Status) is a boolean variable indicating whether the estimated coverage of the non-self space is above a predefined threshold.
- *COVERED* is a boolean variable indicating whether the currently generated candidate detector is an already covered non-self string.
- *SELF* is a boolean variable indicating whether the currently generated candidate detector lies within the space covered by the training self patterns.
- C_0 is the target value for the non-self space coverage.
- C_{self} is the maximum accepted percentage of self space coverage.
- T_{max} is the maximum number of detectors to be generated.
- R_{self} is the detection radius associated with each training self pattern.

The analytical description of V-Detector I involves the following steps:

1. Initialization:

   (a) $m \leftarrow 0$
   (b) $\mathcal{D} \leftarrow []$
   (c) $\mathbf{R} \leftarrow []$
   (d) $SCS \leftarrow FALSE$
   (e) $NCS \leftarrow FALSE$

2. Generation:

   - While $(m \leq T_{max})$ and $(NCS = FALSE)$ and $(SCS = FALSE)$ do:
   (a) $t \leftarrow 0$
   (b) $T \leftarrow 0$
   (c) $COVERED \leftarrow FALSE$
   (d) $SELF \leftarrow FALSE$
   (e) While $(COVERED = TRUE)$ or $(SELF = TRUE)$ do:
      i. Compute vector $\mathbf{x} \in M_{1 \times L}$ as a random point in the $[0, 1]^L$, uniformly sampled.
      ii. Compute matrix $\mathbf{D}_d \in M_{1 \times m}$ according to the following equation:

      $$\mathbf{D}_d(i) = \mathbf{D}(\mathbf{x}, \mathcal{D}_i), \ \forall i \in [m]. \tag{7.185}$$

      iii. Compute vector $\mathbf{I}_d \in M_{1 \times \hat{m}}$ where $\hat{m} \leq m$ according to the following equation:

      $$\mathbf{I}_d = \{i \in [m] : \mathbf{D}_d(i) \leq \mathbf{R}(i)\}. \tag{7.186}$$

      iv. if $(\hat{m} = 0)$ then:
         $COVERED \leftarrow TRUE$ else:
         A. $t \leftarrow t + 1$
         B. if $(t > \frac{1}{1-C_0})$ then:
            $NCS \leftarrow TRUE.$
      v. if $(COVERED = FALSE)$ and $(NCS = FALSE)$ then:
         A. Compute vector $\mathbf{D}_s \in M_{1 \times n}$ according to the following equation:

         $$\mathbf{D}_s(i) = \mathbf{D}(\mathbf{x}, \mathcal{S}_i), \ \forall i \in [n]. \tag{7.187}$$

         B. Compute the candidate detector radius r according to the following equation:

         $$r = \min_{i \in [n]}\{\mathbf{D}_s(i) - R_{self}\}. \tag{7.188}$$

         C. if $(r > R_{self})$ then:
            - $m \leftarrow m + 1$
            - $\mathbf{R} \leftarrow [\mathbf{R}, r]$
            - $\mathcal{D} \leftarrow [\mathcal{D}; \mathbf{x}],$

else:
- $SELF \leftarrow TRUE$
- $T \leftarrow T + 1$
- if $(T > \frac{1}{1-C_{self}})$ then:
  $SELF \leftarrow TRUE$.

The time complexity of V-Detector I may be estimated by taking into consideration that the most time consuming step of the algorithm is the one involving the computation of the distance from the currently generated candidate detector to all the available training self patterns. This individual processing step may be executed in $O(n)$, where n is the size of the given training set. Therefore, the overall time complexity may be given according to the following equation:

$$\textbf{Required Time} = O(mn), \tag{7.189}$$

where m is the total number of detectors to be generated. On the other hand, the required memory to run the algorithm can be estimated by summing up the individual space requirement for each of the most demanding utilized variables $\mathcal{S}, \mathcal{D}, \mathbf{R}, \mathcal{D}_d, \mathcal{D}_s$ and $\mathbf{I}_d$. In particular, the overall space requirement of the algorithm can be given by the following equation:

$$\textbf{Required Space} = (m + n)L + 2m + \hat{m} + n \tag{7.190}$$

References

1. Aickelin, U., Greensmith, J., Twycross, J.: Immune system approaches to intrusion detection - a review. In: Proceedings of the 3rd International Conference on Artificial Immune Systems. LNCS, vol. 3239, pp. 316–329. Springer (2004)
2. Amaral, J.L.M., Amaral, J.F.M., Tanscheit, R.: Real-valued negative selection algorithm with a quasi-monte carlo genetic detector generation. In: ICARIS'07: Proceedings of the 6th International Conference on Artificial Immune Systems, pp. 156–167. Springer, Berlin, Heidelberg (2007)
3. Anchor, K., Zydallis, J., Gunsch, G., Lamont, G.: Extending the computer defense immune system: network intrusion detection with multiobjective evolutionary programming approach. In: ICARIS 2002: 1st International Conference on Artificial Immune Systems Conference Proceedings, pp. 12–21 (2002)
4. Aslantas, V., Ozer, S., Ozturk, S.: A novel clonal selection algorithm based fragile watermarking method. In: ICARIS, pp. 358–369 (2007)
5. Ayara, M., Timmis, J., de Lemos, R., de Castro, L., Duncan, R.: Negative selection: how to generate detectors. In: Timmis, J., Bentley, P. (eds.) 1st International Conference on Artificial Immune Systems, pp. 89–98 (2002)
6. Balthrop, J., Esponda, F., Forrest, S., Glickman, M.: Coverage and generalization in an artificial immune system. In: GECCO'02: Proceedings of the Genetic and Evolutionary Computation Conference, pp. 3–10. Morgan Kaufmann Publishers Inc, San Francisco, CA, USA (2002)
7. Bendiab, E., Meshoul, S., Batouche, M.: An artificial immune system for multimodality image alignment. In: ICARIS, pp. 11–21 (2003)

8. Bezerra, G.B., de Castro, L.N.: Bioinformatics data analysis using an artificial immune network. In: ICARIS, pp. 22–33 (2003)
9. Bezerra, G.B., Barra, T.V., de Castro, L.N., Zuben, F.J.V.: Adaptive radius immune algorithm for data clustering. In: ICARIS, pp. 290–303 (2005)
10. Bezerra, G.B., Barra, T.V., Ferreira, H.M., Knidel, H., de Castro, L.N., Zuben, F.J.V.: An immunological filter for spam. In: ICARIS, pp. 446–458 (2006)
11. Bezerra, G.B., de Castro, L.N., Zuben, F.J.V.: A hierarchical immune network applied to gene expression data. In: ICARIS, pp. 14–27 (2004)
12. Bull, P., Knowles, A., Tedesco, G., Hone, A.: Diophantine benchmarks for the b-cell algorithm. In: ICARIS, pp. 267–279 (2006)
13. Canham, R.O., Tyrrell, A.M.: A hardware artificial immune system and embryonic array for fault tolerant systems. Genet. Program. Evol. Mach. **4**(4), 359–382 (2003)
14. Cayzer, S., Aickelin, U.: On the effects of idiotypic interactions for recommendation communities in artificial immune systems. In: CoRR (2008). abs/0801.3539
15. Ceong, H.T., Kim, Y.-I., Lee, D., Lee, K.H.: Complementary dual detectors for effective classification. In: ICARIS, pp. 242–248 (2003)
16. Chao, D.L., Forrest, S.: Information immune systems. Genet. Program. Evol. Mach. **4**(4), 311–331 (2003)
17. Chen, B., Zang, C.: Unsupervised structure damage classification based on the data clustering and artificial immune pattern recognition. In: ICARIS'09: Proceedings of the 8th International Conference on Artificial Immune Systems, pp. 206–219. Springer, Berlin, Heidelberg (2009)
18. Ciesielski, K., Wierzchon, S.T., Klopotek, M.A.: An immune network for contextual text data clustering. In: ICARIS, pp. 432–445 (2006)
19. Clark, E., Hone, A., Timmis, J.: A markov chain model of the b-cell algorithm. In: Proceedings of the 4th International Conference on Artificial Immune Systems. LNCS, vol. 3627, pp. 318–330. Springer (2005)
20. Coelho, G.P., Zuben, F.J.V.: omni-ainet: An immune-inspired approach for omni optimization. In: ICARIS, pp. 294–308 (2006)
21. Coello, C.A.C., Cortes, N.C.: An Approach to Solve Multiobjective Optimization Problems Based on an Artificial Immune System (2002)
22. Cortés, N.C., Trejo-Pérez, D., Coello, C.A.C.: Handling constraints in global optimization using an artificial immune system. In: ICARIS, pp. 234–247 (2005)
23. Cutello, V., Pavone, M.: Clonal selection algorithms: a comparative case study using effective mutation potentials. In: 4th International Conference on Artificial Immune Systems (ICARIS). LNCS, vol. 4163, pp. 13–28 (2005)
24. DasGupta, D.: An overview of artificial immune systems and their applications. In: Artificial Immune Systems and Their Applications, pp. 3–21. Springer (1993)
25. Dasgupta, D., Krishnakumar, K., Wong, D., Berry, M.: Negative selection algorithm for aircraft fault detection. In: Artificial Immune Systems: Proceedings of ICARIS 2004, pp. 1–14. Springer (2004)
26. de Castro, L., Timmis., J.: Hierarchy and convergence of immune networks: basic ideas and preliminary results. In: ICARIS'01: Proceedings of the 1st International Conference on Artificial Immune Systems, pp. 231–240 (2002)
27. de Castro, L.N.: The immune response of an artificial immune network (ainet). IEEE Congr. Evol. Comput. **1**, 146–153 (2003)
28. De Castro, L.N., Timmis, J.: Artificial Immune Systems: A New Computational Intelligence Approach. Springer Science & Business Media, New York (2002)
29. de Castro, L.N., Zuben, F.J.V.: Learning and optimization using the clonal selection principle. IEEE Trans. Evolutionary Computation **6**(3), 239–251 (2002)
30. de Castro, P.A.D., Coelho, G.P., Caetano, M.F., Zuben, F.J.V.: Designing ensembles of fuzzy classification systems: an immune-inspired approach. In: ICARIS, pp. 469–482 (2005)
31. de Castro, P.A.D., de França, F.O., Ferreira, H.M., Von Zuben, F.J.: Applying biclustering to text mining: an immune-inspired approach. In: ICARIS'07: Proceedings of the 6th International Conference on Artificial Immune Systems, pp. 83–94. Springer, Berlin, Heidelberg (2007)

32. de Mello Honório, L., da Silva, A.M.L., Barbosa, D.A.: A gradient-based artificial immune system applied to optimal power flow problems. In: ICARIS'07: Proceedings of the 6th International Conference on Artificial Immune Systems, pp. 1–12. Springer, Berlin, Heidelberg (2007)

33. D'haeseleer, P.: An immunological approach to change detection: Theoretical results. Comput. Secur. Found. Workshop, IEEE **18** (1996)

34. D'haeseleer, P., Forrest, S., Helman, P.: An immunological approach to change detection: algorithms, analysis and implications. In: SP'96: Proceedings of the 1996 IEEE Symposium on Security and Privacy, p. 110, Washington, DC, USA. IEEE Computer Society (1996)

35. Dilger, W.: Structural properties of shape-spaces. In: ICARIS, pp. 178–192 (2006)

36. Dongmei, F., Deling, Z., Ying, C.: Design and simulation of a biological immune controller based on improved varela immune network model. In: ICARIS, pp. 432–441 (2005)

37. Elberfeld, M., Textor, J.: Efficient algorithms for string-based negative selection. In: ICARIS, pp. 109–121 (2009)

38. Esponda, F., Ackley, E.S., Forrest, S., Helman, P.: On-line negative databases. In: Proceedings of Third International Conference on Artificial Immune Systems (ICARIS 2004), pp. 175–188. Springer (2004)

39. Farmer, J.D., Packard, N.H., Perelson, A.S.: The immune system, adaptation, and machine learning. Physica **22D**, 187–204 (1986)

40. Figueredo, G.P., Ebecken, N.F.F., Barbosa, H.J.C.: The supraic algorithm: a suppression immune based mechanism to find a representative training set in data classification tasks. In: ICARIS, pp. 59–70 (2007)

41. Forrest, S., Perelson, A.S., Allen, L., Cherukuri, R.: Self-nonself discrimination in a computer. In: SP '94: Proceedings of the 1994 IEEE Symposium on Security and Privacy, p. 202, Washington, DC, USA. IEEE Computer Society (1994)

42. Freitas, A.A., Timmis, J.: Revisiting the foundations of artificial immune systems: a problem-oriented perspective. In: ICARIS, pp. 229–241 (2003)

43. Freschi, F., Repetto, M.: Multiobjective optimization by a modified artificial immune system algorithm. In: Proceedings of the 4th International Conference on Artificial Immune Systems, ICARIS 2005. Lecture Notes in Computer Science, vol. 3627, pp. 248–261 (2005)

44. Garain, U., Chakraborty, M.P., Dasgupta, D.: Recognition of handwritten indic script using clonal selection algorithm. In: ICARIS, pp. 256–266 (2006)

45. Goncharova, L.B., Melnikov, Y., Tarakanov, A.O.: Biomolecular immunocomputing. In: ICARIS, pp. 102–110 (2003)

46. Gonza'lez, F., Dasgupta, D., Go'mez, J.: The effect of binary matching rules in negative selection. In: Proceedings of the Genetic and Evolutionary Computation Conference (GECCO)-2003. Lecture Notes in Computer Science, vol. 2723, pp. 195–206. Springer (2003)

47. Gonzalez, F., Dasgupta, D., Niño, L.F.: A randomized real-valued negative selection algorithm. In: Proceedings of Second International Conference on Artificial Immune System (ICARIS 2003), pp. 261–272. Springer (2003)

48. Goodman, D.E., Jr., Boggess, L., Watkins, A.: An investigation into the source of power for airs, an artificial. In: Proceedings of the International Joint Conference on Neural Networks (IJCNN'03), pp. 1678–1683. IEEE (2003)

49. Greensmith, J., Cayzer, S.: An artificial immune system approach to semantic document classification. In: ICARIS, pp. 136–146 (2003)

50. Guzella, T.S., Mota-Santos, T.A., Caminhas, W.M.: A novel immune inspired approach to fault detection. In: ICARIS'07: Proceedings of the 6th International Conference on Artificial Immune Systems, pp.107–118. Springer, Berlin, Heidelberg (2007)

51. Guzella, T.S., Mota-Santos, T.A., Caminhas, W.M.: Towards a novel immune inspired approach to temporal anomaly detection. In: ICARIS'07: Proceedings of the 6th International Conference on Artificial Immune Systems, pp. 119–130. Springer, Berlin, Heidelberg (2007)

52. Haag, C.R., Lamont, G.B., Williams, P.D., Peterson, G.L.: An artificial immune system-inspired multiobjective evolutionary algorithm with application to the detection of distributed computer network intrusions. In: GECCO'07: Proceedings of the 2007 GECCO Conference

Companion on Genetic and Evolutionary Computation, pp. 2717–2724. ACM, New York, NY, USA (2007)
53. Harmer, P.K., Williams, P.D., Gunsch, G.H., Lamont, G.B.: An artificial immune system architecture for computer security applications. IEEE Trans. Evol. Comput. **6**, 252–280 (2002)
54. Hart, E.: Not all balls are round: an investigation of alternative recognition-region shapes. In: ICARIS, pp. 29–42 (2005)
55. Hart, E., Ross, P.: Exploiting the analogy between immunology and sparse distributed memories: A system for clustering non-stationary data. In: ICARIS'01: Proceedings of the 1st International Conference on Artificial Immune Systems, pp. 49–58 (2002)
56. Hart, E., Ross, P.: Studies on the implications of shape-space models for idiotypic networks. In: ICARIS, pp. 413–426 (2004)
57. Hart, E., Ross, P., Webb, A., Lawson, A.: A role for immunology in "next generation" robot controllers. In: ICARIS, pp. 46–56 (2003)
58. Hasegawa, Y., Iba, H.: Multimodal search with immune based genetic programming. In: ICARIS, pp. 330–341 (2004)
59. Haykin, S.: Neural Networks: A Comprehensive Foundation, 2nd edn. Prentice Hall, Upper Saddle River (1999)
60. Hofmeyr, S.A., Forrest, S.: Architecture for an artificial immune system. Evol. Comput. **8**(4), 443–473 (2000)
61. Holland, J.H.: Adaptation in Natural and Artificial Systems. MIT Press, Cambridge (1992)
62. Hone, A., Kelsey, J.: Optima, extrema, and artificial immune systems. In: ICARIS, pp. 80–90 (2004)
63. Hunt, J.E., Cooke, D.E., Holstein, H.: Case memory and retrieval based on the immune system. In: ICCBR '95: Proceedings of the First International Conference on Case-Based Reasoning Research and Development, pp. 205–216. Springer, London, UK (1995)
64. Jansen, T., Zarges, C.: A theoretical analysis of immune inspired somatic contiguous hypermutations for function optimization. In: ICARIS, pp. 80–94 (2009)
65. Jerne, N.K.: Towards a network theory of the immune system. Annales d'immunologie **125C**(1–2), 373–389 (1974)
66. Ji, Z.: Estimating the detector coverage in a negative selection algorithm. In: GECCO 2005: Proceedings of the 2005 Conference on Genetic and Evolutionary Computation, pp. 281–288. ACM Press (2005)
67. Ji, Z., Dasgupta, D.: Real-valued negative selection algorithm with variable-sized detectors. In: LNCS 3102, Proceedings of GECCO, pp. 287–298. Springer (2004)
68. Ji, Z., Dasgupta, D.: V-detector: an efficient negative selection algorithm with probably adequate detector coverage. Inf. Sci. **179**(10), 1390–1406 (2009). Including Special Issue on Artificial Imune Systems
69. Kaers, J., Wheeler, R., Verrelest, H.: The effect of antibody morphology on non-self detection. In: ICARIS, pp. 285–295 (2003)
70. Kalinli, A.: Optimal circuit design using immune algorithm. In: ICARIS, pp. 42–52 (2004)
71. Ko, A., Lau, H.Y.K., Lau, T.L.: An immuno control framework for decentralized mechatronic control. In: ICARIS, pp. 91–105 (2004)
72. Ko, A., Lau, H.Y.K., Lau, T.L.: General suppression control framework: application in self-balancing robots. In: ICARIS, pp. 375–388 (2005)
73. Krautmacher, M., Dilger, W.: Ais based robot navigation in a rescue scenario. In: ICARIS, pp. 106–118 (2004)
74. Lau, H.Y.K., Wong, V.W.K.: Immunologic control framework for automated material handling. In: ICARIS, pp. 57–68 (2003)
75. Lau, H.Y.K., Wong, V.W.K.: Immunologic responses manipulation of ais agents. In: ICARIS, pp. 65–79 (2004)
76. Lau, H., Bate, I., Timmis, J.: An immuno-engineering approach for anomaly detection in swarm robotics. In: ICARIS, pp. 136–150 (2009)
77. Lee, D., Kim, J.-J., Jeong, M., Won, Y., Park, S.H., Lee, K.H.: Immune-based framework for exploratory bio-information retrieval from the semantic web. In: ICARIS, pp. 128–135 (2003)

78. Lee, J., Roh, M., Lee, J., Lee, D.: Clonal selection algorithms for 6-dof pid control of autonomous underwater vehicles. In: ICARIS, pp. 182–190 (2007)
79. Lehmann, M., Dilger, W.: Controlling the heating system of an intelligent home with an artificial immune system. In: ICARIS, pp. 335–348 (2006)
80. Lois, G.M., Boggess, L.: Artificial immune systems for classification: some issues. In: University of Kent at Canterbury, pp. 149–153 (2002)
81. Lu, S.Y.P., Lau, H.Y.K.: An immunity inspired real-time cooperative control framework for networked multi-agent systems. In: ICARIS, pp. 234–247 (2009)
82. Luh, G.-C., Liu, W.-W.: Reactive immune network based mobile robot navigation. In: ICARIS, PP. 119–132 (2004)
83. Luh, G.-C., Wu, C.-Y., Cheng, W.-C.: Artificial immune regulation (air) for model-based fault diagnosis. In: ICARIS, pp. 28–41 (2004)
84. Luo, W., Zhang, Z., Wang, X.: A heuristic detector generation algorithm for negative selection algorithm with hamming distance partial matching rule. In: ICARIS, pp. 229–243 (2006)
85. Luo, W., Wang, X., Wang, X.: A novel fast negative selection algorithm enhanced by state graphs. In: ICARIS, pp. 168–181 (2007)
86. McEwan, C., Hart, E.: On airs and clonal selection for machine learning. In: ICARIS, pp. 67–79 (2009)
87. Morrison, T., Aickelin, U.: An artificial immune system as a recommender system for web sites. In: CoRR (2008). abs/0804.0573
88. Nanas, N., Uren, V.S., Roeck, A.N.D.: Nootropia: a user profiling model based on a self-organising term network. In: ICARIS, pp. 146–160 (2004)
89. Nanas, N., Roeck, A.N.D., Uren, V.S.: Immune-inspired adaptive information filtering. In: ICARIS, pp. 418–431 (2006)
90. Nanas, N., Vavalis, M., Kellis, L.: Immune learning in a dynamic information environment. In: ICARIS, pp. 192–205 (2009)
91. Neal, M.: Meta-stable memory in an artificial immune network. In: Artificial Immune Systems: Proceedings of ICARIS 2003, pp. 168–180. Springer (2003)
92. Oates, R., Greensmith, J., Aickelin, U., Garibaldi, J., Kendall, G.: The application of a dendritic cell algorithm to a robotic classifier. In: ICARIS'07: Proceedings of the 6th International Conference on Artificial Immune Systems, pp. 204–215. Springer, Berlin, Heidelberg (2007)
93. Oda, T., White, T.: Immunity from spam: An analysis of an artificial immune system for junk email detection. In: Artificial Immune Systems, Lecture Notes in Computer Science, pp. 276–289. Springer (2005)
94. Pasek, R.: Theoretical basis of novelty detection in time series using negative selection algorithms. In: ICARIS, pp. 376–389 (2006)
95. Pasti, R., de Castro, L.N.: The influence of diversity in an immune-based algorithm to train mlp networks. In: ICARIS, pp. 71–82 (2007)
96. Percus, J.K., Percus, O., Perelson, A.S.: Predicting the size of the antibody combining region from consideration of efficient self/non-self discrimination. Proceedings of the National Academy of Science 60, 1691–1695 (1993)
97. Percus, J.K., Percus, O.E., Perelson, A.S.: Predicting the size of the t-cell receptor and antibody combining region from consideration of efficient self-nonself discrimination. In: Proceedings of the National Academy of Science, vol. 90 (1993)
98. Perelson, A.S.: Immune network theory. Immunol. Rev. 110, 5–36 (1989)
99. Perelson, A.S., Oster, G.F.: Theoretical studies of clonal selection: Minimal antibody repertoire size and reliability of self- non-self discrimination. J. Theor. Biol. 81, 645–670 (1979)
100. Plett, E., Das, S.: A new algorithm based on negative selection and idiotypic networks for generating parsimonious detector sets for industrial fault detection applications. In: ICARIS, pp. 288–300 (2009)
101. Polat, K., Kara, S., Latifoglu, F., Günes, S.: A novel approach to resource allocation mechanism in artificial immune recognition system: Fuzzy resource allocation mechanism and application to diagnosis of atherosclerosis disease. In: ICARIS, pp. 244–255 (2006)

102. Reche, P.A., Reinherz, E.L.: Definition of mhc supertypes through clustering of mhc peptide binding repertoires. In: ICARIS, pp. 189–196 (2004)
103. Rezende, L.S., da Silva, A.M.L., de Mello Honório, L.: Artificial immune system applied to the multi-stage transmission expansion planning. In: ICARIS, pp. 178–191 (2009)
104. Sahan, S., Polat, K., Kodaz, H., Günes, S.: The medical applications of attribute weighted artificial immune system (awais): diagnosis of heart and diabetes diseases. In: ICARIS, pp. 456–468 (2005)
105. Sarafijanovic, S., Le Boudec, J.-Y: An Artificial Immune System for Misbehavior Detection in Mobile Ad Hoc Networks with Both Innate, Adaptive Subsystems and with Danger Signal (2004)
106. Serapião, A.B.S., Mendes, J.R.P., Miura, K.: Artificial immune systems for classification of petroleum well drilling operations. In: ICARIS'07: Proceedings of the 6th International Conference on Artificial Immune Systems, pp. 47–58. Springer, Berlin, Heidelberg (2007)
107. Shafiq, M.Z., Farooq, M.: Defence against 802.11 dos attacks using artificial immune system. In: ICARIS, pp. 95–106 (2007)
108. Singh, R., Sengupta, R.N.: Bankruptcy prediction using artificial immune systems. In: ICARIS, pp. 131–141 (2007)
109. St, A.T., Tarakanov, A.O., Goncharova, L.B.: Immunocomputing for Bioarrays (2002)
110. Stanfill, C., Waltz, D.: Toward memory-based reasoning. Commun. ACM 29(12), 1213–1228 (1986)
111. Stepney, S., Clark, J.A., Johnson, C.G., Partridge, D., Smith, R.E.: Artificial immune systems and the grand challenge for non-classical computation. In: ICARIS, pp. 204–216 (2003)
112. Stibor, T.: Phase transition and the computational complexity of generating r-contiguous detectors. In: ICARIS'07: Proceedings of the 6th International Conference on Artificial Immune Systems, pp. 142–155. Springer, Berlin, Heidelberg (2007)
113. Stibor, T., Timmis, J., Eckert, C.: A comparative study of real-valued negative selection to statistical anomaly detection techniques. In: Proceedings of the 4th International Conference on Artificial Immune Systems. LNCS, vol. 3627, pp. 262–275. Springer (2005)
114. Stibor, T., Timmis, J., Eckert, C.: On the use of hyperspheres in artificial immune systems as antibody recognition regions. In: Proceedings of 5th International Conference on Artificial Immune Systems. Lecture Notes in Computer Science, pp. 215–228. Springer (2006)
115. Stibor, T., Timmis, J., Eckert, C.: On permutation masks in hamming negative selection. In: Proceedings of 5th International Conference on Artificial Immune Systems. Lecture Notes in Computer Science. Springer (2006)
116. Taylor, D.W., Corne, D.W.: An investigation of the negative selection algorithm for fault detection in refrigeration systems. In: Proceeding of Second International Conference on Artificial Immune Systems (ICARIS), September 1–3, 2003, pp. 34–45. Springer (2003)
117. Tedesco, G., Twycross, J., Aickelin, U.: Integrating innate and adaptive immunity for intrusion detection. In: CoRR (2010). abs/1003.1256
118. Timmis, J.: Assessing the performance of two immune inspired algorithms and a hybrid genetic algorithm for optmisation. In: Proceedings of Genetic and Evolutionary Computation Conference, GECCO 2004, pp. 308–317. Springer (2004)
119. Timmis, J., Hone, A., Stibor, T., Clark, E.: Theoretical advances in artificial immune systems. Theor. Comput. Sci. 403(1), 11–32 (2008)
120. Trapnell, B.C. Jr.: A peer-to-peer blacklisting strategy inspired by leukocyte-endothelium interaction. In: ICARIS, pp. 339–352 (2005)
121. Vargas, P.A., de Castro, L.N., Michelan, R., Zuben, F.J.V.: An immune learning classifier network for autonomous navigation. In: ICARIS, pp. 69–80 (2003)
122. Villalobos-Arias, M., Coello, C.A.C., Hernández-Lerma, O.: Convergence analysis of a multiobjective artificial immune system algorithm. In: ICARIS, pp. 226–235 (2004)
123. Walker, J.H., Garrett, S.M.: Dynamic function optimisation: Comparing the performance of clonal selection and evolution strategies. In: ICARIS, pp. 273–284 (2003)
124. Watkins, A.: Artificial immune recognition system (airs): revisions and refinements. In: Genetic Programming and Evolvable Machines, pp. 173–181 (2002)

125. Watkins, A., Timmis, J.: Exploiting parallelism inherent in airs, an artificial immune classifier. In: Proceedings of the Third International Conference on Artificial Immune Systems. Lecture Notes in Computer Science, vol. 3239, pp. 427–438. Springer (2004)
126. Watkins, A., Timmis, J., Boggess, L.: Artificial immune recognition system (airs): An immune-inspired supervised learning algorithm. Genet. Program. Evol. Mach. **5**(3), 291–317 (2004)
127. White, J.A., Garrett, S.M.: Improved pattern recognition with artificial clonal selection. Proceedings Artificial Immune Systems: Second International Conference, ICARIS **2003**, 181–193 (2003)
128. Wierzchon, S.T.: Discriminative power of receptors: activated by k-contiguous bits rule. J. Comput. Sci. Technol., IEEE **1**(3), 1–13 (2000)
129. Wierzchon, S.T.: Generating optimal repertoire of antibody strings in an artificial immune system. In: Intelligent Information Systems, pp. 119–133 (2000)
130. Wilson, W., Birkin, P., Aickelin, U.: Motif detection inspired by immune memory. In: ICARIS'07: Proceedings of the 6th International Conference on Artificial Immune Systems, pp. 276–287. Springer, Berlin, Heidelberg (2007)
131. Wilson, W.O., Birkin, P., Aickelin, U.: Price trackers inspired by immune memory. In: ICARIS, pp. 362–375 (2006)
132. Woolley, N.C., Milanovic, J.V.: Application of ais based classification algorithms to detect overloaded areas in power system networks. In: ICARIS, pp. 165–177 (2009)

Chapter 8
Experimental Evaluation of Artificial Immune System-Based Learning Algorithms

Abstract In this chapter, we present experimental results to test and compare the performance of Artificial Immune System-based clustering, classification and one-class classification algorithms. The test data are provided via an open access collection of 1000 pieces from 10 classes of western music. This collection has been extensively used in testing algorithms for music signal processing. Specifically, we perform extensive tests on:

- Music Piece Clustering and Music Database Organization,
- Customer Data Clustering in an e-Shopping Application,
- Music Genre Classification, and
- A Music Recommender System.

8.1 Experimentation

This chapter presents contributions of the current monograph to the field of Pattern Recognition, providing experimental justifications concerning the validity of Artificial Immune Systems as an alternative machine learning paradigm. The main effort undertaken in this chapter focuses on addressing the primary problems of Pattern Recognition via developing Artificial Immune System-based machine learning algorithms. Therefore, the relevant research is particularly interested in (i) providing alternative machine learning approaches for the problems of Clustering, Binary Classification and One-Class Classification and (ii) measuring their efficiency against state of the art pattern recognition paradigms such as Support Vector Machines. Pattern classification is specifically studied within the context of the Class Imbalance Problem dealing with extremely skewed training data sets. Specifically, the experimental results presented in this chapter involve degenerated binary classification problems where the class of interest to be recognized is known only through a limited number of positive training instances. In other words, the target class occupies only a negligible volume of the entire pattern space, while the complementary space of negative patterns remains completely unknown during the training process. The effect of the Class Imbalance Problem on the performance of the

© Springer International Publishing AG 2017   237
D.N. Sotiropoulos and G.A. Tsihrintzis, *Machine Learning Paradigms*,
Intelligent Systems Reference Library 118, DOI 10.1007/978-3-319-47194-5_8

utilized Artificial Immune System-based classification algorithm constitutes one of the secondary objectives of this chapter.

The general experimentation framework adopted throughout the current chapter in order to assess the efficiency of the proposed clustering, classification and one class classification algorithms was an open collection of one thousand (1000) pieces from 10 classes of western music. This collection, in particular, has been extensively used in applications concerning music information retrieval and music genre classification [12, 34].

The following list summarizes the pattern recognition problems addressed in the current chapter through the application of specifically designed Artificial Immune System-based machine learning algorithms:

1. Artificial Immune System-Based music piece clustering and Database Organization [26, 27].
2. Artificial Immune System-Based Customer Data Clustering in an e-Shopping application [28].
3. Artificial Immune System-Based Music Genre Classification [29].
4. A Music Recommender Based on Artificial Immune Systems [10].

8.1.1 The Test Data Set

The collection utilized in our experiments contains one thousand (1000) pieces from 10 different classes of western music (Table 8.1). More specifically, this collection contains one hundred (100) pieces, each of thirty second duration, from each of the following ten (10) classes of western music.

The audio signal of each music piece may be represented in multiple ways according to the specific features utilized in order to capture certain aspects of an audio signal. More specifically, there has been a significant amount of work in extracting features that are more appropriate for describing and modelling music signals. We have utilized a specific set of 30 objective features that were originally proposed

Table 8.1 Classes of western music

Class ID	Label
1	Blues
2	Classical
3	Country
4	Disco
5	Hip-Hop
6	Jazz
7	Metal
8	Pop
9	Reggae
10	Rock

Table 8.2 Feature vector constituents

Feature ID	Feature name
1	Mean Centroid
2	Mean Rolloff
3	Mean Flux
4	Mean Zero-crossings
5	STD of Centroid
6	STD of Rolloff
7	STD of Flux
8	STD of Zero-crossings
9	Low Energy
[10...19]	MFCCs
20	Beat A0
21	Beat A1
22	Beat RA
23	Beat P1
24	Beat P2
25	Beat Sum
26	Pitch FA0
27	Pitch UP0
28	Pitch FP0
29	Pitch IP0
30	Pitch Sum

by Tzanetakis and Cook [32, 34] and have dominated the literature in subsequent approaches in this research area. It is worth mentioning that these features not only provide a low-level representation of the statistical properties of the music signal, but also include high-level information extracted by psychoacoustic algorithms. In summary, these features represent rhythmic content (rhythm, beat and tempo information), as well as pitch content describing melody and harmony of a music signal.

Each file, contains an acoustic signal of 30 s duration and is used as input to the feature extraction module. Specifically, short time audio analysis is used in order to break the signal into small, possibly overlapping temporal segments of duration of 50 ms (covering the entire duration of 30 s), and process each segment separately. These segments are called "analysis windows" or "frames" and need to be short enough for the frequency characteristics of the magnitude spectrum to be relatively stable. On the other hand, the term "texture window" describes the shortest window (minimum amount of sound) that is necessary to identify music texture. The texture window is set equal to 30 s in our system.

The actual objective features used in our system are the running mean, median and standard deviation of audio signal characteristics computed over a number of analysis windows. The feature vector constituents appear in Table 8.2. Table 8.2 summarizes the objective feature vector description.[1]

---

[1]STD stands for standard deviation.

Music Surface Features

For the purpose of pattern recognition/classification of music files, we use the statistics of the spectral distribution over time of the corresponding audio signals and represent the "musical surface" [7, 31, 32]. Some of these statistics are defined next.

- **Spectral Centroid:** This feature reflects the *brightness* of the audio signal and is computed as the balancing point (centroid) of the spectrum. It can be calculated as

$$C = \frac{\sum_{n=0}^{N-1} M_t[n] \cdot n}{\sum_{n=0}^{N-1} M_t[n]}, \qquad (8.1)$$

  where $M_t[n]$ is the magnitude of the Fourier transform at frame t and frequency bin n.
- **Spectral Rolloff:** This feature describes the spectral *shape* and is defined as the frequency $R = R(r)$ that corresponds to $r\%$ of the magnitude distribution. It can be seen as a generalization of the spectral centroid, as the spectral centroid is the roll-off value that corresponds to $r = 50\%$ of the magnitude distribution. In our system, we used a roll-off value $r = 95\%$ which has been experimentally determined. N is the length of the discrete signal stored in vector x.

$$\sum_{n=0}^{R} M_t[n] = r \cdot \sum_{n=0}^{N-1} M_t[n]. \qquad (8.2)$$

- **Spectral Flux:** This feature describes the *evolution* of frequency with time and is computed as the difference of the magnitude of the short-time Fourier transform between the current and the previous frame. Therefore, the spectral flux is a measure of local spectral change, given by the equation

$$SF = \sum_{n=0}^{N-1} (N_t[n] - N_{t-1}[n])^2, \qquad (8.3)$$

  where $N_t[n]$ and $N_{t-1}[n]$ is the normalized magnitude of the short-time Fourier transform at window t and $t - 1$, respectively.
- **Zero-Crossings:** A zero-crossing occurs when successive samples in a digital signal have different signs. The corresponding feature is defined as the number of time-domain zero-crossings in the signal. This feature is useful in detecting the *amount of noise* in a signal and can be calculated as

$$Z_n = \sum_m |\mathrm{sgn}[x(m)] - \mathrm{sgn}[x(m - 1)]| \cdot w(n - m), \qquad (8.4)$$

where

$$sgn[x(n)] = \begin{cases} 1, & x(n) \geq 0 \\ 0, & x(n) < 0 \end{cases} \tag{8.5}$$

and

$$w(m) = \begin{cases} \frac{1}{2}, & 0 \leq m \leq N - 1 \\ 0, & \text{otherwise.} \end{cases} \tag{8.6}$$

- **Short-Time Energy Function:** The short-time energy of an audio signal $x(m)$ is defined as

$$E_n = \frac{1}{N} \sum_m [x(m) \cdot w(n - m)]^2, \tag{8.7}$$

where

$$w(m) = \begin{cases} 1, & 0 \leq m \leq N - 1 \\ 0, & \text{otherwise.} \end{cases} \tag{8.8}$$

In Eqs. (8.4) and (8.7), $x(m)$ is the discrete-time audio signal, n is the time index of the short time energy and $w(m)$ is a rectangular window. This feature provides a convenient representation of the *temporal evolution* of the audio signal amplitude variation.

- **Mel-Frequency Cepstral Coefficients (MFCCs):** These coefficients are designed to capture *short-term spectral features*. After taking the logarithm of the amplitude spectrum obtained from the short-time Fourier transform of each frame, the frequency bins are grouped and smoothed according to the Mel-frequency scaling, which has been designed in agreement with human auditory perception. MFCCs are generated by de-correlating the Mel-spectral vectors with a discrete cosine transform.

Rhythm Features and Tempo

Rhythmic features characterize the movement of music signals over time and contain information as *regularity of the tempo*. The feature set for representing rhythm is extracted from a *beat histogram*, that is a curve describing beat strength as a function of tempo values, and can be used to obtain information about the complexity of the beat in the music file. The feature set for representing rhythm structure is based on detecting the most salient periodicities of the signal and is usually extracted from the beat histogram. To construct the beat histogram, the time domain amplitude envelope of each band is first extracted by decomposing the music signal into a number of octave frequency bands. Then, the envelopes of each band are summed together followed by the computation of the autocorrelation of the resulting sum envelop. The dominant peaks of the autocorrelation function, corresponding to the various periodicities of the signals envelope, are accumulated over the whole sound file into a beat histogram where each bin corresponds to the peak lag. The rhythmic content features are then extracted from the beat histogram. In general, they contain the relative amplitude of the first and the second histogram peak, the ratio of the amplitude of the second peak divided by the amplitude of the first peak, the periods of the first and second peak and the overall sum of the histogram.

Pitch Features

The pitch features describe *melody* and *harmony* information in a music signal. A pitch detection algorithm decomposes the signal into two frequency bands and amplitude envelops are extracted for each frequency band where the envelope extraction is performed via half-way rectification and low-pass filtering. The envelopes are summed and an enhanced autocorrelation function is computed so as to reduce the effect of integer multiples of the peak of frequencies to multiple pitch detection. The dominant peaks of the autocorrelation function are accumulated into pitch histograms and the pitch content features extracted from the pitch histograms. The pitch content features typically include the amplitudes and periods of maximum peaks in the histogram, pitch intervals between the two most prominent peaks, and the overall sums of the histograms.

8.1.2 Artificial Immune System-Based Music Piece Clustering and Database Organization

This section presents the development of an Artificial Immune Network for clustering the set of unlabelled multidimensional music feature vectors that are extracted from the utilized music database. The proposed AIN-based clustering approach is assessed for its music data organization and visualization abilities against standard machine learning approaches such as Agglomerative Hierarchical Data Clustering, Fuzzy C-means Clustering and Spectral Clustering. The proposed methodology is combined with traditional agglomerative algorithms based on graph theory so that data clusters may be visualized through the utilization of the minimum spanning tree representation. It is important to note that the identified data clusters, present in the trained Artificial Immune Network, correspond to the intrinsic clusters of the original music dataset.

The original data set representation in three dimensions appears in Fig. 8.1 in which the Principal Component Analysis (PCA) technique has been applied in order to reduce the dimensionality of the input data matrix $\mathbf{Ag} \in M_{1000 \times 30}$. It must be noted that the PCA dimensionality reduction technique is exclusively employed for visualization reasons since the clustering algorithms that have been applied operate on the original thirty-dimensional data set. Moreover, the class label for each data pattern stored in matrix $\mathbf{Ag}$ was completely ignored by each one of the utilized clustering algorithms in order to check their ability to recognize the intrinsic data clusters that are present within the given dataset.

The AIN Learning Algorithm parameters for this particular clustering problem are summarized in Table 8.3,[2] resulting in a total number of 10 memory antibodies

---

[2]GEN corresponds to the number of iterations performed.

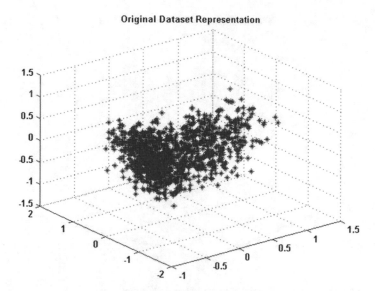

Fig. 8.1 Original music dataset representation in 3 dimensions

Table 8.3 AIN learning algorithm training parameters

AIN parameter	Value
N	10
n	10
ζ	1
σ_d	0.10
σ_s	0.20
GEN	1

directly corresponding to the intrinsic number of classes present within the given dataset. The evolved memory antibodies representation in three dimensions, after application of the PCA dimensionality reduction technique, appears in Fig. 8.2.

Figure 8.4a, b reveal the underlying space distribution of the original dataset by utilizing the minimum spanning tree and hierarchical dendrogram representation of the evolved memory antibodies. The leaves in the AIN-based dendrogram in Fig. 8.3b are significantly fewer than the leaves in the hierarchically-produced dendrogram in Fig. 8.3a, which stems from the fact that the former correspond to cluster representative points in the thirty-dimensional feature space, while the latter correspond directly to music pieces. Thus, the AIN-based dendrogram demonstrates the intrinsic music piece clusters significantly more clearly and compactly than the corresponding hierarchically-produced dendrogram. Moreover, the degree of intra-cluster consistency is clearly significantly higher in the AIN-based rather than the hierarchically-produced clusters, which is of course a direct consequence of a data redundancy reduction achieved by the AIN.

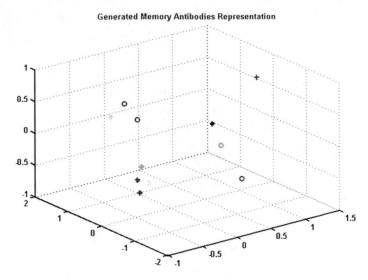

Fig. 8.2 Evolved memory antibodies representation in 3 dimensions

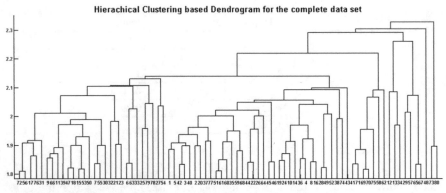

(a) Hierarchical Agglomerative Clustering-Based Dendrogram.

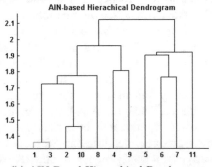

(b) AIN-Based Hierarchical Dendrogram.

Fig. 8.3 Hierarchical AIN-based dendrogram

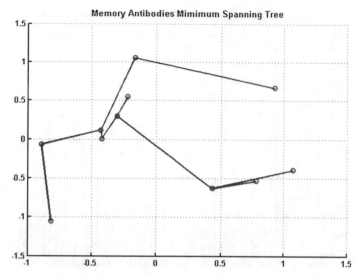

(a) Evolved Memory Antibodies Minimum Spanning Tree Representation.

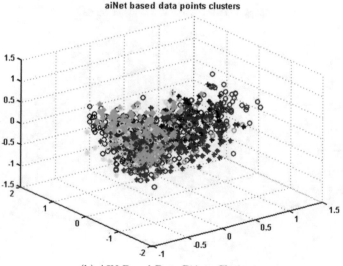

(b) AIN-Based Data Points Clusters.

Fig. 8.4 AIN-based clustering

Figures 8.5a, b and 8.6a, b present the music collection organization obtained by the application of the Spectral Clustering and the Fuzzy C-Means Clustering algorithms respectively. Specifically, Figs. 8.5b and 8.6b present the cluster assignment for each feature vector in the given database according to the particular clustering

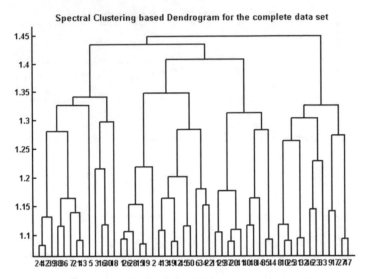

(a) Spectral Clustering Based Dendrogram.

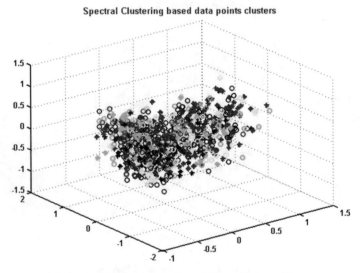

(b) Spectral Clustering Based Data Points Clusters.

Fig. 8.5   Spectral clustering

algorithm. The cluster formation for the AIN-based clustering algorithm appears in
Fig. 8.4b, revealing a more consistent grouping result.

The experimental results presented here reveal that fuzzy c-means clustering
assigned the same degree of cluster membership to all the data points, which implies
that certain intrinsic data *dissimilarities* were *not* captured by the fuzzy c-means

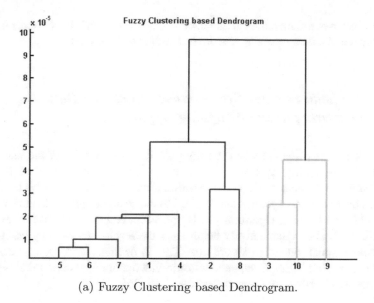

(a) Fuzzy Clustering based Dendrogram.

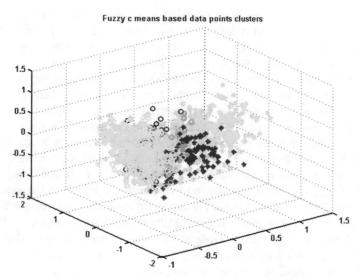

(b) Fuzzy c means based data points clusters.

Fig. 8.6 Fuzzy clustering

clustering algorithm and this makes the clustering result less useful. On the contrary, AIN-based clustering returned significantly higher cluster homogeneity. Moreover, the degree of intra-cluster consistency is clearly significantly higher in the AIN-based rather than the hierarchical and spectral clusters, which is of course a direct consequence of a data redundancy reduction achieved by the AIN.

8.1.3   *Artificial Immune System-Based Customer Data Clustering in an e-Shopping Application*

In this section attention is focused on the problem of adaptivity of the interaction between an e-shopping application and its users. The proposed approach is novel, as it is based on the construction of an Artificial Immune Network (AIN) in which a mutation process is incorporated and applied to the customer profile feature vectors. The AIN-based learning algorithm yields clustering results of users' interests that are qualitatively and quantitatively better than the results achieved by using other more conventional clustering algorithms. This is demonstrated on user data that we collected using *Vision.Com*, an electronic video store application that we have developed as a test-bed for research purposes [28, 36].

Problem Definition

The large quantity of information that exists in an electronic shop, as well as the lack of human salesmen to assist customers, impose a need for adaptivity in e-shops. Adaptivity provides individualized assistance to users, which is dynamically generated. Adaptive responses to users are usually created using the technology of adaptive hypermedia [4]. To create such adaptive behavior of the e-shop to individual users, the system needs information about them, so that it can generate hypotheses about what they need and how they can be assisted in the most appropriate way. This means that an adaptive e-shop needs user modelling components, which monitor the user's actions and generate hypotheses about his/her preferences based on his/her behavior. According to Rich [19] a major technique people use to build models of other people very quickly is the evocation of stereotypes or clusters of characteristics. Given the fact that grouping people can provide quick default assumptions about their preferences and needs [19], the clustering of users' interests has drawn a lot of research energy for purposes of personalization of user interfaces [30].

One solution to the problem of grouping users' behavior can be provided by clustering algorithms that may group users dynamically based on their behavior while they use a system on-line. The main advantage of such an approach is that the categorization of user behavior can be conducted automatically. Clustering algorithms help adaptive systems to categorize users according to their behavior. More specifically adaptive e-commerce systems need not only to acquire information about users' interests in products, but also to have the ability to group users with similar interests. By grouping users together, systems can understand more clearly the user's intention and generate more efficient hypotheses about what a user might need. In this part, clustering algorithms undertake the role of grouping users in an efficient way, thus creating the bone structure of the user model. In this chapter, we present an advanced, immune network-based clustering approach that has been used to provide adaptivity to a test-bed e-shop application constituting a video e-shop.

There exists a significant amount of recommendation mechanisms that have incorporated clustering algorithms (e.g. [1, 8, 9, 11, 14, 24, 37]). All of these applications are recommendation systems, but are primarily concerned with the acquisition of user behavior-related data. In contrast, our work has focused on incorporating an evolutionary clustering algorithm based on the construction of an Artificial Immune System (AIS), in order to personalize the developed e-shop system through grouping similar users' behavior. This means that we have used a very different algorithm that has not been used by these systems. Similarly to our approach, [6, 15] have also utilized an AIS in order to tackle the task of film and web site recommendation respectively by identifying a neighborhood of user preferences. Although, their systems have used a similar algorithm to the one presented in here, the algorithm that we have used is significantly more sophisticated. In particular, we have based our research on the construction of an AIN which incorporates a mutation process that applies to the customer profile feature vectors. More specifically, our work has focused on how a significant amount of redundancy within the customer profile dataset can be revealed and reduced, how many clusters intrinsic to the customer dataset can be identified and what the spatial structure of the data within the identified clusters is.

The proposed approach consists of developing an AIN for the problem of clustering a set of unlabelled multidimensional customer profile feature vectors generated in an electronic video store application in which customer preferences are quantified and groups of customer profiles are maintained. The utilized algorithm has been examined thoroughly by comparing it with three other clustering algorithms, namely agglomerative hierarchical clustering, fuzzy c-means clustering and spectral clustering. A video e-shop application has been developed for this purpose so that the customer data collected could be fed into each of the four clustering algorithm in order to check their relative performance in recognizing valid customer profiles.

Operation of e-Shop System and Customer Data Description

Vision.Com is an adaptive e-commerce video store that learns from customers preferences. Its aim is to provide help to customers, choosing the best movie for them. The web-based system ran on a local network with IIS playing the role of the web server. We used this technique in order to avoid network problems during peak hours. For every user the system creates a different record at the database. In Vision.Com, every customer may browse a large number of movies by navigating through four movie categories, namely social, action, thriller and comedy movies. Every customer has a personal shopping cart. A customer intending to buy a movie must simply move the movie into his/her cart by pressing the specific button. S/He may also remove one or more movies from his/her cart by choosing to delete them. After deciding on which movies to buy, a customer can easily purchase them by pressing the button "buy".

All navigational moves of a customer are recorded by the system in the statistics database. In this way Vision.Com maintains statistics regarding visits to various movie categories and individual movies. Same type statistics are maintained for every customer and every movie that was moved to a shopping cart. Moreover, the same procedure is followed for those movies that are eventually bought by every customer. All of these statistical results are scaled to the unit interval [0, 1].

More specifically, Vision.Com interprets user actions in a way that results in estimates/predictions of user interests in individual movies and movie categories. Each user action contributes to the individual user profile by indicating relevant degrees of interest into a movie category or individual movie. If a user browses into a movie, this indicates interest of this user for the particular movie and its category. If the user puts this movie into the shopping cart, this implies *increased* interest in the particular movie and its category. Finally, if the user decides to buy the movie, then this shows highest interest of this user in the specific movie and its category and is recorded as an increase in an *interest counter*. On the other hand, if the user removes the movie from the shopping cart without buying it, the interest counter is unchanged. Apart from movie categories that are already presented, other movie features taken into consideration are *price range*, *leading actor* and *director*. The price of every movie belongs to one of the five price ranges in euro: 20–25, 26–30, 31–35, 36–40 and over 41. As a consequence, every customer's interest in one of the above features is recorded as a percentage of his/her visits to movie pages. For example, interest of the customer at a particular movie category is calculated as in Eq. (8.9). Every result can also be considered as a probability of a customer's intention to buy a movie.

$$InterestInMovieCategory = \frac{VisitsInSpecificCategory}{VisitsInAllCategories} \tag{8.9}$$

Vision.Com was used by 150 users to buy movies. The system collected data about the user's behavior. The data collected consisted of three similar parts. The first part contains statistical data of the visits that every user made to specific movies. The second part contains data of the cart moves (i.e. which movies the user moved into his/her cart). The last part consists of statistical data concerning the preferences on the movies bought by every user. Every record in every part is a vector of the same 80 features that were extracted from movie features and represents the preferences of a user. The 80 features of these vectors are computed as the movie features we described above. Every 80 featured vector consists of the four movie categories, the five price ranges, all the leading actors and all the directors. The value of each feature is the percentage of interest of every individual customer in this particular feature (Eq. (8.9)).

In the context of Vision.Com, antibodies correspond to neighboring user profiles that are represented by the feature vectors. Specifically, the immune network system clustered users interests as well as movies and represented each resulting cluster with corresponding antibodies.

Comparison of Customer Data Clustering Algorithms and Conclusions

The proposed clustering approach was realized on the basis of the real-valued 80-dimensional vector space $\mathbb{R}^{80}$ so that the affinity level between any given pair of feature vectors may be computed by the normalized Euclidean distance given by Eq. (7.57). Therefore, $\mathbf{S} = [0, 1]^{80}$ is the resulting shape-space, quantifying all interactions between elements of the utilized AIN. In this context, the antigenic pattern

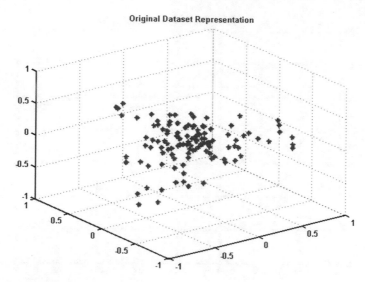

Original Dataset Representation

Fig. 8.7 Evolved memory antibodies representation in 3 dimensions

set to be recognized by the evolved AIN is composed of a set of 150 80-dimensional feature vectors such that $\mathbf{Ag} \in M_{150 \times 80}$. The three-dimensional representation of the data points stored in matrix $\mathbf{Ag}$ appears in Fig. 8.7 after the application of the PCA dimensionality reduction technique.

This section evaluates the clustering efficiency of the proposed AIS-based algorithm against widely used machine learning paradigms such as (a) *agglomerative hierarchical clustering*, (b) *fuzzy c-means clustering* and (c) *spectral clustering*. The ultimate purpose behind this comparison process is to determine which clustering methodology is more efficient in identifying the number and constituent elements of the most general customer profiles present within the given dataset. Specifically, we applied these four clustering methodologies on the 150 customer profile feature vectors collected as previously described.

The memory antibodies produced by the evolved AIN may be considered as providing an alternative compact representation of the original customer profile feature vectors set. This is true since the set minimal set of the 6 generated memory antibodies maintains the spatial structure of the original dataset. The ability of the produced memory antibodies in revealing the underlying space distribution of the given data set is clearly illustrated through the utilization of the minimum spanning tree representation appearing in Fig. 8.8, indicating significant data compression, combined with clear revelation and visualization of the intrinsic data clusters.

Also, in Fig. 8.9, we show the partitioning of the complete dataset into six clusters by the spectral (center right), fuzzy c-means (bottom left), and AIN-based (bottom right) clustering algorithms, respectively. We observe that spectral clustering does not result in cluster homogeneity, while fuzzy c-means clustering results in higher cluster homogeneity, but in only four clusters rather than six required.

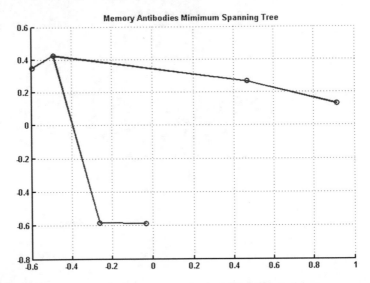

Fig. 8.8 Evolved memory antibodies representation in 3 dimensions

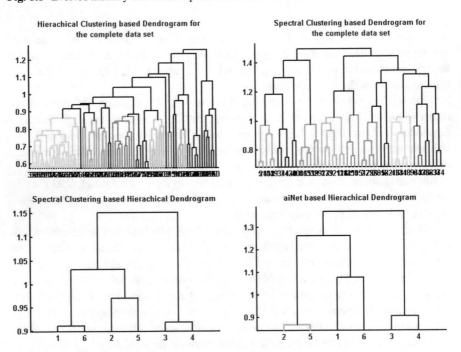

Fig. 8.9 Evolved memory antibodies representation in 3 dimensions

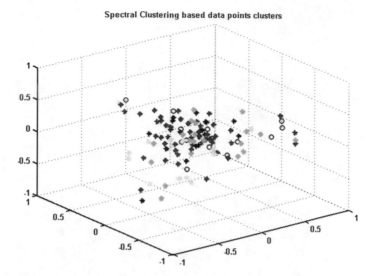

Fig. 8.10 Evolved memory antibodies representation in 3 dimensions

Specifically, we observed that fuzzy c-means clustering assigned the same degree of cluster membership to all the data points, which implies that certain intrinsic data *dissimilarities* were not captured by the fuzzy c-means clustering algorithm and this makes the clustering result less useful. On the contrary, AIN-based clustering returned significantly higher cluster homogeneity. Moreover, the degree of intra-cluster consistency is clearly significantly higher in the AIN-based rather than the hierarchical and spectral clusters, which is of course a direct consequence of a data redundancy reduction achieved by the AIN. These conclusions are highly supported by the spectral clustering, fuzzy c-means clustering and AIN clustering-based data clusters formations appearing in Figs. 8.10, 8.11 and 8.12.

Figures 8.8 and 8.9, lead to the conclusion that *Vision.Com* customers exhibit certain patterns of behavior when shopping and tend to group themselves into six clusters. The 22 antibodies that arose via the AIN-based clustering algorithm correspond to customer behavior representatives and, thus, can be seen as important customer profiles, which eventually correspond to stereotypes in user models. This process is promising because it can provide recommendations based on the users' interests and the characteristics of a movie irrespective of whether this movie has ever been selected by a user before. Thus, the recommendation system can recommend new movies or movies newly acquired by the e-shop as efficiently as previously stored movies. This approach forms the basis of work which is currently in progress and will be reported on a future occasion.

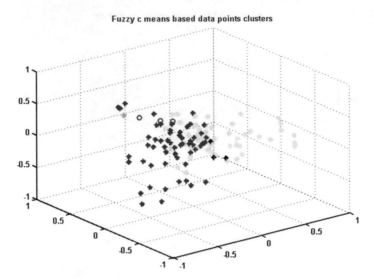

Fig. 8.11 Evolved memory antibodies representation in 3 dimensions

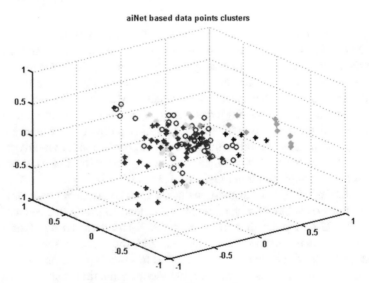

Fig. 8.12 Evolved memory antibodies representation in 3 dimensions

8.1.4   AIS-Based Music Genre Classification

In this section the problem of automated music genre classification is addressed through the utilization of a novel approach that is focused on the implementation of an Artificial Immune System-based classifier. The proposed methodology relies upon the AIRS learning algorithm which exploits the inherent pattern recognition capabilities of the adaptive immune system. Automated music genre classification constitutes a non-trivial multi-class classification problem since boundaries between genres are extremely overlapping and fuzzy. Therefore, it may serve as an ideal framework in order to assess the validity of the alternative machine learning paradigm proposed by Artificial Immune Systems. The AIS-based classifier is initially compared against the state of the art machine learning paradigm of Support Vector Machines in a variety of classification settings with gradually increasing complexity. The primary domain of interest, however, concerns the investigation of highly unbalanced classification problems in the context of Artificial Immune Systems. The relevant research is motivated by the observation that the natural immune system has the intrinsic property of self/non-self cell discrimination, especially when the non-self (complementary) space of cells is significantly larger than the class of self cells. In other words, the inherent ability of the adaptive immune system to persistently address a classification problem that involves highly skewed pattern spaces, provides the fundamental inspiration in order to investigate the Class Imbalance Problem within the framework of Artificial Immune Systems.

The experimental results presented in this section are organized in three groups of increasingly unbalanced classification problems appearing in the following list:

1. **Balanced Multi Class Classification Problems**: Tables 8.4, 8.5, 8.6, 8.7, 8.8, 8.9, 8.10, 8.11, 8.12, 8.13, 8.14, 8.15, 8.16, 8.17, 8.18, 8.19, 8.20, 8.21, 8.22, 8.23, 8.24, 8.25, 8.26, 8.27, 8.28, 8.29, 8.30, 8.31, 8.32, 8.33, 8.34, 8.35, 8.36, 8.37, 8.38, 8.39, 8.40, 8.41, 8.42, 8.43, 8.44, 8.45, 8.46, 8.47 and 8.48 summarize the relative classification performance of the AIS-based classifier in a series of classification settings with gradually increasing complexity, against the SVM classifier. Specifically, this group involves the full range between a simple binary classification problem to a very hard 10-class classification problem. The complete range of problems addressed in this session is described within the following list:

   - $C1$ versus $C2$;
   - $C1$ versus $C2$ versus $C3$;
   - $C1$ versus $C2$ versus $C3$ versus $C4$;
   - $C1$ versus $C2$ versus $C3$ versus $C4$ versus $C5$;
   - $C1$ versus $C2$ versus $C3$ versus $C4$ versus $C5$ versus $C6$
   - $C1$ versus $C2$ versus $C3$ versus $C4$ versus $C5$ versus $C6$ versus $C7$;
   - $C1$ versus $C2$ versus $C3$ versus $C4$ versus $C5$ versus $C6$ versus $C7$ versus $C8$;
   - $C1$ versus $C2$ versus $C3$ versus $C4$ versus $C5$ versus $C6$ versus $C7$ versus $C8$ versus $C9$;
   - $C1$ versus $C2$ versus $C3$ versus $C4$ versus $C5$ versus $C6$ versus $C7$ versus $C8$ versus $C9$ versus $C10$;

2. **One Against All Balanced Classification Problems**: Tables 8.49, 8.50, 8.51, 8.52, 8.53, 8.54, 8.55, 8.56, 8.57, 8.58, 8.59, 8.60, 8.61, 8.62, 8.63, 8.64, 8.65, 8.66, 8.67, 8.68, 8.69, 8.70, 8.71, 8.72, 8.73, 8.74, 8.75, 8.76, 8.77 and 8.78 summarize the classification performance of the AIS-based classifier in a special class of balanced binary classification problems, where a particular class of positive/target patterns is to be recognized against the complementary pattern space. In other words, the complementary space of negative/outlier patterns is systematically under-sampled so that both classes are equally represented during the training process. The complete range of classification problems addressed in this session is described within the following list:

   - $C1$ versus $\{C2 \cup C3 \cup C4 \cup C5 \cup C6 \cup C7 \cup C8 \cup C9 \cup C10\}$;
   - $C2$ versus $\{C1 \cup C3 \cup C4 \cup C5 \cup C6 \cup C7 \cup C8 \cup C9 \cup C10\}$;
   - $C3$ versus $\{C1 \cup C2 \cup C4 \cup C5 \cup C6 \cup C7 \cup C8 \cup C9 \cup C10\}$;
   - $C4$ versus $\{C1 \cup C2 \cup C3 \cup C5 \cup C6 \cup C7 \cup C8 \cup C9 \cup C10\}$;
   - $C5$ versus $\{C1 \cup C2 \cup C3 \cup C4 \cup C6 \cup C7 \cup C8 \cup C9 \cup C10\}$;
   - $C6$ versus $\{C1 \cup C2 \cup C3 \cup C4 \cup C5 \cup C7 \cup C8 \cup C9 \cup C10\}$;
   - $C7$ versus $\{C1 \cup C2 \cup C3 \cup C4 \cup C5 \cup C6 \cup C8 \cup C9 \cup C10\}$;
   - $C8$ versus $\{C1 \cup C2 \cup C3 \cup C4 \cup C5 \cup C6 \cup C7 \cup C9 \cup C10\}$;
   - $C9$ versus $\{C1 \cup C2 \cup C3 \cup C4 \cup C5 \cup C6 \cup C7 \cup C8 \cup C10\}$;
   - $C10$ versus $\{C1 \cup C2 \cup C3 \cup C4 \cup C5 \cup C6 \cup C7 \cup C8 \cup C9\}$;

3. **One Against All Unbalanced Classification Problems**: Tables 8.79, 8.80, 8.81, 8.82, 8.83, 8.84, 8.85, 8.86, 8.87, 8.88, 8.89, 8.90, 8.91, 8.92, 8.93, 8.94, 8.95, 8.96, 8.97, 8.98, 8.99, 8.100, 8.101, 8.102, 8.103, 8.104, 8.105, 8.106, 8.107, 8.108, 8.109, 8.110, 8.111, 8.112, 8.113, 8.114, 8.115, 8.116, 8.117, 8.118, 8.119, 8.120, 8.121, 8.122, 8.123, 8.124, 8.125, 8.126, 8.127 and 8.128 summarize the classification performance of the AIS-based classifier in a series of extremely unbalanced binary classification problems, where the negative class of outlier patterns is not under-sampled so that it is equally represented with the positive class of target patterns. This experimentation session, in particular, involves the classification problems with the highest degree of unbalance. This is true, since the positive class of patterns to be recognized occupies only a negligible volume of the complete pattern space so that the corresponding training process is biased towards the majority class of patterns. The complete range of *unbalanced* classification problems addressed in this session is described by the following list:

   - $C1$ versus $\{C2 \cup C3 \cup C4 \cup C5 \cup C6 \cup C7 \cup C8 \cup C9 \cup C10\}$;
   - $C2$ versus $\{C1 \cup C3 \cup C4 \cup C5 \cup C6 \cup C7 \cup C8 \cup C9 \cup C10\}$;
   - $C3$ versus $\{C1 \cup C2 \cup C4 \cup C5 \cup C6 \cup C7 \cup C8 \cup C9 \cup C10\}$;
   - $C4$ versus $\{C1 \cup C2 \cup C3 \cup C5 \cup C6 \cup C7 \cup C8 \cup C9 \cup C10\}$;
   - $C5$ versus $\{C1 \cup C2 \cup C3 \cup C4 \cup C6 \cup C7 \cup C8 \cup C9 \cup C10\}$;
   - $C6$ versus $\{C1 \cup C2 \cup C3 \cup C4 \cup C5 \cup C7 \cup C8 \cup C9 \cup C10\}$;

- $C7$ versus $\{C1 \cup C2 \cup C3 \cup C4 \cup C5 \cup C6 \cup C8 \cup C9 \cup C10\}$;
- $C8$ versus $\{C1 \cup C2 \cup C3 \cup C4 \cup C5 \cup C6 \cup C7 \cup C9 \cup C10\}$;
- $C9$ versus $\{C1 \cup C2 \cup C3 \cup C4 \cup C5 \cup C6 \cup C7 \cup C8 \cup C10\}$;
- $C10$ versus $\{C1 \cup C2 \cup C3 \cup C4 \cup C5 \cup C6 \cup C7 \cup C8 \cup C9\}$;

The experimental results presented in the following sections reveal the validity of the alternative machine learning paradigm provided by Artificial Immune Systems. The performance of the proposed classification methodology, in particular, was found to be similar to that of the Support Vector Machines when tested in balanced multi-class classification problems. The most important findings, however, relate to the special nature of the adaptive immune system which is particulary evolved in order to deal with an extremely unbalanced classification problem, namely the self/non-self discrimination process. Self/Non-self discrimination within the adaptive immune system is an essential biological process involving a severely imbalanced classification problem since the subspace of non-self cells occupies the vast majority of the complete molecular space. Therefore, the identification of any given non-self molecule constitutes a very hard pattern recognition problem that the adaptive immune system resolves remarkably efficient. This fact was the primary source of inspiration that led to the application of the proposed AIS-based classification algorithm on a series of gradually imbalanced classification problems. Specifically, the classification accuracy of the AIS-based classifier is significantly improved against the SVM classifier for the balanced One versus All classification problems where the class of outlier is appropriately down-sampled so that both classes are equally represented during the training process.

The most interesting behavior of the AIS-based classifier was observed during the third experimentation session involving a series of 10 severely imbalanced pattern recognition problems. Specifically, each class of patterns pertaining to the original dataset was treated as the target class to be recognized against the rest of the classes of the complementary space. In this context, the AIS-based classifier exhibited superior classification efficiency especially in recognizing the minority class of patterns. The true positive rate of recognition for the minority class of patterns is significantly higher for the complete set of the utilized experimentation datasets. More importantly, the proposed classification scheme based on the principles of the adaptive immune system demonstrates an inherent ability in dealing with the class imbalance problem.

Balanced Multi Class Classification Problems

Table 8.4 AIRS run information

C1 versus C2	
Airs classifier run parameter	Value
Affinity threshold scalar	0.2
Clonal rate	20
Hyper mutation rate	2
Stimulation threshold	0.99
Total resources	150
Nearest neighbors number	10
Affinity threshold	0.377
Total training instances	200
Total memory cell replacements	5
Mean antibody clones per refinement iteration	150.162
Mean total resources per refinement iteration	150
Mean pool size per refinement iteration	159.12
Mean memory cell clones per antigen	29.865
Mean antibody refinement iterations per antigen	19.555
Mean antibody prunings per refinement iteration	151.335
Data reduction percentage	2.5 %

Table 8.5 AIRS classification results

C1 versus C2	
Airs classifier classification parameter	Value
Correctly classified instances	**94 %**
Incorrectly classified instances	**6 %**
Kappa statistic	0.88
Mean absolute error	0.06
Root mean squared error	0.2449
Relative absolute error	12 %
Root relative squared error	48.9898 %
Total number of instances	200

Table 8.6 AIRS classification results by class

C1 versus C2					
AIRS detailed accuracy by class					
Class ID	TP rate	FP rate	Preclion	Recall	F-measure
1	0.96	0.08	0.923	0.96	0.941
2	0.92	0.04	0.958	0.92	0.939

Table 8.7 SVM classification results

C1 versus C2	
SVM classifier classification parameter	Value
Correctly classified instances	**92.5**%
Incorrectly classified instances	**7.5**%
Kappa statistic	0.85
Mean absolute error	0.075
Root mean squared error	0.2739
Relative absolute error	15%
Root relative squared error	54.7723%
Total number of instances	200

Table 8.8 SVM classification results by class

C1 versus C2

SVM detailed accuracy by class

Class ID	TP rate	FP rate	Precision	Recall	F-measure
1	0.93	0.08	0.921	0.93	0.925
2	0.92	0.07	0.9294	0.92	0.925

Table 8.9 AIRS run information

C1 versus C2 versus C3	
Airs classifier run parameter	Value
Affinity threshold scalar	0.2
Clonal rate	20
Hyper mutation rate	8
Stimulation threshold	0.99
Total resources	150
Nearest neighbors number	5
Affinity threshold	0.369
Total training instances	300
Total memory cell replacements	1
Mean antibody clones per refinement iteration	151.486
Mean total resources per refinement iteration	150
Mean pool size per refinement iteration	160.392
Mean memory cell clones per antigen	119.797
Mean antibody refinement iterations per antigen	22.613
Mean antibody prunings per refinement iteration	156.477
Data reduction percentage	0.333%

Table 8.10 AIRS Classification Results

C1 versus C2 versus C3	
Airs classifier classification parameter	Value
Correctly classified instances	**72.3333** %
Incorrectly classified instances	**27.6667** %
Kappa statistic	0.585
Mean absolute error	0.1844
Root mean squared error	0.4295
Relative absolute error	41.5
Root relative squared error	91.1043
Total Number of Instances	300

Table 8.11 AIRS classification results by class

C1 versus C2 versus C3					
AIRS detailed accuracy by class					
Class ID	TP rate	FP rate	Precision	Recall	F-measure
1	0.68	0.205	0.624	0.68	0.651
2	0.87	0.045	0.906	0.87	0.888
3	0.62	0.165	0.653	0.62	0.636

Table 8.12 SVM classification results

C1 versus C2 versus C3	
SVM classifier classification parameter	Value
Correctly classified instances	**75.3333** %
Incorrectly classified instances	**24.66674** %
Kappa statistic	0.63
Mean absolute error	0.2874
Root mean squared error	0.3728
Relative absolute error	64.6667 %
Root relative squared error	79.0921 %
Total number of instances	300

Table 8.13 SVM classification results by class

C1 versus C2 versus C3					
SVM detailed accuracy by class					
Class ID	TP rate	FP rate	Precision	Recall	F-measure
1	0.7	0.185	0.654	0.7	0.676
2	0.87	0.02	0.956	0.87	0.911
3	0.69	0.165	0.676	0.69	0.683

Table 8.14 AIRS run information

C1 versus C2 versus C3 versus C4

Airs classifier run parameter	Value
Affinity threshold scalar	0.2
Clonal rate	20
Hyper mutation rate	6
Stimulation threshold	0.99
Total resources	150
Nearest neighbors number	10
Affinity threshold	0.365
Total training instances	400
Total memory cell replacements	2
Mean antibody clones per refinement iteration	151.331
Mean total resources per refinement iteration	150
Mean pool size per refinement iteration	160.229
Mean memory cell clones per antigen	90.7
Mean antibody refinement iterations per antigen	22.152
Mean antibody prunings per refinement iteration	155.113
Data reduction percentage	0.5 %

Table 8.15 AIRS classification results

C1 versus C2 versus C3 versus C4

Airs classifier classification parameter	Value
Correctly classified instances	**68.5** %
Incorrectly classified instances	**31.5** %
Kappa statistic	0.58
Mean absolute error	0.1575
Root mean squared error	0.3969
Relative absolute error	42 %
Root relative squared error	91.6515 %
Total number of instances	400

Table 8.16 AIRS classification results by class

C1 versus C2 versus C3 versus C4

AIRS detailed accuracy by class

Class ID	TP rate	FP rate	Precision	Recall	F-measure
1	0.58	0.143	0.574	0.58	0.577
2	0.87	0.017	0.946	0.87	0.906
3	0.58	0.16	0.547	0.58	0.563
4	0.71	0.1	0.703	0.71	0.706

Table 8.17 SVM classification results

C1 versus C2 versus C3 versus C4

SVM classifier classification parameter	Value
Correctly classified instances	**71** %
Incorrectly classified instances	**29** %
Kappa statistic	0.6133
Mean absolute error	0.285
Root mean squared error	0.3635
Relative absolute error	76 %
Root relative squared error	83.9422 %
Total number of instances	400

Table 8.18 SVM classification results by class

C1 versus C2 versus C3 versus C4

SVM detailed accuracy by class

Class ID	TP rate	FP rate	Precision	Recall	F-measure
1	0.54	0.127	0.587	0.54	0.563
2	0.89	0.017	0.947	0.89	0.918
3	0.62	0.113	0.646	0.62	0.633
4	0.79	0.13	0.669	0.79	0.725

Table 8.19 AIRS run information

C1 versus C2 versus C3 versus C4 versus C5

Airs classifier run parameter	Value
Affinity threshold scalar	0.2
Clonal rate	15.0
Hyper mutation rate	8
Stimulation threshold	0.99
Total resources	150
Nearest neighbors number	10
Affinity threshold	0.362
Total training instances	500
Total memory cell replacements	5
Mean antibody clones per refinement iteration	148.348
Mean total resources per refinement iteration	150
Mean pool size per refinement iteration	160.23
Mean memory cell clones per antigen	91.032
Mean antibody refinement iterations per antigen	19.634
Mean antibody prunings per refinement iteration	152.501
Data reduction percentage	1 %

Table 8.20 AIRS classification results

C1 versus C2 versus C3 versus C4 versus C5

Airs classifier classification parameter	Value
Correctly classified instances	**61** %
Incorrectly classified instances	**39** %
Kappa statistic	0.5125
Mean absolute error	0.156
Root mean squared error	0.395
Relative absolute error	48.75 %
Root relative squared error	98.7421 %
Total number of instances	500

Table 8.21 AIRS classification results by class

C1 versus C2 versus C3 versus C4 versus C5

AIRS detailed accuracy by class

Class ID	TP rate	FP rate	Precision	Recall	F-measure
1	0.57	0.168	0.46	0.57	0.509
2	0.84	0.023	0.903	0.84	0.87
3	0.5	0.108	0.538	0.5	0.518
4	0.6	0.128	0.541	0.6	0.569
5	0.54	0.063	0.684	0.54	0.603

Table 8.22 SVM classification results

C1 versus C2 versus C3 versus C4 versus C5

SVM classifier classification parameter	Value
Correctly classified instances	**63.4**%
Incorrectly classified instances	**36.6**%
Kappa statistic	0.5425
Mean absolute error	0.2637
Root mean squared error	0.3515
Relative absolute error	82.4%
Root relative squared error	87.8721%
Total number of instances	500

Table 8.23 SVM classification results by class

C1 versus C2 versus C3 versus C4 versus C5

SVM detailed accuracy by class

Class ID	TP rate	FP rate	Precision	Recall	F-measure
1	0.47	0.108	0.522	0.47	0.495
2	0.88	0.015	0.936	0.88	0.907
3	0.63	0.105	0.6	0.63	0.615
4	0.55	0.143	0.491	0.55	0.519
5	0.64	0.088	0.646	0.64	0.643

Table 8.24 AIRS run information

C1 versus C2 versus C3 versus C4 versus C5 versus C6

Airs classifier run parameter	Value
Affinity threshold scalar	0.2
Clonal rate	20.0
Hyper mutation rate	8
Stimulation threshold	0.99
Total resources	150
Nearest neighbors number	10
Affinity threshold	0.363
Total training instances	600
Total memory cell replacements	9
Mean antibody clones per refinement iteration	151.042
Mean total resources per refinement iteration	150
Mean pool size per refinement iteration	159.941
Mean memory cell clones per antigen	121.275
Mean antibody refinement iterations per antigen	21.235
Mean antibody prunings per refinement iteration	156.427
Data reduction percentage	1.5%

Table 8.25 AIRS classification results

C1 versus C2 versus C3 versus C4 versus C5 versus C6	
Airs classifier classification parameter	Value
Correctly classified instances	**57.6667** %
Incorrectly classified instances	**42.3333** %
Kappa statistic	0.492
Mean absolute error	0.1411
Root mean squared error	0.3756
Relative absolute error	50.8 %
Root relative squared error	100.7968 %
Total number of instances	600

Table 8.26 AIRS classification results by class

C1 versus C2 versus C3 versus C4 versus C5 versus versus C6

AIRS detailed accuracy by class

Class ID	TP rate	FP rate	Precision	Recall	F-measure
1	0.55	0.13	0.458	0.55	0.5
2	0.71	0.046	0.755	0.71	0.732
3	0.44	0.118	0.427	0.44	0.433
4	0.59	0.102	0.536	0.59	0.562
5	0.57	0.05	0.695	0.57	0.626
6	0.6	0.062	0.659	0.6	0.628

Table 8.27 SVM classification results

C1 versus C2 versus C3 versus C4 versus C5 versus C6	
SVM classifier classification parameter	Value
Correctly classified instances	**61.8333** %
Incorrectly classified instances	**38.1667** %
Kappa statistic	0.542
Mean absolute error	0.2375
Root mean squared error	0.3339
Relative absolute error	85.5067 %
Root relative squared error	89.5877 %
Total number of instances	600

Table 8.28 SVM classification results by class

C1 versus C2 versus C3 versus C4 versus C5 versus C6

SVM detailed accuracy by class

Class ID	TP rate	FP rate	Precision	Recall	F-measure
1	0.48	0.112	0.462	0.48	0.471
2	0.75	0.036	0.806	0.75	0.777
3	0.58	0.074	0.611	0.58	0.595
4	0.6	0.1	0.545	0.6	0.571
5	0.6	0.066	0.645	0.6	0.622
6	0.7	0.07	0.667	0.7	0.683

Table 8.29 AIRS run information

C1 versus C2 versus C3 versus C4 versus C5 versus C6 versus C7

Airs classifier run parameter	Value
Affinity threshold scalar	0.2
Clonal rate	20.0
Hyper mutation rate	4
Stimulation threshold	0.99
Total resources	150
Nearest neighbors number	10
Affinity threshold	0.369
Total training instances	700
Total memory cell replacements	18
Mean antibody clones per refinement iteration	150.955
Mean total resources per refinement iteration	150
Mean pool size per refinement iteration	159.874
Mean memory cell clones per antigen	60.483
Mean antibody refinement iterations per antigen	21.214
Mean antibody prunings per refinement iteration	153.48
Data reduction percentage	2.571 %

Table 8.30 AIRS classification results

C1 versus C2 versus C3 versus C4 versus C5 versus C6 versus C7

Airs classifier classification parameter	Value
Correctly classified instances	**57.8571** %
Incorrectly classified instances	**42.1429** %
Kappa statistic	0.5083
Mean absolute error	0.1204
Root mean squared error	0.347
Relative absolute error	49.1667 %
Root relative squared error	99.1632 %
Total number of instances	700

Table 8.31 AIRS classification results by class

C1 versus C2 versus C3 versus C4 versus C5 versus C6 versus C7

AIRS detailed accuracy by class

Class ID	TP rate	FP rate	Precision	Recall	F-measure
1	0.53	0.135	0.396	0.53	0.453
2	0.78	0.043	0.75	0.78	0.765
3	0.54	0.09	0.5	0.54	0.519
4	0.58	0.088	0.523	0.58	0.55
5	0.55	0.055	0.625	0.55	0.585
6	0.57	0.055	0.633	0.57	0.6
7	0.5	0.025	0.769	0.5	0.606

Table 8.32 SVM classification results

C1 versus C2 versus C3 versus C4 versus C5 versus C6 versus C7

SVM classifier classification parameter	Value
Correctly classified instances	**59.7143** %
Incorrectly classified instances	**40.2857** %
Kappa statistic	0.53
Mean absolute error	0.2145
Root mean squared error	0.3179
Relative absolute error	87.6032 %
Root relative squared error	90.8443 %
Total number of instances	700

Table 8.33 SVM classification results by class

C1 versus C2 versus C3 versus C4 versus C5 versus C6 versus C7

SVM detailed accuracy by class

Class ID	TP rate	FP rate	Precision	Recall	F-measure
1	0.44	0.073	0.5	0.44	0.468
2	0.72	0.032	0.791	0.72	0.754
3	0.56	0.073	0.56	0.56	0.56
4	0.54	0.122	0.425	0.54	0.476
5	0.61	0.055	0.649	0.61	0.629
6	0.62	0.052	0.667	0.62	0.642
7	0.69	0.063	0.645	0.69	0.667

Table 8.34 AIRS run information

C1 versus C2 versus C3 versus C4 versus C5 versus C6 versus C7 versus C8

Airs classifier run parameter	Value
Affinity threshold scalar	0.2
Clonal rate	10.0
Hyper mutation rate	8
Stimulation threshold	0.99
Total resources	150
Nearest neighbors number	10
Affinity threshold	0.365
Total training instances	800
Total memory cell replacements	25
Mean antibody clones per refinement iteration	148.391
Mean total resources per refinement iteration	150
Mean pool size per refinement iteration	166.172
Mean memory cell clones per antigen	60.746
Mean antibody refinement iterations per antigen	25.686
Mean antibody prunings per refinement iteration	150.189
Data reduction percentage	3.125 %

Table 8.35 AIRS classification results

C1 versus C2 versus C3 versus C4 versus C5 versus C6 versus C7 versus C8

Airs classifier classification parameter	Value
Correctly classified instances	**52.5** %
Incorrectly classified instances	**47.5** %
Kappa statistic	0.4571
Mean absolute error	0.1188
Root mean squared error	0.3446
Relative absolute error	54.2857 %
Root relative squared error	104.1976 %
Total number of instances	800

Table 8.36 AIRS classification results by class

C1 versus C2 versus C3 versus C4 versus C5 versus C6 versus C7 versus C8

AIRS detailed accuracy by class

Class ID	TP rate	FP rate	Precision	Recall	F-measure
1	0.47	0.114	0.37	0.47	0.414
2	0.71	0.023	0.816	0.71	0.759
3	0.49	0.093	0.43	0.49	0.458
4	0.49	0.089	0.441	0.49	0.464
5	0.41	0.051	0.532	0.41	0.463
6	0.58	0.059	0.586	0.58	0.583
7	0.48	0.031	0.686	0.48	0.565
8	0.57	0.083	0.496	0.57	0.53

Table 8.37 SVM classification results

C1 versus C2 versus C3 versus C4 versus C5 versus C6 versus C7 versus C8

SVM classifier classification parameter	Value
Correctly classified instances	**57.125** %
Incorrectly classified instances	**42.875** %
Kappa statistic	0.51
Mean absolute error	0.1961
Root mean squared error	0.3055
Relative absolute error	89.6429 %
Root relative squared error	92.373 %
Total number of instances	800

Table 8.38 SVM classification results by class

C1 versus C2 versus C3 versus C4 versus C5 versus C6 versus C7 versus C8

SVM detailed accuracy by class

Class ID	TP rate	FP rate	Precision	Recall	F-measure
1	0.47	0.084	0.443	0.47	0.456
2	0.73	0.024	0.811	0.73	0.768
3	0.52	0.07	0.515	0.52	0.517
4	0.49	0.097	0.419	0.49	0.452
5	0.53	0.043	0.639	0.53	0.579
6	0.65	0.054	0.631	0.65	0.64
7	0.66	0.064	0.595	0.66	0.626
8	0.52	0.053	0.584	0.52	0.55

Table 8.39 AIRS run information

C1 versus C2 versus C3 versus C4 versus C5 versus C6 versus C7 versus C8 versus C9

Airs classifier run parameter	Value
Affinity threshold scalar	0.2
Clonal rate	20.0
Hyper mutation rate	6
Stimulation threshold	0.99
Total resources	150
Nearest neighbors number	10
Affinity threshold	0.362
Total training instances	900
Total memory cell replacements	16
Mean antibody clones per refinement iteration	150.856
Mean total resources per refinement iteration	150
Mean pool size per refinement iteration	159.747
Mean memory cell clones per antigen	91.594
Mean antibody refinement iterations per antigen	21.087
Mean antibody prunings per refinement iteration	154.871
Data reduction percentage	1.778 %

Table 8.40 AIRS classification results

C1 versus C2 versus C3 versus C4 versus C5 versus C6 versus C7 versus C8 versus C9

Airs classifier classification parameter	Value
Correctly classified instances	**48.6667** %
Incorrectly classified instances	**51.3333** %
Kappa statistic	0.4225
Mean absolute error	0.1141
Root mean squared error	0.3377
Relative absolute error	57.75 %
Root relative squared error	107.4709 %
Total number of instances	900

Table 8.41 AIRS classification results by class

C1 versus C2 versus C3 versus C4 versus C5 versus C6 versus C7 versus C8 versus C9

AIRS detailed accuracy by class

Class ID	TP rate	FP rate	Precision	Recall	F-measure
1	0.52	0.106	0.38	0.52	0.439
2	0.79	0.034	0.745	0.79	0.767
3	0.49	0.088	0.412	0.49	0.447
4	0.49	0.096	0.389	0.49	0.434
5	0.21	0.058	0.313	0.21	0.251
6	0.57	0.046	0.606	0.57	0.588
7	0.45	0.02	0.738	0.45	0.559
8	0.55	0.068	0.505	0.55	0.526
9	0.31	0.063	0.383	0.31	0.343

Table 8.42 SVM classification results

C1 versus C2 versus C3 versus C4 versus C5 versus C6 versus C7 versus C8 versus C9

SVM classifier classification parameter	Value
Correctly classified instances	**53** %
Incorrectly classified instances	**47** %
Kappa statistic	0.4713
Mean absolute error	0.1798
Root mean squared error	0.2934
Relative absolute error	91.0243 %
Root relative squared error	93.3676 %
Total number of instances	900

Table 8.43 SVM classification results by class

C1 versus C2 versus C3 versus C4 versus C5 versus C6 versus C7 versus C8 versus C9

SVM detailed accuracy by class

Class ID	TP rate	FP rate	Precision	Recall	F-measure
1	0.44	0.075	0.423	0.44	0.431
2	0.72	0.02	0.818	0.72	0.766
3	0.54	0.058	0.54	0.54	0.54
4	0.43	0.1	0.35	0.43	0.386
5	0.34	0.054	0.442	0.34	0.384
6	0.62	0.054	0.59	0.62	0.605
7	0.64	0.051	0.61	0.64	0.624
8	0.55	0.051	0.573	0.55	0.561
9	0.49	0.066	0.48	0.49	0.485

Table 8.44 AIRS run information

C1 versus C2 versus C3 versus C4 versus C5 versus C6 versus C7 versus C8 versus C9 versus C10

Airs classifier run parameter	Value
Affinity threshold scalar	0.2
Clonal rate	20.0
Hyper mutation rate	10
Stimulation threshold	0.99
Total resources	150
Nearest neighbors number	10
Affinity threshold	0.362
Total training instances	1000
Total memory cell replacements	19
Mean antibody clones per refinement iteration	150.977
Mean total resources per refinement iteration	150
Mean pool size per refinement iteration	159.864
Mean memory cell clones per antigen	152.033
Mean antibody refinement iterations per antigen	21.264
Mean antibody prunings per refinement iteration	157.801
Data reduction percentage	1.9 %

Table 8.45 AIRS classification results

C1 versus C2 versus C3 versus C4 versus C5 versus C6 versus C7 versus C8 versus C9 versus C10

Airs classifier classification parameter	Value
Correctly classified instances	**44.3** %
Incorrectly classified instances	**55.7** %
Kappa statistic	0.3811
Mean absolute error	0.1114
Root mean squared error	0.3338
Relative absolute error	61.8889 %
Root relative squared error	111.2555 %
Total number of instances	1000

Table 8.46 AIRS classification results by class

C1 versus C2 versus C3 versus C4 versus C5 versus C6 versus C7 versus C8 versus C9 versus C10

AIRS detailed accuracy by class

Class ID	TP rate	FP rate	Precision	Recall	F-measure
1	0.48	0.1	0.348	0.48	0.403
2	0.76	0.033	0.717	0.76	0.738
3	0.43	0.092	0.341	0.43	0.381
4	0.36	0.088	0.313	0.36	0.335
5	0.26	0.047	0.382	0.26	0.31
6	0.53	0.043	0.576	0.53	0.552
7	0.47	0.031	0.627	0.47	0.537
8	0.54	0.073	0.45	0.54	0.491
9	0.39	0.061	0.415	0.39	0.402
10	0.21	0.05	0.318	0.21	0.253

Table 8.47 SVM classification results

C1 versus C2 versus C3 versus C4 versus C5 versus C6 versus C7 versus C8 versus C9 versus C10

SVM classifier classification parameter	Value
Correctly classified instances	**50 %**
Incorrectly classified instances	**50 %**
Kappa statistic	0.4444
Mean absolute error	0.1659
Root mean squared error	0.2827
Relative absolute error	92.1852 %
Root relative squared error	94.2452 %
Total number of instances	1000

Table 8.48 SVM classification results by class

C1 versus C2 versus C3 versus C4 versus C5 versus C6 versus C7 versus C8 versus C9 versus C10

SVM detailed accuracy by class

Class ID	TP rate	FP rate	Precision	Recall	F-measure
1	0.4	0.061	0.421	0.4	0.41
2	0.72	0.026	0.758	0.72	0.738
3	0.51	0.059	0.49	0.51	0.5
4	0.44	0.083	0.37	0.44	0.402
5	0.42	0.05	0.483	0.42	0.449
6	0.61	0.051	0.57	0.61	0.589
7	0.63	0.053	0.568	0.63	0.597
8	0.52	0.039	0.598	0.52	0.556
9	0.48	0.06	0.471	0.48	0.475
10	0.27	0.073	0.29	0.27	0.28

One Versus All Balanced Classification Problems

Table 8.49 AIRS run information

C1 versus all balanced	
Airs classifier run parameter	Value
Affinity threshold scalar	0.2
Clonal rate	20.0
Hyper mutation rate	8
Stimulation threshold	0.99
Total resources	150
Nearest neighbors number	3

Table 8.50 AIRS classification results

Airs classifier		
C1 versus all balanced		
Fold ID	Accuracy	Error rate
1	75	25
2	85	15
3	85	15
4	55	45
5	65	35
6	65	35
7	65	35
8	50	50
9	75	25
10	85	15
Mean value	**70.5**	**29.5**

Table 8.51 SVM classification results

SVM classifier

C1 versus all balanced

Fold ID	Accuracy	Error rate
1	50	50
2	50	50
3	0	100
4	100	0
5	100	0
6	100	0
7	50	50
8	0	100
9	50	50
10	50	50
Mean value	**55**	**45**

Table 8.52 AIRS run information

C2 versus all balanced

Airs classifier run parameter	Value
Affinity threshold scalar	0.2
Clonal rate	10.0
Hyper mutation rate	8
Stimulation threshold	0.99
Total resources	150
Nearest neighbors number	10

Table 8.53 AIRS classification results

Airs classifier

C2 versus all balanced

Fold ID	Accuracy	Error rate
1	100	0
2	85	15
3	85	15
4	80	20
5	95	5
6	95	5
7	80	20
8	95	5
9	95	5
10	90	10
Mean value	**90**	**10**

Table 8.54   SVM classification results

SVM classifier		
C2 versus all balanced		
Fold ID	Accuracy	Error rate
1	100	0
2	100	0
3	100	0
4	100	0
5	100	0
6	100	0
7	100	0
8	100	0
9	100	0
10	100	0
Mean value	**100**	**0**

Table 8.55   AIRS run information

C3 versus all balanced	
Airs classifier run parameter	Value
Affinity threshold scalar	0.2
Clonal rate	20.0
Hyper mutation rate	8
Stimulation threshold	0.99
Total resources	150
Nearest neighbors number	3

Table 8.56   AIRS classification results

Airs classifier		
C3 versus all balanced		
Fold ID	Accuracy	Error rate
1	60	40
2	65	35
3	75	25
4	70	30
5	75	25
6	70	30
7	65	35
8	75	25
9	85	15
10	65	35
Mean value	**70.5**	**29.5**

Table 8.57 SVM classification results

SVM classifier

C3 versus all balanced

Fold ID	Accuracy	Error rate
1	100	0
2	50	50
3	50	50
4	50	50
5	50	50
6	100	0
7	50	50
8	50	50
9	50	50
10	100	0
Mean value	**65**	**35**

Table 8.58 AIRS run information

C4 versus all balanced

Airs classifier run parameter	Value
Affinity threshold scalar	0.2
Clonal rate	20.0
Hyper mutation rate	8
Stimulation threshold	0.99
Total resources	150
Nearest neighbors number	3

Table 8.59 AIRS classification results

Airs classifier

C4 versus all balanced

Fold ID	Accuracy	Error rate
1	70	30
2	85	15
3	70	30
4	70	30
5	80	20
6	65	35
7	90	10
8	65	35
9	60	40
10	90	10
Mean value	**74.5**	**25.5**

Table 8.60 SVM classification results

SVM classifier		
C4 versus all balanced		
Fold ID	Accuracy	Error rate
1	50	50
2	100	0
3	100	0
4	100	0
5	100	0
6	50	50
7	100	0
8	100	0
9	50	50
10	50	50
Mean value	**80**	**20**

Table 8.61 AIRS run information

C6 versus all balanced	
Airs classifier run parameter	Value
Affinity threshold scalar	0.2
Clonal rate	20.0
Hyper mutation rate	8
Stimulation threshold	0.99
Total resources	150
Nearest neighbors number	10

Table 8.62 AIRS classification results

Airs classifier		
C5 versus all balanced		
Fold ID	Accuracy	Error rate
1	75	25
2	85	15
3	75	25
4	80	20
5	70	30
6	75	25
7	70	30
8	70	30
9	50	50
10	80	20
Mean value	**73**	**27**

Table 8.63 SVM classification results

SVM classifier		
C5 versus all balanced		
Fold ID	Accuracy	Error rate
1	0	100
2	100	0
3	50	50
4	50	50
5	0	100
6	0	100
7	100	0
8	100	0
9	0	100
10	50	50
Mean value	**45**	**55**

Table 8.64 AIRS run information

C6 versus all balanced	
Airs classifier run parameter	Value
Affinity threshold scalar	0.2
Clonal rate	10.0
Hyper mutation rate	8
Stimulation threshold	0.99
Total resources	150
Nearest neighbors number	10

Table 8.65 AIRS classification results

Airs classifier		
C6 versus all balanced		
Fold ID	Accuracy	Error rate
1	80	20
2	80	20
3	80	20
4	70	30
5	85	15
6	80	20
7	80	20
8	85	15
9	90	10
10	80	20
Mean value	**81**	**19**

Table 8.66 SVM classification results

SVM Classifier		
C6 versus all balanced		
Fold ID	Accuracy	Error rate
1	100	0
2	50	50
3	100	0
4	100	0
5	100	0
6	100	0
7	100	0
8	50	50
9	100	0
10	100	0
Mean value	**90**	**10**

Table 8.67 AIRS run information

C7 versus all balanced	
Airs classifier run parameter	Value
Affinity threshold scalar	0.2
Clonal rate	20.0
Hyper mutation rate	10
Stimulation threshold	0.99
Total resources	150
Nearest neighbors number	10

Table 8.68 AIRS classification results

Airs classifier		
C7 versus all balanced		
Fold ID	Accuracy	Error rate
1	90	10
2	100	0
3	90	10
4	75	25
5	85	15
6	75	25
7	95	5
8	65	35
9	75	25
10	65	35
Mean value	**81.5**	**18.5**

Table 8.69 SVM classification results

SVM classifier		
C7 versus all balanced		
Fold ID	Accuracy	Error rate
1	100	0
2	100	0
3	100	0
4	50	50
5	100	0
6	50	50
7	100	0
8	100	0
9	50	50
10	50	50
Mean value	**80**	**20**

Table 8.70 AIRS run information

C8 versus all balanced	
Airs classifier run parameter	Value
Affinity threshold scalar	0.2
Clonal rate	20.0
Hyper mutation rate	4
Stimulation threshold	0.99
Total resources	150
Nearest neighbors number	10

Table 8.71 AIRS classification results

Airs classifier		
C8 versus all balanced		
Fold ID	Accuracy	Error rate
1	50	50
2	65	35
3	75	25
4	65	35
5	70	30
6	80	20
7	75	25
8	70	30
9	55	45
10	75	25
Mean value	**68**	**32**

Table 8.72 SVM classification results

SVM classifier		
C8 versus all balanced		
Fold ID	Accuracy	Error rate
1	50	50
2	100	0
3	50	50
4	50	50
5	50	50
6	50	50
7	100	0
8	50	50
9	100	0
10	50	50
Mean value	**65**	**35**

Table 8.73 AIRS run information

C9 versus all balanced	
Airs classifier run parameter	Value
Affinity threshold scalar	0.2
Clonal rate	20.0
Hyper mutation rate	2
Stimulation threshold	0.99
Total resources	150
Nearest neighbors number	10

Table 8.74 AIRS classification results

Airs classifier		
C9 versus all balanced		
Fold ID	Accuracy	Error rate
1	65	35
2	75	25
3	75	25
4	90	10
5	70	30
6	80	20
7	75	25
8	70	30
9	65	35
10	65	35
Mean value	**73**	**27**

Table 8.75 SVM classification results

SVM classifier		
C9 versus all balanced		
Fold ID	Accuracy	Error rate
1	100	0
2	100	0
3	50	50
4	100	0
5	100	0
6	100	0
7	50	50
8	50	50
9	50	50
10	100	0
Mean value	**80**	**20**

Table 8.76 AIRS run information

C10 versus all balanced	
Airs classifier run parameter	Value
Affinity threshold scalar	0.2
Clonal rate	20.0
Hyper mutation rate	6
Stimulation threshold	0.99
Total resources	150
Nearest neighbors number	10

Table 8.77 AIRS classification results

Airs classifier		
C10 versus all balanced		
Fold ID	Accuracy	Error rate
1	65	35
2	60	40
3	65	35
4	65	35
5	55	45
6	75	25
7	55	45
8	55	45
9	80	20
10	60	40
Mean value	**63.5**	**36.5**

Table 8.78 SVM classification results

SVM classifier		
C10 versus all balanced		
Fold ID	Accuracy	Error rate
1	50	50
2	0	100
3	100	0
4	100	0
5	100	0
6	100	0
7	50	50
8	100	0
9	100	0
10	100	0
Mean value	**80**	**20**

One Versus All Unbalanced Classification Problems

Table 8.79 AIRS run information

C1 versus all unbalanced	
Airs classifier run parameter	Value
Affinity threshold scalar	0.2
Clonal rate	20
Hyper mutation rate	8
Stimulation threshold	0.99
Total resources	150
Nearest neighbors number	4
Affinity threshold	0.362
Total training instances	1000
Total memory cell replacements	25
Mean antibody clones per refinement iteration	150.278
Mean total resources per refinement iteration	150
Mean pool size per refinement iteration	159.166
Mean memory cell clones per antigen	125.851
Mean antibody refinement iterations per antigen	1.546
Mean antibody prunings per refinement iteration	156.36
Data reduction percentage	2.5 %

Table 8.80 AIRS classification results

C1 versus all unbalanced

Airs classifier classification parameter	Value
Correctly classified instances	**87.2**%
Incorrectly classified instances	**12.8**%
Kappa statistic	0.2558
Mean absolute error	0.128
Root mean squared error	0.3578
Relative absolute error	70.8319%
Root relative squared error	119.2564%
Total number of instances	1000

Table 8.81 AIRS classification results by class

C1 versus all unbalanced

AIRS detailed accuracy by class

Class	TP rate	FP rate	Precision	Recall	F-measure
Minority	**0.31**	0.066	0.344	0.31	0.326
Majority	**0.934**	0.69	0.924	0.934	0.929

Table 8.82 SVM classification results

C1 versus all unbalanced

SVM classifier classification parameter	Value
Correctly classified instances	**90.1**%
Incorrectly classified instances	**9.9**%
Kappa statistic	0.0179
Mean absolute error	0.099
Root mean squared error	0.3146
Relative absolute error	54.784%
Root relative squared error	104.8804%
Total number of Instances	1000

Table 8.83 SVM classification results by class

C1 versus all unbalanced

SVM detailed accuracy by class

Class	TP rate	FP rate	Precision	Recall	F-measure
Minority	**0.01**	0	1	0.01	0.02
Majority	1	0.99	0.901	1	0.948

Table 8.84 AIRS run information

C2 versus all unbalanced	
Airs classifier run parameter	Value
Affinity threshold scalar	0.2
Clonal rate	15
Hyper mutation rate	8
Stimulation threshold	0.99
Total resources	150
Nearest neighbors number	4
Affinity threshold	0.362
Total training instances	1000
Total memory cell replacements	26
Mean antibody clones per refinement iteration	150.333
Mean total resources per refinement iteration	150
Mean pool size per refinement iteration	159.204
Mean memory cell clones per antigen	126.304
Mean antibody refinement iterations per antigen	19.883
Mean antibody prunings per refinement iteration	156.334
Data reduction percentage	2.6 %

Table 8.85 AIRS classification results

C2 versus all unbalanced	
Airs classifier classification parameter	Value
Correctly classified instances	**94.6 %**
Incorrectly classified instances	**5.4 %**
Kappa statistic	0.6973
Mean absolute error	0.054
Root mean squared error	0.2324
Relative absolute error	29.8822 %
Root relative squared error	77.4593 %
Total number of Instances	1000

Table 8.86 AIRS classification results by class

C2 versus all unbalanced

AIRS detailed accuracy by class

Class	TP rate	FP rate	Precision	Recall	F-measure
Minority	**0.72**	0.029	0.735	0.72	0.727
Majority	**0.971**	0.28	0.969	0.971	0.97

Table 8.87 SVM classification results

C2 versus all unbalanced	
SVM classifier classification parameter	Value
Correctly classified instances	**95.2**%
Incorrectly classified instances	**4.8**%
Kappa statistic	0.6923
Mean absolute error	0.048
Root mean squared error	0.2191
Relative absolute error	26.562%
Root relative squared error	73.0294%
Total number of Instances	1000

Table 8.88 SVM classification results by class

C2 versus all unbalanced					
SVM detailed accuracy by class					
Class	TP rate	FP rate	Precision	Recall	F-measure
Minority	**0.61**	0.01	0.871	0.61	0.718
Majority	**0.99**	0.39	0.958	0.99	0.974

Table 8.89 AIRS run information

C3 versus all unbalanced	
Airs classifier run parameter	Value
Affinity threshold scalar	0.2
Clonal rate	20
Hyper mutation rate	2
Stimulation threshold	0.99
Total resources	150
Nearest neighbors number	10
Affinity threshold	0.362
Total training instances	1000
Total memory cell replacements	38
Mean antibody clones per refinement iteration	149.589
Mean total resources per refinement iteration	150
Mean pool size per refinement iteration	158.526
Mean memory cell clones per antigen	31.399
Mean antibody refinement iterations per antigen	18.309
Mean antibody prunings per refinement iteration	150.923
Data reduction percentage	3.8%

Table 8.90 AIRS classification results

C3 versus all unbalanced	
Airs classifier classification parameter	Value
Correctly classified instances	**90 %**
Incorrectly classified instances	**10 %**
Kappa statistic	0.2775
Mean absolute error	0.1
Root mean squared error	0.3162
Relative absolute error	55.3374 %
Root relative squared error	105.4088 %
Total number of Instances	1000

Table 8.91 AIRS classification results by class

C3 versus all unbalanced					
AIRS detailed accuracy by class					
Class	TP rate	FP rate	Precision	Recall	F-measure
Minority	**0.24**	0.027	0.5	0.24	0.324
Majority	**0.973**	0.76	0.92	0.973	0.946

Table 8.92 SVM classification results

C3 versus all unbalanced	
SVM classifier classification parameter	Value
Correctly classified instances	**90.1 %**
Incorrectly classified instances	**9.9 %**
Kappa statistic	0.0481
Mean absolute error	0.099
Root mean squared error	0.3146
Relative absolute error	54.784 %
Root relative squared error	104.8804 %
Total number of Instances	1000

Table 8.93 SVM classification results by class

C3 versus all unbalanced					
SVM detailed accuracy by class					
Class	TP rate	FP rate	Precision	Recall	F-measure
Minority	**0.03**	0.002	0.6	0.03	0.057
Majority	**0.998**	0.97	0.903	0.998	0.948

Table 8.94 AIRS run information

C4 versus all unbalanced	
Airs classifier run parameter	Value
Affinity threshold scalar	0.2
Clonal rate	20
Hyper mutation rate	8
Stimulation threshold	0.99
Total resources	150
Nearest neighbors number	4
Affinity threshold	0.362
Total training instances	1000
Total memory cell replacements	27
Mean antibody clones per refinement iteration	150.475
Mean total resources per refinement iteration	150
Mean pool size per refinement iteration	159.356
Mean memory cell clones per antigen	125.899
Mean antibody refinement iterations per antigen	19.992
Mean antibody prunings per refinement iteration	156.423
Data reduction percentage	2.7 %

Table 8.95 AIRS classification results

C4 versus all unbalanced	
Airs classifier classification parameter	Value
Correctly classified instances	**86 %**
Incorrectly classified instances	**14 %**
Kappa statistic	0.2616
Mean absolute error	0.14
Root mean squared error	0.3742
Relative absolute error	77.4724 %
Root relative squared error	124.7214 %
Total number of Instances	1000

Table 8.96 AIRS classification results by class

C4 versus all unbalanced

AIRS detailed accuracy by class

Class	TP rate	FP rate	Precision	Recall	F-measure
Minority	**0.36**	0.084	0.321	0.36	0.34
Majority	**0.916**	0.64	0.928	0.916	0.922

Table 8.97 SVM classification results

C4 versus all unbalanced	
SVM classifier classification parameter	Value
Correctly classified instances	**90.1**%
Incorrectly classified instances	**9.9**%
Kappa statistic	0.0179
Mean absolute error	0.099
Root mean squared error	0.3146
Relative absolute error	54.784%
Root relative squared error	104.8804%
Total number of instances	1000

Table 8.98 SVM classification results by class

C4 versus all unbalanced					
SVM detailed accuracy by class					
Class	TP rate	FP rate	Precision	Recall	F-measure
Minority	**0.01**	0	1	0.01	0.02
Majority	**1**	0.99	0.901	1	0.948

Table 8.99 AIRS run information

C5 versus all unbalanced	
Airs classifier run parameter	Value
Affinity threshold scalar	0.2
Clonal rate	20
Hyper mutation rate	8
Stimulation threshold	0.99
Total resources	150
Nearest neighbors number	4
Affinity threshold	0.362
Total training instances	1000
Total memory cell replacements	19
Mean antibody clones per refinement iteration	150.415
Mean total resources per refinement iteration	150
Mean pool size per refinement iteration	159.296
Mean memory cell clones per antigen	125.858
Mean antibody refinement iterations per antigen	19.907
Mean antibody prunings per refinement iteration	156.387
Data reduction percentage	1.9%

Table 8.100 AIRS classification results

C5 versus all unbalanced

Airs classifier classification parameter	Value
Correctly classified instances	**87** %
Incorrectly classified instances	**13** %
Kappa statistic	0.2225
Mean absolute error	0.13
Root mean squared error	0.3606
Relative absolute error	71.9387 %
Root relative squared error	120.1845 %
Total number of instances	1000

Table 8.101 AIRS classification results by class

C5 versus all unbalanced

AIRS detailed accuracy by class

Class	TP rate	FP rate	Precision	Recall	F-measure
Minority	**0.27**	0.063	0.321	0.27	0.293
Majority	**0.937**	0.73	0.92	0.937	0.928

Table 8.102 SVM classification results

C5 versus all unbalanced

SVM classifier classification parameter	Value
Correctly classified instances	**90** %
Incorrectly classified instances	**10** %
Kappa statistic	0
Mean absolute error	0.1
Root mean squared error	0.3162
Relative absolute error	55.3374 %
Root relative squared error	105.4088 %
Total number of instances	1000

Table 8.103 SVM classification results by class

C5 versus all unbalanced

SVM detailed accuracy by class

Class	TP rate	FP rate	Precision	Recall	F-measure
Minority	**0**	0	0	0	0
Majority	**1**	1	0.9	1	0.947

Table 8.104 AIRS run information

C6 versus all unbalanced	
Airs classifier run parameter	Value
Affinity threshold scalar	0.2
Clonal rate	20
Hyper mutation rate	4
Stimulation threshold	0.99
Total resources	150
Nearest neighbors number	4
Affinity threshold	0.362
Total training instances	1000
Total memory cell replacements	24
Mean antibody clones per refinement iteration	150.293
Mean total resources per refinement iteration	150
Mean pool size per refinement iteration	159.191
Mean memory cell clones per antigen	62.976
Mean antibody refinement iterations per antigen	19.699
Mean antibody prunings per refinement iteration	153.135
Data reduction percentage	2.4 %

Table 8.105 AIRS classification results

C6 versus all unbalanced	
Airs classifier classification parameter	Value
Correctly classified instances	**92.2 %**
Incorrectly classified instances	**7.8 %**
Kappa statistic	0.4786
Mean absolute error	0.078
Root mean squared error	0.2793
Relative absolute error	43.1632 %
Root relative squared error	93.0945 %
Total number of instances	1000

Table 8.106 AIRS classification results by class

C6 versus all unbalanced					
AIRS detailed accuracy by class					
Class	TP rate	FP rate	Precision	Recall	F-measure
Minority	**0.42**	0.022	0.677	0.42	0.519
Majority	**0.978**	0.58	0.938	0.978	0.958

Table 8.107 SVM classification results

C6 versus all unbalanced

SVM classifier classification parameter	Value
Correctly classified instances	**91.9** %
Incorrectly classified instances	**8.1** %
Kappa statistic	0.3425
Mean absolute error	0.081
Root mean squared error	0.2846
Relative absolute error	44.8233 %
Root relative squared error	94.8679 %
Total number of instances	1000

Table 8.108 SVM classification results by class

C6 versus all unbalanced

SVM detailed accuracy by class

Class	TP rate	FP rate	Precision	Recall	F-measure
Minority	**0.24**	0.006	0.828	0.24	0.372
Majority	**0.994**	0.76	0.922	0.994	0.957

Table 8.109 AIRS run information

C7 versus all unbalanced

Airs classifier run parameter	Value
Affinity threshold scalar	0.2
Clonal rate	20
Hyper mutation rate	8
Stimulation threshold	0.99
Total resources	150
Nearest neighbors number	4
Affinity threshold	0.362
Total training instances	1000
Total memory cell replacements	20
Mean antibody clones per refinement iteration	150.432
Mean total resources per refinement iteration	150
Mean pool size per refinement iteration	159.315
Mean memory cell clones per antigen	125.815
Mean antibody refinement iterations per antigen	19.973
Mean antibody prunings per refinement iteration	156.381
Data reduction percentage	2 %

Table 8.110 AIRS classification results

C7 versus all unbalanced	
Airs classifier classification parameter	Value
Correctly classified instances	**90.8** %
Incorrectly classified instances	**9.2** %
Kappa statistic	0.4498
Mean absolute error	0.092
Root mean squared error	0.3033
Relative absolute error	50.9104 %
Root relative squared error	101.1046 %
Total number of instances	1000

Table 8.111 AIRS classification results by class

C7 versus all unbalanced					
AIRS detailed accuracy by class					
Class	TP rate	FP rate	Precision	Recall	F-measure
Minority	**0.46**	0.042	0.548	0.46	0.5
Majority	**0.958**	0.54	0.941	0.958	0.949

Table 8.112 SVM classification results

C7 versus all unbalanced	
SVM classifier classification parameter	Value
Correctly classified instances	**92.1** %
Incorrectly classified instances	**7.9** %
Kappa statistic	0.4051
Mean absolute error	0.079
Root mean squared error	0.2811
Relative absolute error	43.7166 %
Root relative squared error	93.6894 %
Total number of instances	1000

Table 8.113 SVM classification results by class

C7 versus all unbalanced					
SVM detailed accuracy by class					
Class	TP rate	FP rate	Precision	Recall	F-measure
Minority	**0.31**	0.011	0.756	0.31	0.44
Majority	**0.989**	0.69	0.928	0.989	0.958

Table 8.114 AIRS run information

C8 versus all unbalanced

Airs classifier run parameter	Value
Affinity threshold scalar	0.2
Clonal rate	10
Hyper mutation rate	8
Stimulation threshold	0.99
Total resources	150
Nearest neighbors number	4
Affinity threshold	0.362
Total training instances	1000
Total memory cell replacements	32
Mean antibody clones per refinement iteration	147.913
Mean total resources per refinement iteration	150
Mean pool size per refinement iteration	165.661
Mean memory cell clones per antigen	62.822
Mean antibody refinement iterations per antigen	24.011
Mean antibody prunings per refinement iteration	149.919
Data reduction percentage	3.2 %

Table 8.115 AIRS classification results

C8 versus all unbalanced

Airs classifier classification parameter	Value
Correctly classified instances	**91.4 %**
Incorrectly classified instances	**8.6 %**
Kappa statistic	0.4704
Mean absolute error	0.086
Root mean squared error	0.2933
Relative absolute error	47.5902 %
Root relative squared error	97.7521 %
Total number of instances	1000

Table 8.116 AIRS classification results by class

C8 versus all unbalanced

AIRS detailed accuracy by class

Class	TP rate	FP rate	Precision	Recall	F-measure
Minority	**0.46**	0.036	0.59	0.46	0.517
Majority	**0.964**	0.54	0.941	0.964	0.953

Table 8.117 SVM classification results

C8 versus all unbalanced	
SVM classifier classification parameter	Value
Correctly classified instances	**92 %**
Incorrectly classified instances	**8 %**
Kappa statistic	0.3865
Mean absolute error	0.08
Root mean squared error	0.2828
Relative absolute error	44.2699 %
Root relative squared error	94.2805 %
Total number of instances	1000

Table 8.118 SVM classification results by class

C8 versus all unbalanced					
SVM detailed accuracy by class					
Class	TP rate	FP rate	Precision	Recall	F-measure
Minority	**0.29**	0.01	0.763	0.29	0.42
Majority	**0.99**	0.71	0.926	0.99	0.957

Table 8.119 AIRS run information

C9 versus all unbalanced	
Airs classifier run parameter	Value
Affinity threshold scalar	0.2
Clonal rate	20
Hyper mutation rate	2
Stimulation threshold	0.99
Total resources	150
Nearest neighbors number	10
Affinity threshold	0.362
Total training instances	1000
Total memory cell replacements	34
Mean antibody clones per refinement iteration	149.756
Mean total resources per refinement iteration	150
Mean pool size per refinement iteration	158.687
Mean memory cell clones per antigen	31.431
Mean antibody refinement iterations per antigen	18.573
Mean antibody prunings per refinement iteration	151.072
Data reduction percentage	3.4 %

Table 8.120 AIRS classification results

C9 versus all unbalanced	
Airs classifier classification parameter	Value
Correctly classified instances	**90.1 %**
Incorrectly classified instances	**9.9 %**
Kappa statistic	0.2969
Mean absolute error	0.099
Root mean squared error	0.3146
Relative absolute error	54.784 %
Root relative squared error	104.8804 %
Total number of instances	1000

Table 8.121 AIRS classification results by class

C9 versus all unbalanced					
AIRS detailed accuracy by class					
Class	TP rate	FP rate	Precision	Recall	F-measure
Minority	**0.26**	0.028	0.51	0.26	0.344
Majority	**0.972**	0.74	0.922	0.972	0.946

Table 8.122 SVM classification results

C9 versus all unbalanced	
SVM classifier classification parameter	Value
Correctly classified instances	**90 %**
Incorrectly classified instances	**10 %**
Kappa statistic	0.0157
Mean absolute error	0.1
Root mean squared error	0.3162
Relative absolute error	55.3374 %
Root relative squared error	105.4088 %
Total number of instances	1000

Table 8.123 SVM classification results by class

C9 versus all unbalanced					
SVM detailed accuracy by class					
Class	TP rate	FP rate	Precision	Recall	F-measure
Minority	**0.01**	0.001	0.5	0.01	0.02
Majority	**0.999**	0.99	0.901	0.999	0.947

Table 8.124 AIRS run information

C10 versus all unbalanced	
Airs classifier run parameter	Value
Affinity threshold scalar	0.2
Clonal rate	20
Hyper mutation rate	10
Stimulation threshold	0.99
Total resources	150
Nearest neighbors number	10
Affinity threshold	0.362
Total training instances	1000
Total memory cell replacements	24
Mean antibody clones per refinement iteration	150.428
Mean total resources per refinement iteration	150
Mean pool size per refinement iteration	159.298
Mean memory cell clones per antigen	157.128
Mean antibody refinement iterations per antigen	20.005
Mean antibody prunings per refinement iteration	157.933
Data reduction percentage	2.4 %

Table 8.125 AIRS classification results

C10 versus all unbalanced	
Airs classifier classification parameter	Value
Correctly classified instances	**89.8 %**
Incorrectly classified instances	**10.2 %**
Kappa statistic	0.0556
Mean absolute error	0.102
Root mean squared error	0.3194
Relative absolute error	56.4442 %
Root relative squared error	106.4577 %
Total number of instances	1000

Table 8.126 AIRS classification results by class

C10 versus all unbalanced					
AIRS detailed accuracy by class					
Class	TP rate	FP rate	Precision	Recall	F-measure
Minority	**0.04**	0.007	0.4	0.04	0.073
Majority	**0.993**	0.96	0.903	0.993	0.946

Table 8.127 SVM classification results

C10 versus all unbalanced

SVM classifier classification parameter	Value
Correctly classified instances	**90** %
Incorrectly classified instances	**10** %
Kappa statistic	0
Mean absolute error	0.1
Root mean squared error	0.3162
Relative absolute error	55.3374 %
Root relative squared error	105.4088 %
Total number of instances	1000

Table 8.128 SVM classification results by class

C10 versus all unbalanced

SVM detailed accuracy by class

Class	TP rate	FP rate	Precision	Recall	F-measure
Minority	**0**	0	0	0	0
Majority	**1**	1	0.9	1	0.947

8.1.5 Music Recommendation Based on Artificial Immune Systems

In this section, the music recommendation process is addressed as a one-class classification problem by developing an AIS-based Negative Selection (NS) algorithm. The primary objective of the proposed methodology is to exploit the inherent ability of the NS-based classifier in handling severely unbalanced classification problems in order to capture user preferences. The main idea that motivated the recommendation approach presented in this section stems from the fact that users' interests occupy only a small fraction of a given multimedia collection. That is, the music preferences of a particular user tend to be contained within a negligible volume of the complete pattern space. Therefore, the problem of identifying multimedia instances that a specific user would evaluate as preferable, constitutes an extremely unbalanced pattern recognition problem that could be addressed within the context of one-class classification. This is true, since in most real-life situations the user supplied feedback to a recommendation system is exclusively given in the form of positive examples from the target class to be recognized.

Specifically, the adapted approach decomposes the music recommendation problem into a two-level cascading recommendation scheme. The first recommendation level incorporates the AIS-based one-class classification algorithm in order to discriminate between positive and negative patterns on the basis of zero knowledge from the subspace of outliers. The second level, on the other hand, is responsible for assigning a particular degree of preference according to past user ratings. For this purpose, the second recommendation level applies either a content-based approach or a collaborative filtering technique. The implementation and evaluation of the proposed Cascade Hybrid recommender approach, enhanced by the one class classifier in the first level and the collaborative filtering in the second level, demonstrates the efficiency of the proposed recommendation scheme. The presented technique benefits from both content-based and collaborative filtering methodologies. The content-based level eliminates the drawbacks of the pure collaborative filtering that do not take into account the subjective preferences of an individual user, as they are biased towards the items that are most preferred by the rest of the users. On the other hand, the collaborative filtering level eliminates the drawbacks of the pure content-based recommender which ignore any beneficial information related to users with similar preferences. The combination of two approaches in a cascade form, mimics the social process when someone has selected some items according to his preferences and asks for opinions about these by others, in order to achieve the best selection.

Fundamental Problems of Recommender Systems

In this section we refer the fundamental problems of recommender system and the solutions that each of the above techniques offer to these.

The *cold-start problem* [23] related with the learning rate curve of the recommender system. This problem could be analyzed into two different sub-problems:

- *New-User problem*, is the problem of making recommendations to new user [18], as it is related with the situation where almost nothing is known about his/her preferences.
- *New-Item problem*, is the problem where ratings are required for items that have not been rated by users. Therefore, until the new item is rated by a satisfactory number of users, the recommender system would not be able to recommend this item. This problem it appears mostly to collaborative approaches, as it could be eliminated with the use of content-based or hybrid approaches where content information is used to infer similarities among items.

This problem is also related, with the *coverage* of a recommender which is a measure for the domain of items over which the system could produce recommendations. For example low coverage of the domain means that it is used a limited space of items in results of recommender and these results usually could be biased by preferences of other users. This is also known as the problem of *Over-specialization*. When the system can only recommend items that score highly against a users profile, the user is limited to being recommended items that are similar to those already rated. This problem, which has also been studied in other domains, is often addressed by introducing some randomness. For example, the use of genetic algorithms has been proposed as a possible solution in the context of information filtering [25].

Novelty Detection - Quality of Recommendations. From those items that a recommender system recommend to users, there are items that already known and items that are new (novel) unknown to them. Therefore, there is a competitiveness between the desire for novelty and the desire for high quality recommendation. One the one hand quality of the recommendations [20] is related with "trust" that users express for the recommendations. This means that recommender should minimize false positive errors, more specifically the recommender should not recommend items that are not desirable. On the other hand, novelty is related with the "timestamp - age" of items, the older items should be treated as less relevant than the new ones which means increase to the novelty rate. Thus, a high novelty rate will produce a poor quality recommendations because the users would not be able to identify most of the items in the list of recommendations.

The *sparsity problem* [2, 13] is related to the unavailability of a large number of rated item for each active user. The number of items that are rated by users is usually a very small subset of those items that are totally available. For example in Amazon if the active users may have purchased 1 % of the items and the total amount of items is approximately 1 million of books this means that there are only 10,000 of books which are rated. Consequently, such sparsity in ratings affects in an accurate selection of the neighbors in the step of neighborhood formation, and leads in poor recommendation results.

A number of possible solutions have been proposed to overcome the sparsity problem such as content-based similarities, item-based collaborative filtering methods, use of demographic data and a number of hybrid approaches [5]. A different approach to deal with this problem is proposed in [21] where it is utilized dimensional reduction technique, such as Singular Value Decomposition, in order to transform the sparse user-item matrix R into a dense matrix. The SVD is a method for matrix factorization that produce the best lower rank approximations of the original matrix [16].

Recommender systems, especially with large electronic sites, have to deal with the problem of *scalability* as there is a constantly growing number of users and items [3, 35]. Therefore, it required an increasing amount of computational resources as the amount of data grows. A recommendation method that could be efficient when the number of data is limited could be very time-consuming, scales poorly and unable to generate a satisfactory number of recommendations in a large amount of data. Thus it is important the recommendation approach to be capable of scaling up in a successful manner [22].

Recommendation as a One-Class Classification Problem

The main problem dominating the design of an efficient multimedia recommender system is the difficulty faced by its users when attempting to articulate their needs. However, users are extremely good at characterizing a specific instance of multimedia information as preferable or not. This entails that it is possible to obtain a sufficient number of positive and negative examples from the user in order to employ an appropriate machine learning methodology to acquire a user preference profile. Positive and negative evidence concerning the preferences of a specific user are utilized by the machine learning methodology so as to derive a model of how that particular user valuates the information content of a multimedia file. Such a model could enable a recommender system to classify unseen multimedia files as desirable or non-desirable according to the acquired model of the user preferences. Thus, the problem of recommendation is formulated as a binary classification problem where the set of probable classes would include two class instances, $C_+ = $ prefer/like and $C_- = $ don't prefer/dislike. However, the burden of obtaining a sufficient number of positive and negative examples from a user is not negligible. Additionally, users find it sensibly hard to explicitly express what they consider as non desirable since the reward they will eventually receive does not outweigh the cost undertaken in terms of time and effort. It is also very important to mention that the class of desirable patterns occupies only a negligible volume of the patterns space since the multimedia instances that a particular user would characterize as preferable are only few compared to the vast majority of the non - desirable patterns.

This fact justifies the *highly unbalanced nature of the recommendation problem*, since non - targets occur only occasionally and their measurements are very costly. Moreover, if there were available patterns from the non - target class they could not be trusted in principle as they would be badly sampled, with unknown priors and ill - defined distributions. In essence, non - targets are weakly defined since they appear as any kind of deviation or anomaly from the target objects.Since samples from both classes are not available, machine learning models based on defining a boundary between the two classes are not applicable. Therefore, a natural choice in order to overcome this problem is building a model that either provides a statistical description for the class of the available patterns or a description concerning the shape/structure of the class that generated the training samples. *This insight has led us to reformulate the problem of recommendation as a one-class classification problem where the only available patterns originate from the target class to be learned.* Specifically, the one-class to be considered as the target class is the class of desirable patterns while the complementary space of the universe of discourse corresponds to the class on non-desirable patterns. Otherwise stated, our primary concern is to derive an inductive bias which will form the basis for the classification of unseen patterns as preferable or not. In the context of building a music piece recommender system, available training patterns correspond to those multimedia instances that a particular user assigned to the class of preferable patterns. The recommendation of new music pieces is then performed by utilizing the one-class classifier for assigning unseen music pieces in the database either in the class of desirable patterns or in the complementary class of non-desirable patterns.

The general setting of the recommendation problem where there is a unique class of interest and everything else belongs to another class manifests its extremely non - symmetrical nature. Additionally, the probability density for the class of target patterns may be scattered along the different intrinsic classes of the data. For example the universe of discourse for our music piece recommender system is a music database which is intrinsically partitioned into 10 disjoint classes of musical genres. Thus, the target class of preferable patterns for a particular may be formed as a mixing of the various musical genres in arbitrary proportions. The non - symmetrical nature of the recommendation problem is an additional fact validating its formulation as a one - class learning problem.

Another important factor that leads toward the selection of one - class learning as a valid paradigm for the problem of recommendation is that the related misclassification costs are analogously unbalanced. The quantities of interest are the false positive rate and the false negative rate. The false positive rate expresses how often a classifier falsely predicts that a specific pattern belongs to the target class of patterns while it originated from the complementary class. The false negative rate expresses how often a classifier falsely predicts that a specific pattern belongs to the complementary class of patterns while it originated from the target class. In the context of designing a music piece recommender system the cost related to the false positive rate is of greater impact than the cost related to the false negative rate. False positives result in recommending items that a particular user would classify as non - desirable thus effecting the quality of recommendation. In contrast, false negatives result in not recommending items that a particular user would classify as desirable. Thus, it is of vital importance in minimizing the false positive rate which results in improving the accuracy of recommendation.

Cascade Classification Architecture

The problem of recommendation is addressed by developing a two-level cascade classification architecture. The first level involves the incorporation of a one-class classifier which is trained exclusively on positive patterns. The one-class learning component at the first level serves the purpose of recognizing instances from the class of desirable patterns against those patterns originating from the class of non-desirable ones. On the other hand, the second level is based on a multi-class classifier, which is also trained exclusively on positive data, but in order to discriminate amongst the various classes of positive patterns.

It must be mentioned that the class of negative/non-desirable patterns occupies a significantly larger volume within the pattern space V in comparison with the volume occupied by the positive/desirable patterns. This is a reasonable fact since user preferences are concentrated within a small fraction of the universe of discourse. Thus, in the context of building an efficient music recommendation system it is a matter of crucial importance to be able to recognize the majority of instances that belong to the complementary space of non-desirable data. This particular problem is addressed by the first level component of our cascade classification architecture by developing a one-class classifier trained exclusively on samples from the minority class of desirable patterns. In other words, the first level classifier is designated to confront the extremely unbalanced nature of the underlying machine learning problem. This problem arises when a content-based, item oriented approach is adopted in order to address the problem of recommendation. The second level component addresses the problem of discriminating amongst the classes of desirable patterns, which may be formulated as a balanced multi-class machine learning problem since the users' preferences are evenly distributed over the classes of desirable patterns.

In order to formulate the problem of recommendation as a two-level machine learning problem, it is necessary to precisely define the training and testing procedures followed for both classifiers. It must be explicitly mentioned that the training procedure of classifiers at both levels is conducted exclusively on positive data. This constitutes the most important aspect of our approach since we completely ignore the majority class of non-desirable patterns during the training process. Negative patterns are only used within the testing procedure where we need to accurately measure the efficiency of the two-level classifier in predicting the class of unseen patterns.

Cascade Content Based Recommendation (First Level One Class - Second Level Multi Class)

Let $U = \{u_1, u_2, \ldots, u_m\}$ be the set of users and $I = \{i_1, i_2, \ldots, i_n\}$ be the set of items pertaining to the music database which was used for the implementation of our music recommendation system. Each music file in the database corresponds to a feature vector in the high-dimensional Euclidean vector space V, defined in Sect. 8.1.1. Each user that participated in our experiments was asked to assign a unique rating value for each item in the database within the range of $\{0, 1, 2, 3\}$. Thus, user ratings define four disjoint classes of increasing degree of interest, namely C_0, C_1, C_2 and C_3. C_0 corresponds to the class of non-desirable/negative patterns, while the class of desirable/positive patterns may be defined as the union $(C_1 \cup C_2 \cup C_3)$ of C_1, C_2 and C_3. In order to indicate the user involvement in defining the four classes of interest, we may write that

$$\forall u \in U, V = C_0(u) \cup C_1(u) \cup C_2(u) \cup C_3(u) \text{ where}$$
$$C_0(u) \cap C_1(u) \cap C_2(u) \cap C_3(u) = \emptyset \tag{8.10}$$

More specifically, letting $R(u, i)$ be the rating value that the user u assigned to item i, the four classes of interest may be defined by the following equations:

$$C_0(u) = \{i \in I : R(u, i) = 0\}$$
$$C_1(u) = \{i \in I : R(u, i) = 1\}$$
$$C_2(u) = \{i \in I : R(u, i) = 2\}$$
$$C_3(u) = \{i \in I : R(u, i) = 3\} \tag{8.11}$$

At this point, we need to mention that if $I(u)$ denotes the subset of items for which user u provided a rating, it follows that $\forall u \in U, I(u) = I$. Thus, the positive/desirable and negative/non-desirable classes of patterns for each user may be defined as follows:

$$\forall u \in U, \mathbf{P}(u) = C_1(u) \cup C_2(u) \cup C_3(u)$$
$$\forall u \in U, \mathbf{N}(u) = C_0(u) \tag{8.12}$$

The training/testing procedure for the classifiers at both levels involves the partitioning of each class of desirable patterns for each user into K disjoint subsets such that:

$$\forall u \in U, j \in \{1, 2, 3\}, \ C_j(u) = \bigcup_{k \in [K]} C_j(u, k) \text{ where}$$
$$\forall k \in [K], |C_j(u, k)| = \frac{1}{K}|C_j(u)| \text{ such that} \tag{8.13}$$
$$\bigcap_{k \in [K]} C_j(u, k) = \emptyset$$

Letting $C_j(u, k)$ be the set of patterns from the positive class j that is used throughout the testing procedure then the corresponding set of training patterns will be denoted as $\widehat{C}_j(u, k)$ so that the following equation holds:

$$\forall j \in \{1, 2, 3\}, \forall k \in [K], \ \widehat{C}_j(u, k) \cup C_j(u, k) = C_j(u) \tag{8.14}$$

In other words, Eq. (8.14) defines the K-fold cross validation partitioning that is utilized in order to measure the performance accuracy of the cascade classification scheme which lies within the core of our music recommendation system. If $P(u, k), N(u, k)$ are the sets of positive and negative patterns respectively, as they are presented to the first level classifier during the testing stage at fold k for a particular user u, we may write that:

$$\begin{aligned} P(u, k) &= C_1(u, k) \cup C_2(u, k) \cup C_3(u, k) \\ N(u, k) &= C_0(u, k) = C_0(u) = N(u) \end{aligned} \tag{8.15}$$

In case the K-fold cross validation partitioning is not taken into consideration the set of positive patterns for a particular user may be addressed as $P(u)$ so that $P(u) = C_1(u) \cup C_2(u) \cup C_3(u)$.

The training procedure concerning the first level of our cascade classification architecture involves developing a one-class classifier for each particular user. These one-class classifiers are trained in order to recognize those data instances that originated from the positive class of patterns. In other words, each one-class classifier realizes a discrimination function denoted by $f_u(\mathbf{v})$, where $\mathbf{v}$ is vector in V, which is learned from the fraction of training positive patterns. More specifically, if $f_{u,k}(\mathbf{v})$ is the discrimination function that corresponds to user u at fold k, then this function would be the result of training the one-class classifier on $\widehat{C}_1(u, k) \cup \widehat{C}_2(u, k) \cup \widehat{C}_3(u, k)$. The purpose of each discrimination function $f_u(\mathbf{v})$ constitutes in recognizing the testing positive patterns $P(u)$ against the complete set of negative patterns $N(u)$.

On the other hand, the training procedure that concerns the second level of our cascade classification architecture involves developing a multi-class classifier for each particular user. This entails training the multi-class classifier on the same set of positive data, $\widehat{C}_1(u, k) \cup \widehat{C}_2(u, k) \cup \widehat{C}_3(u, k)$, but this time in order to discriminate amongst the data pertaining to the set $P(u, k)$. In other words, each second level classifier realizes a discrimination function denoted by $g_u(\mathbf{v})$ whose purpose consists in partitioning the space of testing positive data $P(u)$ into the 3 corresponding subspaces, $C_1(u)$, $C_2(u)$ and $C_3(u)$, of desirable patterns. If we need to explicitly address the discrimination function concerning user u at fold k then we can write $g_{u,k}(\mathbf{v})$.

The recommendation ability of our system is based on its efficiency in predicting the rating value that a particular user assigned to a music file which was not included within the training set. Having in mind that $P(u, k) \cup N(u, k)$ is the set of testing data that are presented to the fist level of our cascade classification mechanism, the one-class component operates as a filter that distinguishes those music pieces that a user

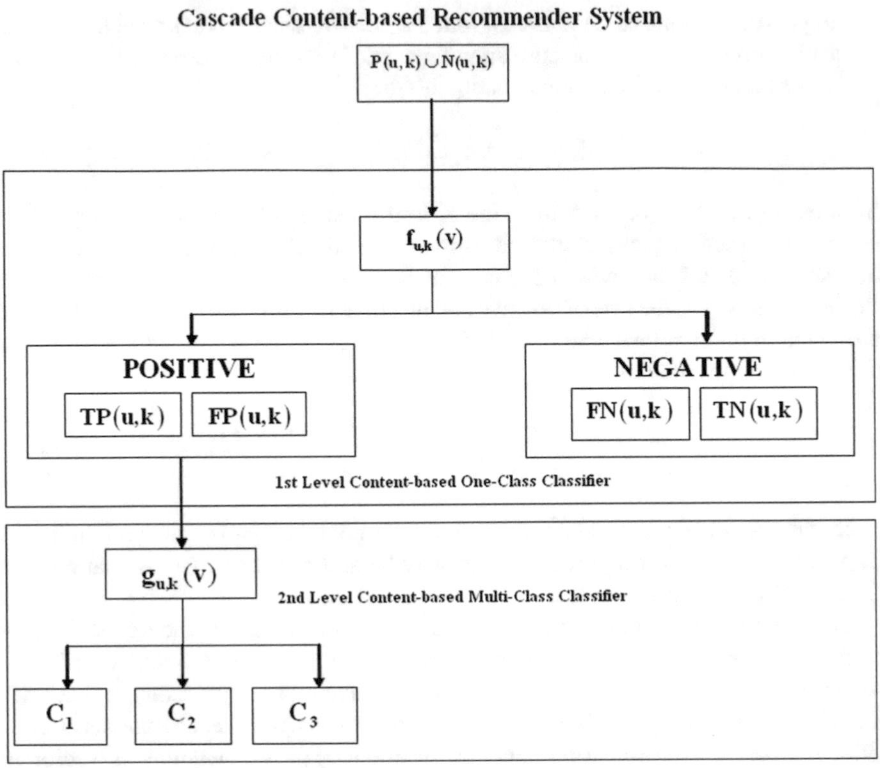

Cascade Content-based Recommender System

Fig. 8.13 Cascade content-based recommender

assigned to the class of desirable patterns. Specifically, the first level discrimination function $f_{u,k}(\mathbf{v})$ for user u at fold k partitions the set of testing data into positive and negative patterns as it is depicted in Fig. 8.13. In other words, the testing procedure concerning the first level of our cascade classifier involves the assignment of a unique value within the set $\{-1, 1\}$ for each input element such that:

$$\forall u \in U, \ \forall k \in [K], \ \forall \mathbf{v} \in P(u, k) \cup N(u, k), \ f_{u,k}(\mathbf{v}) \in \{-1, +1\} \qquad (8.16)$$

The subset of testing instances that are assigned to the class of desirable patterns are subsequently fed to the second level classifier which assigns a particular rating value within the range of $\{1, 2, 3\}$. Specifically, we may write that:

$$\forall u \in U, \forall k \in [K], \forall \mathbf{v} \in P(u, k) \cup N(u, k) : f_{u,k}(\mathbf{v}) = +1, \ g_{u,k}(\mathbf{v}) \in \{1, 2, 3\}$$
$$(8.17)$$

Let the true rating value concerning an object $\mathbf{v} \in V$ for a particular user u at fold k be $R_{u,k}(\mathbf{v})$ so that the following equation holds:

$$\forall u \in U, \ \forall k \in [K], \ \forall j \in \{0, 1, 2, 3\}, R_{u,k}(\mathbf{v}) = j \Leftrightarrow \mathbf{v} \in C_j(u, k) \qquad (8.18)$$

the estimated rating value assigned by our system will be addressed as $\widehat{R}_{u,k}(\mathbf{v})$. Thus, the estimated rating can be computed by the following equation:

$$\widehat{R}_{u,k}(\mathbf{v}) = \begin{cases} 0, \ \forall \mathbf{v} \in P(u, k) \cup N(u, k) : f_{u,k}(\mathbf{v}) = +1; \\ 1, \ \forall \mathbf{v} \in P(u, k) \cup N(u, k) : f_{u,k}(\mathbf{v}) = -1 \text{ and } g_{u,k}(\mathbf{v}) = 1; \\ 2, \ \forall \mathbf{v} \in P(u, k) \cup N(u, k) : f_{u,k}(\mathbf{v}) = -1 \text{ and } g_{u,k}(\mathbf{v}) = 2; \\ 3, \ \forall \mathbf{v} \in P(u, k) \cup N(u, k) : f_{u,k}(\mathbf{v}) = -1 \text{ and } g_{u,k}(\mathbf{v}) = 3. \end{cases} \qquad (8.19)$$

Cascade Hybrid Recommendation (First Level One Class - Second Level Collaborative Filtering)

The previous discussion manifests that the major problem addressed in this chapter is building an efficient music recommendation system in the complete absence of negative examples. Since negative examples are in general extremely difficult to find we employed classification paradigms that operate exclusively on the basis of positive patterns. This justifies the incorporation of the one-class classification component within the first level our cascade classification architecture. Thus, the first classification level serves the purpose of filtering out the majority of the non-desirable patterns. However, our aim is to provide the user with high quality recommendations which involves predicting the true class of unseen patterns with high accuracy. This is the rationale behind the second classification level which takes as input the set of patterns that were assigned to the minority class by the first classification level. In order to provide high quality recommendations it is vital to correctly discriminate amongst the various classes of desirable patterns. This effort is undertaken within the second multi-class classification level of our cascade classification architecture.

A natural modification of our cascade classification architecture consists in replacing the second multi-class classification level with a collaborative filtering component as it is illustrated in Fig. 8.14. Having in mind that the first classification level realizes the broader distinction between positive and negative patterns, the subsequent collaborative filtering component produces specific rating values within the range of $\{1, 2, 3\}$. Specifically, the collaborative filtering methods that we utilized were:

- Pearson Correlation [3]
- Vector Similarity [3] and
- Personality Diagnosis [17].

Personality Diagnosis (PD) may be thought of as a hybrid between memory—and model-based approaches. The main characteristic is that predictions have a meaningful probabilistic semantics. Moreover, this approach assumes that preferences constitute a manifestation of their underlying personality type for each user. Therefore, taking into consideration the active users known ratings of items, it is possible to estimate the probability that he or she has the same personality type with another user [17].

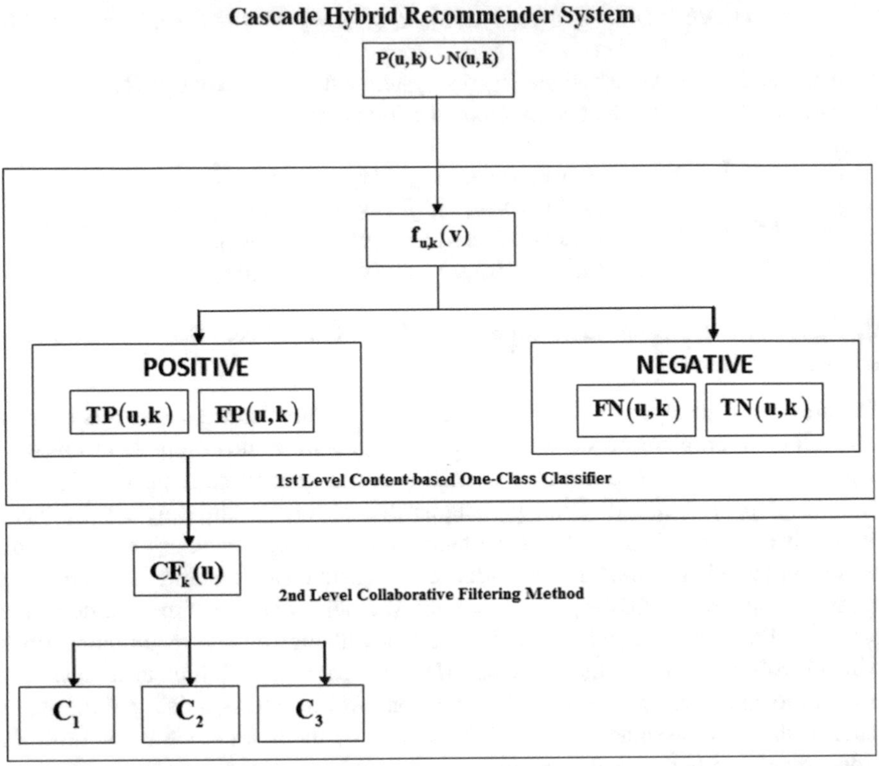

Cascade Hybrid Recommender System

Fig. 8.14   Cascade hybrid recommender

The training and testing procedures concerning the second level collaborative filtering component are identical to the ones used for the multi-class classification component. Specifically, training was conducted on the ratings that correspond to thecst of pattern $\widehat{C}_1(u, k) \cup \widehat{C}_2(u, k) \cup \widehat{C}_3(u, k)$. Accordingly, the testing procedure was conducted on the rating values that correspond to the set of testing patterns $C_1(u, k) \cup C_2(u, k) \cup C_3(u, k)$.

Measuring the Efficiency of the Cascade Classification Scheme

The efficiency of the adapted cascade classification scheme was measured in terms of the *Mean Absolute Error* and the *Ranked Evaluation* measure. The Mean Absolute Error (MAE) constitutes the most commonly used measure in order to evaluate the efficiency of recommendation systems. More formally, MAE concerning user u at fold k may be defined as:

$$MAE(u, k) = \frac{1}{|P(u, k)| + |N(u, k)|} \sum_{v \in P(u,k) \cup N(u,k)} |R_{u,k}(\mathbf{v}) - \widehat{R}_{u,k}(\mathbf{v})| \qquad (8.20)$$

The Ranked Scoring (RS)[Needs Reference] assumes that the recommendation is presented to the user as a list of items ranked by their predicted ratings. Specifically, RS assesses the expected utility of a ranked list of items, by multiplying the utility of an item for the user by the probability that the item will be viewed by the user. The utility of an item is computed as the difference between its observed rating and the default or neutral rating d in the domain (which can be either the midpoint of the rating scale or the average rating in the dataset), while the probability of viewing decays exponentially as the rank of items increases. Formally, the RS of a ranked list of items $\mathbf{v_j} \in P(u_i, k) \cup N(u_i, k)$ sorted according to the index j in order of declining $R_{u_i,k}(\mathbf{v_j})$ for a particular user u_i at fold k is given by:

$$RS_{u_i,k} = \sum_{\mathbf{v_j} \in P(u_i,k) \cup N(u_i,k)} \max \{R_{u_i,k}(\mathbf{v_j}) - d), 0\} \times \frac{1}{2^{(j-1)(i-1)}} \qquad (8.21)$$

Having in mind that the set of testing patterns for the first level classifier at fold k is formed by the patterns pertaining to the sets $C_0(u)$, $C_1(u, k)$, $C_2(u, k)$ and $C_3(u, k)$, we may write that

$$|P(u, k)| = |C_1(u, k)| + |C_2(u, k)| + |C_3(u, k)| \qquad (8.22)$$

and

$$|N(u, k)| = |C_0(u)| \qquad (8.23)$$

According to Eqs. (8.22) and (8.23) we may define the quantities true positive rate (TPR), false negative rate (FNR), true negative rate (TNR) and false positive rate (FPR) concerning user u for the f-th fold of the testing stage as follows:

$$TPR(u, k) = \frac{TP(u, k)}{|P(u, k)|} \qquad (8.24)$$

$$FNR(u, k) = \frac{FP(u, k)}{|P(u, k)|} \qquad (8.25)$$

$$TNR(u, k) = \frac{TN(u, k)}{|N(u, k)|} \qquad (8.26)$$

and

$$FPR(u, k) = \frac{FP(u, k)}{|N(u, k)|} \qquad (8.27)$$

It is important to note that the quantities defined by the Eqs. (8.24), (8.25), (8.26), and (8.27) refer to the classification performance of the first level classifier in the adapted cascade classification scheme. True Positive $TP(u, k)$ is the number of positive/desirable patterns that were correctly assigned to the positive class of patterns while False Negative $TN(u, k)$ is the number of positive/desirable patterns that were incorrectly assigned to the negative class of patterns. True Negative $TN(u, k)$ is the number of negative/non - desirable patterns that were correctly assigned to the negative class of patterns while False Positive $FP(u, k)$ is the number of negative/non - desirable patterns that were incorrectly assigned to the positive class. More formally, having in mind Eq. (8.19) the above quantities may be described by the following equations:

$$TP(u, k) = \{\mathbf{v} \in P(u, k) : f_{u,k}(\mathbf{v}) = +1\} \tag{8.28}$$

$$FP(u, k) = \{\mathbf{v} \in N(u, k) : f_{u,k}(\mathbf{v}) = +1\} \tag{8.29}$$

$$TN(u, k) = \{\mathbf{v} \in N(u, k) : f_{u,k}(\mathbf{v}) = -1\} \tag{8.30}$$

$$FN(u, k) = \{\mathbf{v} \in P(u, k) : f_{u,k}(\mathbf{v}) = -1\} \tag{8.31}$$

Computing the mean value for the above quantities through the different folds results in the following equations:

$$\overline{TPR}(u) = \frac{1}{K} \sum_{f \in F} TPR(u, k) \tag{8.32}$$

$$\overline{FNR}(u) = \frac{1}{K} \sum_{f \in F} FNR(u, k) \tag{8.33}$$

$$\overline{TNR}(u) = \frac{1}{K} \sum_{f \in F} TNR(u, k) \tag{8.34}$$

$$\overline{FPR}(u) = \frac{1}{K} \sum_{f \in F} FPR(u, k) \tag{8.35}$$

It is possible to bound the mean absolute error for the complete two level classifier according to its performance during the second stage of the multi-class classification scheme. The best case scenario concerning the classification performance of the second level multi-class classifier suggests that all the true positive patterns, which are passed to the second classification level, are correctly classified. Moreover, the best case scenario involves that all the false negative patterns of the first classification level originated from C_1. Thus, the following inequality holds:

$$\forall u \in U \; \forall k \in [K] \; MAE(u, k) \geq \frac{FN(u, k) + FP(u, k)}{|P(u, k)| + |N(u, k)|} \tag{8.36}$$

Given the Eqs. (8.24), (8.25), (8.26), and (8.27) and letting

$$\lambda(u, k) = \frac{|P(u, k)|}{|N(u, k)|} = \frac{|P(u)|}{|N(u)|} = \lambda(u) \tag{8.37}$$

since the number of positive and negative patterns used during the testing stage do not change for each fold and for each user, inequality (8.36) may be written as

$$MAE(u, k) \geq FNR(u, k) \times \frac{\lambda(u)}{\lambda(u) + 1} + FPR(u, k) \times \frac{1}{\lambda(u) + 1} \tag{8.38}$$

Given that

$$\overline{MAE}(u) = \frac{1}{K} \sum_{k \in [K]} MAE(u, k) \tag{8.39}$$

Inequality (8.38) may be written as:

$$\overline{MAE}(u) \geq \overline{FNR}(u) \times \frac{\lambda(u)}{\lambda(u) + 1} + \overline{FPR}(u) \times \frac{1}{\lambda(u) + 1} \tag{8.40}$$

If we consider the average value for the mean absolute error over all users then we may write that:

$$\overline{MAE} = \frac{1}{|U|} \sum_{u \in U} \overline{MAE}(u) \tag{8.41}$$

which yields that:

$$\overline{MAE} \geq \frac{1}{|U|} \sum_{u \in U} \overline{FNR}(u) \times \frac{\lambda(u)}{\lambda(u) + 1} + \overline{FPR}(u) \times \frac{1}{\lambda(u) + 1} \tag{8.42}$$

The worst case case scenario concerning the classification performance of the second level is that the multi-class second level classifier incorrectly assigned all true positive patterns to C_3 while they originated from C_1. In addition all the false negative patterns originated from C_3 and all the false positive patterns were assigned to C_3. Thus, we may write the following inequality:

$$\forall u \in U \; \forall f \in [K] \; MAE(u, k) \leq \frac{3 \times FN(u, k) + 2 \times TP(u, k) + 3 \times FP(u, k)}{P(u, k) + N(u, k)} \tag{8.43}$$

Given the Eqs. (8.24), (8.25), (8.26), (8.27) and (8.37) then Eq. (8.43) may be written as:

$$MAE(u, k) \leq \frac{3 \times FNR(u, k) \times \lambda(u)}{\lambda(u) + 1} + \frac{2 \times TPR(u, k) \times \lambda(u)}{\lambda(u) + 1} + \frac{3 \times FPR(u, k)}{\lambda(u) + 1} \tag{8.44}$$

Given Eq. (8.39) inequality (8.44) yields that:

$$\overline{MAE(u)} \leq \frac{3 \times \overline{FNR}(u) \times \lambda(u)}{\lambda(u) + 1} + \frac{2 \times \overline{TPR}(u) \times \lambda(u)}{\lambda(u) + 1} + \frac{3 \times \overline{FPR}(u)}{\lambda(u) + 1} \quad (8.45)$$

Thus, the average value for the mean absolute error has an upper bound given by the following inequality:

$$\overline{MAE} \leq \frac{1}{|U|} \sum_{u \in U} \frac{3 \times \overline{FNR}(u) \times \lambda(u)}{\lambda(u) + 1} + \frac{2 \times \overline{TPR}(u) \times \lambda(u)}{\lambda(u) + 1} + \frac{3 \times \overline{FPR}(u)}{\lambda(u) + 1}$$
$$(8.46)$$

Inequalities (8.42), (8.46) yield that the minimum and maximum values for the average mean absolute error over all users are described by the following equations:

$$\min_{u \in U} \overline{MAE} = \frac{1}{|U|} \sum_{u \in U} \frac{\overline{FNR}(u) \times \lambda(u)}{\lambda(u)+1} + \frac{\overline{FPR}(u)}{\lambda(u)+1} \quad (8.47)$$

$$\max_{u \in U} \overline{MAE} = \frac{1}{|U|} \sum_{u \in U} \frac{3 \times \overline{FNR}(u) \times \lambda(u)}{\lambda(u)+1} + \frac{2 \times \overline{TPR}(u) \times \lambda(u)}{\lambda(u)+1} + \frac{1}{|U|} \sum_{u \in U} \frac{3 \times \overline{FPR}(u)}{\lambda(u)+1} \quad (8.48)$$

System Evaluation

Data Description

The audio signal may be represented in multiple ways according to the specific features utilized in order to capture certain aspects of an audio signal. More specifically, there has been a significant amount of work in extracting features that are more appropriate for describing and modeling the music signal. In this chapter we have utilized a specific set of 30 objective features that was originally proposed by Tzanetakis and Cook [33] which dominated many subsequent approaches in the same research field. It is worth mentioning that these features not only provide a low level representation of the statistical properties of the music signal but also include high level information, extracted by psychoacoustic algorithms in order to represent rhythmic content (rhythm, beat and tempo information) and pitch content describing melody and harmony of a music signal. The music database that we utilized in order to evaluate the accuracy of our recommendation system involved 1085 music pieces and a set of 16 users. Each user has provided an average value of 360 positive ratings so that the corresponding sparsity level, sparsity level $= 1 - \frac{\text{non zero entries}}{\text{total entries}}$, was found to be 0.67.

Experimental Setup

The experimental results provided in this section correspond to the testing stage of our system which was thoroughly described in Sect. 6. Specifically, the evaluation process involved three recommendation approaches:

1. The first approach corresponds to the standard collaborative filtering methodologies, namely the Pearson Correlation, the Vector Similarity and the Personality Diagnosis.
2. The second approach corresponds to the Cascade Content-based Recommendation methodology which was realized on the basis of a two-level classification scheme. Specifically, we tested two types of one-class classifiers for the first level, namely the One-Class SVM and the V-Detector. On the other hand, the second classification level was realized as a multi-class SVM.
3. Finally, the third approach corresponds to the Cascade Hybrid Recommendation methodology which was implemented by a one-class SVM classification component at the first level and a collaborative filtering counterpart at the second level. Thus, the third recommendation approach involves three different recommenders according to the different collaborative filtering methodologies that can be embedded within the second level.

The following subsections provide a detailed description concerning the three types of experiments that were conducted in order to evaluate the efficiency of our cascade recommendation architecture.

- The first type of experiments demonstrates the contribution of the one-class classification component at the first level of our cascade recommendation system. Specifically, we provide MAE/RS measurements concerning the mean overall performance of the standard collaborative filtering methodologies in comparison to the hybrid recommendation approach for the complete set of users. Additionally, we measure the relative performance of the Cascade Content-based Recommender against the rest recommendation approaches in order to indicate the recommendation system that exhibited the best overall performance.
- The second type of experiments demonstrates the contribution of the second classification level within the framework of the Cascade Content-based Recommendation methodology. The main purpose of this experimentation session was to reveal the classification benefit gained by the second multi-class classification level.
- The third type of experiments constitutes a comparative study concerning the classification performance of the Cascade Content-based Recommendation approaches. Specifically, we explicitly measure the relative performance of the utilized one-class classifiers (One-Class SVM and V-Detector) in recognizing true positive and true negative patterns.

Comparative Study: Collaborative Filtering Methods, One Class Content-Based Methods, Hybrid Methods

In this section we provide a detailed description concerning the first type of experiments. Our primary concern focused on conducting a comparative study of the various recommendation approaches that were implemented in this chapter. It is very important to assess the recommendation ability of each individual system in order to distinguish the one that exhibited the best overall performance. Specifically, the recommendation accuracy was measured in terms of the average MAE over all folds for the complete set of users. The details concerning the exact testing procedure for

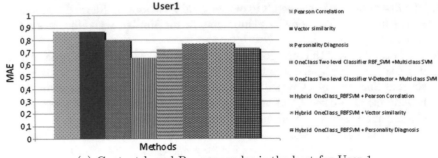

(a) Content-based Recommender is the best for User 1

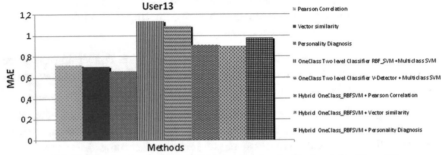

(b) Collaborative Filtering Recommender is the best for User 13

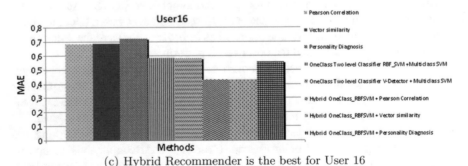

(c) Hybrid Recommender is the best for User 16

Fig. 8.15 MAE boundaries for one class SVM

each user are thoroughly described in Sect. 6. Our findings indicate that there was
no recommendation approach that outperformed the others for the complete set of
users. This means that there were occasions for which the best recommendations, for
a particular user, were given by the standard Collaborative Filtering approach. On the
other hand, there were occasions for which either the Cascade Content-based Recommender or the Cascade Hybrid Recommender provided more accurate predictions
concerning the true user ratings.

Typical examples of the previously mentioned situations are illustrated in
Fig. 8.15a–c. Specifically, Fig. 8.15a demonstrates that the best recommendation

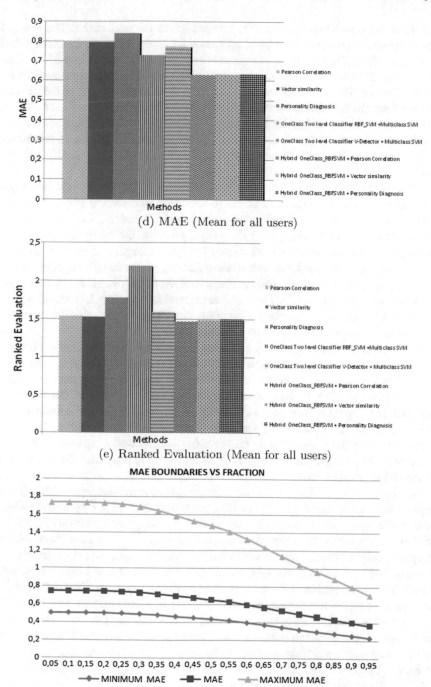

(d) MAE (Mean for all users)

(e) Ranked Evaluation (Mean for all users)

Fig. 8.15 (continued)

approach for User1 was the Cascade Content-based Recommender. The rest recommendation approaches in order of efficiency for User1 were the Cascade Hybrid Recommender and standard Collaborative Filtering. Furthermore, Fig. 8.15b demonstrates that the best recommendation approach for User13 was the standard Collaborative Filtering. The rest recommendation approaches for this user were the Cascade Hybrid Recommender at the second place and the Cascade Content-based Recommender at the third. Finally, Fig. 8.15c demonstrates that the ranked list of efficiency for the set of the utilized recommendation approaches presents the Cascade Content-based Recommender at the second place and the standard Collaborative Filtering at the third.

The most important finding that we need to convey as the result of the first experimentation session concerns the fact that the best overall recommendation approach over all users and folds was the Cascade Hybrid Recommender. This fact is explicitly illustrated in Fig. 8.15d where the hybrid approach presents the lowest mean MAE value taken over all users and folds during the testing stage. It is worth to mention that the pure content-based and collaborative filtering methodologies occupy the second and third positions respectively in the ranked list of the overall recommendation accuracy. We postulate that this is not an accidental fact but it is an immediate consequence that follows the incorporation of the one-class classification component at the first level of the cascade recommendation scheme.

The recommendation approaches that rely exclusively on Collaborative Filtering, estimate the rating value that a particular user would assign to an unseen item on the basis of the ratings that the rest users provided for the given item. In other words, the pure Collaborative filtering approaches do not take into account the subjective preferences of an individual user, as they are biased towards the items that are most preferred by the rest of the users. The major drawback of the standard collaborative filtering approaches is that they disorientate the user by operating exclusively on a basis formed by the preferences of the rest users, ignoring the particular penchant an individual user might have.

On the other hand, the pure content-based recommendation approaches fail to exploit the neighborhood information for a particular user. They operate exclusively on classifiers which are trained to be user specific, ignoring any beneficial information related to users with similar preferences. A natural solution to the problems related to the previously mentioned recommendation approaches would be the formation of a hybrid recommendation system. Such a system would incorporate the classification power of the Content-based Recommenders and the ability of standard collaborative filtering approaches to estimate user ratings on the basis of similar users' profiles.

The Cascade Hybrid Recommendation approach presented in this chapter mimics the social process when someone has selected some items according to his preferences and asks for opinions about these by others, in order to achieve the best selection. In other words, the one-class classification component, at the first level, provides specialized recommendations by filtering out those music files that a particular user would characterize as non-desirable. This is achieved through the user-specific training process of the one-class classifiers which are explicitly trained on user-defined positive classes of patterns. On the other hand, the second level of recommendation exploits the neighborhood of preferences formed by users with similar opinions. We strongly claim that the recommendation superiority exhibited by the Cascade Hybrid Recommender lays its foundations on the more efficient utilization of the Collaborative Filtering component. This is achieved by constraining its operation only on the subset of patterns that are already recognized as desirable. Therefore, this approach resolves the problem of user disorientation by asking for the opinions of the rest users only for the items that a particular user assigns to the positive class of patterns.

One Class SVM - Fraction: Analysis

This experimentation session reveals the contribution of the second multi-class classification level in the overall recommendation ability of the Cascade Content-based Recommender. Equations (8.47) and (8.48) provide the minimum and maximum values for the average mean absolute error over all users, given the classification performance of the first (one-class) classification level. Having in mind that the previously mentioned lower and upper bounds on the average MAE concern the overall performance of the cascade recommender at both levels, they reflect the impact of the second multi-class classification component. The lower bound on the average MAE indicates the best case scenario which involves that the second classification level performs inerrably. On the other hand, the upper bound on the average MAE indicates the worst case scenario, suggesting that the performance of the second classification level is a total failure. In this context, if we measure the actual value for the average MAE (over all users) we can asses the classification influence of the second level in the overall recommendation accuracy of our system. Thus, if the actual value of the average MAE is closer to the lower bound, this implies that the second classification level operated nearly to the best possible way. In the contrary, if the actual value of the average MAE is closer to the upper bound, we can infer that the second classification level yielded almost nothing to the overall performance of our recommendation system. Figure 8.15 presents a curve of actual values for the average MAE relative to the corresponding lower and upper bound curves. The set of values for each curve is generated by parameterizing the one-class SVM classifier with respect to the fraction of the positive data that should be rejected during the training process.

Comparison Study of One Class Methods

Finally, the third experimentation session involves a comparison of the utilized one class classifiers, namely the one-class SVM and the V-Detector, concerning the classification performance of the corresponding cascade content-based recommendation approaches. The relative performance of both classifiers was measured in terms of the precision, recall, F1-measure and MAE. The precision is defined by the following equation

$$\overline{Precision} = \frac{\overline{TP}}{\overline{TP} + \overline{FP}} \tag{8.49}$$

which provides the average precision over all users and folds in relation to the average values for the true positives and the false positives. The recall is defined by the following equation

$$\overline{Recall} = \frac{\overline{TP}}{\overline{TP} + \overline{FN}} \tag{8.50}$$

which provides the average recall over all users and folds in relation to the average values for the true positives and the false negatives. Finally, the F1-measure is defined by the following equation

$$\overline{F1} = \frac{2 \times \overline{Precision} \times \overline{Recall}}{\overline{Precision} + \overline{Recall}} \tag{8.51}$$

which provides the average value for the F1-measure over all users and folds. Precision quantifies the amount of information you aren't loosing while recall expresses the amount of data you aren't loosing. Higher values for precision and recall indicate superior classification performance. The F1-measure is a combination of precision and recall which ranges within the [0, 1] interval. The minimum value (0) indicates the worst possible performance while the maximum value (1) indicates performance of the highest efficiency. Figure 8.20 demonstrates the average value of the MAE for both one-class classification approaches over all users and folds where the one-class SVM classifier presents a lower MAE value in comparison to the V-Detector. MAE is a measure related to the overall classification performance of the Cascade Content-based Recommender where values closer to zero indicate higher classification accuracy. It is very important to note that in the context of the highly unbalanced classification problem related to recommendation, the quality that dominates the value of MAE is the number of the the correctly classified negative patterns (True Negatives) (Fig. 8.17). Since the vast majority of patterns belong to the negative class, correctly identifying them reduces the overall classification inaccuracy. Thus, the lower MAE value for the one-class SVM indicates that this classifier performs better in filtering out the non-desirable patterns. On the other hand, the F1-measure, that specifically relates to precision and recall according to Eq. (8.51), is dominated by the amount of positive patterns that are correctly classified (True Positives), according to Eqs. (8.49) and (8.50). The F1-measure quantifies the amount of valuable positive recommendations that the system provides to the user. Figures 8.18, 8.19 and 8.21 demonstrate that the V-Detector performs better in the context of providing valuable positive recommendations. This constitutes a major concern of recommendation systems that specifically relates to the ability of the utilized classifiers in correctly identifying the true positive patterns. The V-Detector classifier provides significantly more true positive patterns as it is illustrated in Fig. 8.16 when it is compared against the one-class SVM. Thus, if we consider the ratio

$$QualityRate = \frac{\overline{F1}}{MAE} \tag{8.52}$$

Fig. 8.16 True positives

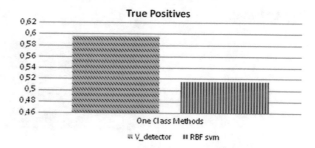

Fig. 8.17 True negatives

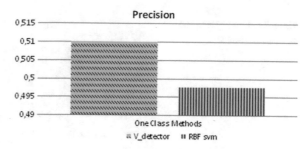

Fig. 8.18 Precision

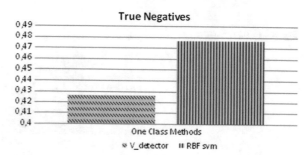

Fig. 8.19 Recall

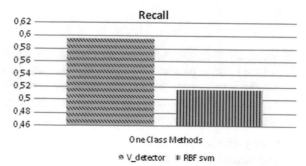

as an overall measure that evaluates the quality of recommendation for both classifiers, V-Detectors performs slightly better as is illustrated in Fig. 8.22. It is obvious that high quality recommendations involve increasing the F1-measure values and decreasing MAE which results in higher Quality Rate values. We need to mention

Fig. 8.20 MAE

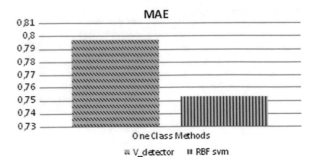

Fig. 8.21 F1 measure

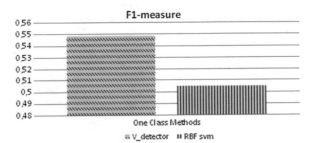

Fig. 8.22 Quality rate for
one class methods

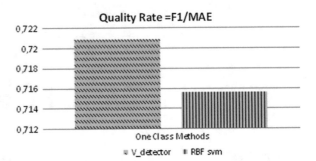

the measurement concerning the one-class SVM where conducted by rejected the
5 % of the positive patterns during the training procedure.

Finally, a remarkable finding concerns the behavior of the one-class classifiers
with respect to the fraction of positive and negative patterns that they identify during
the testing process. Our experiments indicate that:

- the classification performance of the one-class SVM classifier involves increasing
  True Negative Rates as the fraction of rejected positive patterns during training
  approaches the 95 %,
- while the classification performance of the the V-Detector involves increasing True
  Positive Rates as the parameter C_o, which affects the estimated coverage of the
  non-self (negative) space, decreases.

Thus, it is a matter of choice whether the recommendation process will focus on increasing the true positive rate or the true negative rate. Increasing the true negative rate results in lower MAE values while increasing the true positive rate results in higher F1-measure values. Specifically, the fact that the non-desirable patterns occupy the vast majority of the universe of discourse suggests that the quality of recommendation is crucially influenced by the number of the correctly identified negative patterns. In other words, constraining the amount of the false positive patterns that pass to the second level of the recommendation system increases the reliability related to the quality of the recommended items. The most appropriate measure concerning the quality of recommendation is given by the Ranked Evaluation Score since it presents the amount of true positive items that are at the top of the ranked list. This fact is clearly demonstrated in Fig. 8.15e where the Ranked Evaluation Score for the Cascade Content-based recommender of the one-class SVM outperforms the rest recommendation approaches.

References

1. Abraham, A.: Business intelligence from web usage mining. JIKM **2**(4), 375–390 (2003)
2. Adomavicius, G., Tuzhilin, E.: Toward the next generation of recommender systems: a survey of the state-of-the-art and possible extensions. IEEE Trans. Knowl. Data Eng. **17**, 734–749 (2005)
3. Breese, J.S., Heckerman, D., Kadie, C.: Empirical analysis of predictive algorithms for collaborative filtering. In: Proceedings of the Fourteenth Conference on Uncertainty in Artificial Intelligence, pp. 43–52. Morgan Kaufmann (1998)
4. Brusilovsky, P.: Adaptive hypermedia. User Model. User-Adapt. Interact. **11**(1–2), 87–110 (2001)
5. Burke, R.: Hybrid recommender systems: survey and experiments. User Model. User-Adapt. Interact. **12**(4), 331–370 (2002)
6. Cayzer, S., Aickelin, U.: A recommender system based on idiotypic artificial immune networks. J. Math. Model. Algorithms **4**, 181–198 (2005)
7. Foote, J.T.: Content-based retrieval of music and audio. In: PRoceedings. Of Spiemultimedia Storage and Archiving Systems II, pp. 138–147 (1997)
8. Jin, X., Zhou, Y., Mobasher, B.:. A unified approach to personalization based on probabilistic latent semantic models of web usage and content. In: Proceedings of the AAAI 2004 Workshop on Semantic Web Personalization (SWP'04) (2004)
9. Jin, X., Zhou, Y., Mobasher, B.: Web usage mining based on probabilistic latent semantic analysis. In: KDD, pp. 197–205 (2004)
10. Lampropoulos, A.S., Tsihrintzis, G.A.: A music recommender based on artificial immune systems. Multimedia Systems (2010)
11. Lampropoulos, A.S., Tsihrintzis, G.A.: Machine Learning Paradigms - Applications in Recommender Systems. Volume 92 of Intelligent Systems Reference Library. Springer, Berlin (2015)
12. Li, T., Ogihara, M., Li, Q.: A comparative study on content-based music genre classification. In: SIGIR '03: Proceedings of the 26th Annual International ACM SIGIR Conference on Research and Development in Informaion Retrieval, pp. 282–289. ACM, New York (2003)
13. Linden, G., Smith, B., York, J.: Amazon.com recommendations: Item-to-item collaborative filtering. IEEE Int. Comput. **7**(1), 76–80 (2003)
14. Menczer, F., Monge, A.E., Street, W.N.: Adaptive assistants for customized e-shopping. IEEE Intell. Syst. **17**(6), 12–19 (2002)

15. Morrison, T., Aickelin, U.: An artificial immune system as a recommender system for web sites. CoRR. arXiv:0804.0573 (2008)
16. Pazzani, M.J., Billsus, D.: Adaptive web site agents. In: Proceedings of the 3rd Annual Conference on Autonomous Agents: AGENTS'99, pp. 394–395. ACM, New York (1999)
17. Pennock, D.M., Horvitz, E., Lawrence, S., Giles, C.L.: Collaborative filtering by personality diagnosis: a hybrid memory and model-based approach. In: UAI'00: Proceedings of the 16th Conference on Uncertainty in Artificial Intelligence, pp. 473–480. Morgan Kaufmann Publishers Inc., San Francisco (2000)
18. Rashid, A.M., Albert, I., Cosley, D., Lam, S.K., McNee, S.M., Konstan, J.A., Riedl, J.: Getting to know you: learning new user preferences in recommender systems. In: IUI'02: Proceedings of the 7th International Conference on Intelligent User Interfaces, pp. 127–134. ACM, New York (2002)
19. Rich, E.: User modeling via stereotypes. Cogn. Sci. 3(4), 329–354 (1979)
20. Sarwar, B., Karypis, G., Konstan, J., Riedl, J.: Analysis of recommendation algorithms for e-commerce. In: Proceedings of the 2nd ACM Conference on Electronic Commerce, pp. 158–167. ACM, New York (2000)
21. Sarwar, B.M., Karypis, G., Konstan, J.A., Riedl, J.T.: Application of dimensionality reduction in recommender system - a case study. In: In ACM WebKDD Workshop (2000)
22. Sarwar, B., Karypis, G., Konstan, J., Reidl, J.: Item-based collaborative filtering recommendation algorithms. In: Proceedings of the 10th International Conference on World Wide Web, pp. 285–295. ACM, New York (2001)
23. Schein, A.I., Popescul, A., Ungar, L.H., Pennock, D.M.: Methods and metrics for cold-start recommendations. In: SIGIR '02: Proceedings of the 25th Annual International ACM SIGIR Conference on Research and Development in Information Retrieval, pp. 253–260. ACM, New York (2002)
24. Shahabi, C., Faisal, A., Kashani, F.B., Faruque, J.: Insite: a tool for real-time knowledge discovery from users web navigation. In: VLDB, pp. 635–638 (2000)
25. Sheth, B., Maes, P.: Evolving agents for personalized information filtering. In: In Proceedings of the 19th Conference on Artificial Intelligence for Applications, pp. 345–352 (1993)
26. Sotiropoulos, D., Tsihrintzis, G.: Immune system-based clustering and classification algorithms. Mathematical Methods (2006)
27. Sotiropoulos, D., Lampropoulos, A., Tsihrintzis, G.: Artificial immune system-based music piece similarity measures and database organization. In: Proceedings of the 5th EURASIP Conference (2005)
28. Sotiropoulos, D.N., Tsihrintzis, G.A., Savvopoulos, A., Virvou, M.: Artificial immune system-based customer data clustering in an e-shopping application. In: Knowledge-Based Intelligent Information and Engineering Systems, 10th International Conference, KES 2006, Bournemouth, UK, October 9–11, 2006, Proceedings, Part I, pp. 960–967 (2006)
29. Sotiropoulos, D., Lampropoulos, A., Tsihrintzis, G.: Artificial immune system-based music genre classification. In: New Directions in Intelligent Interactive Multimedia. Volume 142 of Studies in Computational Intelligence, pp. 191–200 (2008)
30. Stathopoulou, I., Tsihrintzis, G.A.: Visual Affect Recognition, Volume 214 of Frontiers in Artificial Intelligence and Applications. IOS Press, Amsterdam (2010)
31. Tzanetakis, G.: Manipulation, analysis and retrieval systems for audio signals. Ph.D. thesis (2002)
32. Tzanetakis, G., Cook, P.: Marsyas: a framework for audio analysis. Org. Sound 4(3), 169–175 (1999)
33. Tzanetakis, G., Cook, P.: Musical genre classification of audio signals. IEEE Trans. Speech Audio Process. 10(5), 293–302 (2002)
34. Tzanetakis, G., Cook, P.R.: Music genre classification of audio signals. IEEE Trans. Speech Audio Process. 10(5), 293–302 (2002)
35. Ungar, L., Foster, D., Andre, E., Wars, S., Wars, F.S., Wars, D.S., Whispers, J.H.: Clustering methods for collaborative filtering. In: Proceedings of the AAAI Workshop on Recommendation Systems. AAAI Press (1998)

36. Virvou, M., Savvopoulos, A., Tsihrintzis, G.A., Sotiropoulos, D.N.: Constructing stereotypes for an adaptive e-shop using ain-based clustering. In: Adaptive and Natural Computing Algorithms, 8th International Conference, ICANNGA 2007, Warsaw, Poland, Proceedings, Part I, April 11—14, 2007, pp. 837–845 (2007)
37. Wang, Q., Makaroff, D.J., Edwards, H.K.: Characterizing customer groups for an e-commerce website. In: *EC '04: Proceedings of the 5th ACM Conference on Electronic Commerce*, pp. 218–227 (2004)

Chapter 9
Conclusions and Future Work

Abstract In this chapter, we summarize the research monograph, presenting its major findings and suggesting avenues of future research.

The primary effort undertaken in this monograph focused on addressing the fundamental problems of Pattern Recognition by developing Artificial Immune System-based machine learning algorithms. Therefore, the relevant research is particularly interested in providing alternative machine learning approaches for the problems of *Clustering*, *Classification* and *One-Class Classification*, measuring their efficiency against state of the art pattern recognition paradigms such as the Support Vector Machines. The main source of inspiration stemmed from the fact that the Adaptive Immune System constitutes one of the most sophisticated biological systems that is particularly evolved in order to continuously address an extremely unbalanced pattern classification problem. Its task is to perform the self/non-self discrimination process. Pattern classification was specifically studied within the context of the Class Imbalance Problem dealing with extremely skewed training data sets. Specifically, the experimental results presented in this monograph involve degenerated binary classification problems where the class of interest to be recognized is known through a limited number of positive training instances. In other words, the target class occupies only a negligible volume of the entire pattern space while the complementary space of negative patterns remains completely unknown during the training process. Therefore, the effect of the Class Imbalance Problem on the performance of the proposed Artificial Immune System-based classification algorithm constitutes one of the secondary objectives of this monograph.

The proposed Artificial Immune System-based clustering algorithm was found to be an excellent tool for data analysis demonstrating the following properties:

- reveals redundancy within a given data set,
- identifies he intrinsic clusters present within a given data set and
- unravels the spatial structure of a given data set by providing a more compact representation.

The performance of the proposed Artificial Immune System-based classification algorithm, on the other hand, was found to be similar to that of the Support Vector Machines when tested in balanced multi-class classification problems. The most

© Springer International Publishing AG 2017

D.N. Sotiropoulos and G.A. Tsihrintzis, *Machine Learning Paradigms*,
Intelligent Systems Reference Library 118, DOI 10.1007/978-3-319-47194-5_9

important findings, however, relate to the special nature of the adaptive immune system which is particularly evolved in order to deal with an extremely unbalanced classification problem, namely the self/non-self discrimination process. Self/Non-self discrimination within the adaptive immune system is an essential biological process involving a severely imbalanced classification problem since the subspace of non-self cells occupies the vast majority of the complete molecular space. Therefore, the identification of any given non-self molecule constitutes a very hard pattern recognition problem that the adaptive immune system resolves remarkably efficient. This fact was the primary source of inspiration that led to the application of the proposed AIS-based classification algorithm on a series of gradually imbalanced classification problems. Specifically, the classification accuracy of the AIS-based classifier is significantly improved against the SVM classifier for the balanced One vs All classification problems where the class of outlier is appropriately down-sampled so that both classes are equally represented during the training process.

The most interesting behavior of the AIS-based classifier was observed during the third experimentation session involving a series of 10 severely imbalanced pattern recognition problems. Specifically, each class of patterns pertaining to the original data set was treated as the target class to be recognized against the rest of the classes of the complementary space. In this context, the AIS-based classifier exhibited superior classification efficiency especially in recognizing the minority class of patterns. The true positive rate of recognition for the minority class of patterns is significantly higher for the complete set of the utilized experimentation data sets. More importantly, the proposed classification scheme based on the principles of the adaptive immune system demonstrates an inherent ability in dealing with the class imbalance problem.

Finally, in this monograph the music recommendation process was addressed as a one-class classification problem by developing an AIS-based Negative Selection (NS) algorithm. The primary objective of the proposed methodology was to exploit the inherent ability of the NS-based classifier in handling severely unbalanced classification problems in order to capture user preferences. The main idea that motivated the recommendation approach presented in this section stems from the fact that users' interests occupy only a small fraction of a given multimedia collection. That is, the music preferences of a particular user tend to be contained within a negligible volume of the complete pattern space. Therefore, the problem of identifying multimedia instances that a specific user would evaluate as preferable, constitutes an extremely unbalanced pattern recognition problem that could be addressed within the context of one-class classification. This is true, since in most real-life situations the user supplied feedback to a recommendation system is exclusively given in the form of positive examples from the target class to be recognized.

Specifically, the adapted approach decomposes the music recommendation problem into a two-level cascading recommendation scheme. The first recommendation level incorporates the AIS-based one-class classification algorithm in order to discriminate between positive and negative patterns on the basis of zero knowledge from the subspace of outliers. The second level, on the other hand, is responsible for assigning a particular degree of preference according to past user ratings. For this purpose, the second recommendation level applies either a content-based approach or a collaborative filtering technique. The implementation and evaluation of the proposed Cascade Hybrid recommender approach, enhanced by the one class classifier in the first level and the collaborative filtering in the second level, demonstrates the efficiency of the proposed recommendation scheme. The presented technique benefits from both content-based and collaborative filtering methodologies. The content-based level eliminates the drawbacks of the pure collaborative filtering that do not take into account the subjective preferences of an individual user, as they are biased towards the items that are most preferred by the rest of the users. On the other hand, the collaborative filtering level eliminates the drawbacks of the pure content-based recommender which ignore any beneficial information related to users with similar preferences. The combination of two approaches in a cascade form, mimics the social process when someone has selected some items according to his preferences and asks for opinions about these by others, in order to achieve the best selection. In particular, the Negative Selection-based One-Class Classifier at the first level of the proposed methodology demonstrates superior recommendation quality. This is true, since the corresponding true positive rate, is significantly higher than the one achieved through the utilization of the One-Class Support Vector Machines.

A natural extension of the work presented in this monograph involves the incorporation of the proposed Artificial Immune System-based classification algorithms within the general framework of combining pattern classifiers. The relevant experimentation indicates that Artificial Immune System-based classification algorithms can be significantly improved by adapting a pairwise combination technique. However, a very interesting approach would be to exploit the multi-class classification ability of an ensemble of Artificial Immune System-based One Class Classifiers. Moreover, a very promising avenue of future research can be drawn by utilizing a game theoretic approach in order to devise more efficient strategies for the combination of the individual classifiers.

This and other related research work is currently underway and will be reported elsewhere in the future.

Printed in the United States
By Bookmasters